Aphid Ecology

Aphid Ecology
An optimization approach

Second edition

A.F.G. Dixon

School of Biological Sciences,
University of East Anglia

SPRINGER SCIENCE+BUSINESS MEDIA, LLC

First edition 1985

Second edition 1998

Originally published by Blackie and Son Ltd; 1998 A. F. G. Dixon in 1985
Softcover reprint of the hardcover 2nd edition 1985

Typeset in 10/12pt Times New Roman by Type Study, Scarborough

ISBN 978-94-010-6480-4 ISBN 978-94-011-5868-8 (eBook)
DOI 10.1007/978-94-011-5868-8

A catalogue record for this book is available from the British Library

Library of Congress Catalog Card Number: 97-69057

Printed on permanent acid-free text paper, manufactured in accordance with ANSI/NISO Z39.48–1992 and ANSI/NISO Z39.48–1984 (Permanence of Paper).

Contents

Preface

This book is about the ecology of a group of small, sap sucking, insects, the aphids, some of which are serious pests of agricultural crops and forest trees. The association of aphids with plants is considered in relation to certain of the features that aphids developed early in their evolution. An optimization approach is used to explain how the modes of feeding and reproduction of aphids have led to a very close and specific association with plants and have affected their size and population structure, and account for the paucity of species in the tropics.

This book is written for specialists, postgraduates and advanced undergraduates in entomology and ecology, but I hope that other people will find it interesting.

I am grateful for the opportunity to acknowledge the help of colleagues and past students, in particular Pavel Kindlmann and Seamus Ward, in developing the ideas presented here. A number of people have helped in the preparation of this book. I am especially grateful to June, whose support made the task possible, and to Drs Aoki, Bromley, Dussourd, Fukatsu, Gibson, Jukes and Tjallingii for supplying photographs used in the text. I also wish to thank Diane Alden for preparing the figures and Philippa Riley for typing the manuscript.

A.F.G.D.

1 Introduction

Aphids have fascinated and frustrated man for a very long time. This is mainly because of their intricate life cycles and close association with their host plants, their polyphenism, and their ability to reproduce both asexually and sexually. Thus aphids are ideal for studying many of the topical issues in ecology. The term 'ecology' is used here in a broad sense to include aspects of the functional biology of the group necessary for an understanding of the population and community levels of organization. The modes of feeding and reproduction that developed early in aphid evolutionary history led to a close and specific association with certain plants, and affected aphid size and population structure. In turn, these consequences have influenced aphid life-history patterns, dispersal behaviour, population dynamics, species diversity and world-wide distribution.

There are about 4000 species of aphid. Compared with 10 000 grasshoppers, 12 000 geometrid moths and 60 000 weevils, there are relatively few aphids. However, their diversity is expressed as polyphenism as well as speciation. The greatest number of aphid species occurs in the temperate regions and there one plant species in four is infested. However, the relatively few species of aphids among insects is in itself intriguing, because like so many other aspects of aphid ecology, the explanation is to be found in the functional biology of the group.

Unfortunately there are few biological studies on aphids indigenous to Africa, South America and India. Until such gaps in our knowledge are filled, an account of aphid ecology will be unbalanced because it is based mainly on information about species that live in the northern temperate region. The results of future studies of aphid biology in the equatorial and the hot dry regions of the world are likely to challenge, as well as complement, the generalizations about aphid biology that are founded mainly on studies of one subfamily of aphids, the Aphididae, in temperate and markedly seasonal environments.

1.1 Distinguishing features of aphids

They are small (1–10 mm), soft-bodied, plant-sucking insects. Several or all generations comprise parthenogenetic females, which do not require

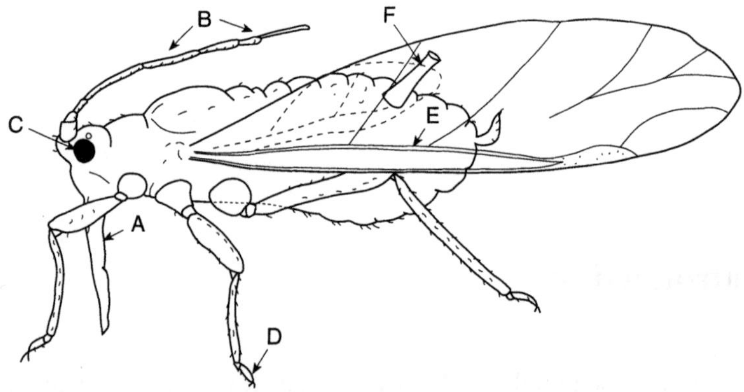

Figure 1.1 Diagnostic morphological features of aphids. (After Heie, 1980.)

fertilization and are viviparous. Species in which periods of asexual reproduction alternate with sexual reproduction are said to show cyclical pathenogenesis. As the eggs of parthenogenetic females commence development immediately after ovulation, a nymph can have embryos developing within itself that also have embryos. This parthenogenesis and telescoping of generations enable aphids to achieve very high rates of increase. The winged aphids are known as alatae and wingless aphids as apterae. Polyphenism, which is the occurrence within a species of different forms or morphs, is also a characteristic of aphids.

The more obvious diagnostic morphological features are illustrated in Figure 1.1. (A) The base of the proboscis lies between and behind the fore coxae; (B) the antennae have two short thick basal segments and a thinner flagellum, of at most four segments, the ultimate one of which consists of a proximal part and a thinner distal part, the process terminalis; (C) there is an ocular tubercle made up of three lenses (a triommatidium) situated behind each compound eye; (D) there are two tarsal segments; (E) the wings have only one prominent longitudinal vein; and (F) there is a pair of siphunculi on the dorsum of the fifth abdominal segment.

1.2 Origin of aphids

Aphids, adelgids and phylloxerids are very closely related and are either placed in the insect super family Aphidoidea (Blackman and Eastop, 1994), or into two super families the Phylloxeroidea and Aphidoidea, within the order Homoptera, the plant-sucking bugs. They are all small insects; wings, if present, are membranous. Although the oldest fossil aphid, the wing of *Triassoaphis cubitus*, is from the Triassic (Figure 1.2), it is likely that the Aphidoidea evolved 280 million years ago, in the Carboniferous, from the same stock that gave rise to the now extinct Archescytinidae and

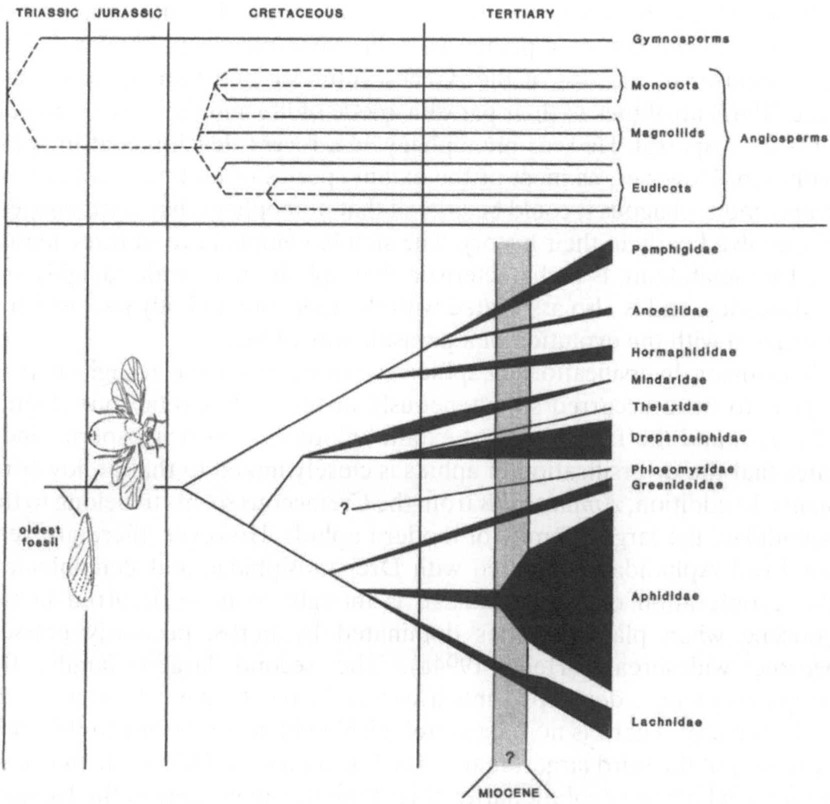

Figure 1.2 Simplified phylogenetic trees and the relative stratigraphic ranges of aphids and plants. (After Crane *et al.*, 1995; Heie, 1967.)

Permaphidopsidae, which are thought to have lived on primitive gymnosperms like the Cordaitales and Cycadophyta. Reproduction by means of unfertilized eggs is common to all three groups, and as the adelgids and phylloxerids appeared in the late Carboniferous or early Permian, parthenogenesis possibly evolved before these three families separated, over 200 million years ago. The resultant cyclical parthenogenesis, in which periods of parthenogenetic reproduction alternate with sexual reproduction to give a holocycle, is thought to have evolved in a seasonal climate, possibly that associated with the glacial period in the Lower Permian. However, viviparity, another characteristic feature of aphids, must have evolved later, as the Archescytinidae and the modern adelgids and phylloxerids are oviparous (Heie, 1967). The characteristic shape and veining of their wings, and the structure of their proboscis and legs, had evolved by the Jurassic (e.g. *Juraphis crassipes*; *cf.* Figures 1.1 and 1.2), whereas the cauda and siphunculi appeared later, in the Cretaceous (Shaposhnikov, 1977).

Primeval aphids were probably polyphagous, and fed on the parenchymatous and phloem tissues of plants. That the Aphidoidea and Phylloxeroidea are a quarter of the size of the Archescyntinidae and Permaphidopsidae, Heie (1967) attributes to their parasitic mode of life and their use of air currents for dispersal. He sees monophagy as a recent development in aphid evolution. However, as most of the extant species of all three families are mainly monophagous it could be argued that monophagy, like parthenogenesis, evolved early in their history. The simple nymphal eye of three lenses, the triommatidium, is a characteristic that aphids share with adelgids and phylloxerids, and is also associated with the reduction in body size, which is associated with the evolution of a parasitic way of life.

The major diversifications in aphids and flowering plants (Angiosperms) appear to have occurred simultaneously in the early Cretaceous (Figure 1.2). This and the fact that most extant aphids live on Angiosperms indicates that the diversification of aphids is closely linked to that of flowering plants. In addition, *Aphidocallis* from the Cretaceous seems to belong to the Aphididae, the largest family of modern aphids. However, there are very few fossil Aphididae compared with Drepanosiphidae and Pemphigidae. The proliferation of the Aphididae is thought to have occurred in the Miocene when plant societies dominated by herbs, primarily grasses, became widespread (Heie, 1994a). The second largest family, the Drepanosiphidae, developed much earlier in the Upper Cretaceous and early Tertiary. There is no palaeontological evidence relevant to the early evolution of the third largest family, the Lachnidae. As 80% of them live on conifers, which are evolutionarily older than the angiosperms, the lachnids have generally been regarded as primitive. However, two genera, *Aphis* in the Aphididae and *Cinara* in the Lachnidae, each have a large number of species that are difficult to distinguish from one another. This and the absence of early fossil lachnids led Heie (1967) to suggest that the Lachnidae is not a particularly old family.

1.3 Host plant associations

The numbers of aphid species that have colonized the various families of plants is clearly not correlated with their species diversity. Of the plant groups, the Compositae hosts the most species (600) of aphids and is the third most diverse plant group. The Leguminosae is ranked fourth in terms of species diversity but only ninth in terms of hosting aphids, and the Orchidaceae (20 000 species), the second most diverse plant group, only hosts nine species of aphids. In contrast, other groups of plants, like the Coniferae, which is ranked 58th in terms of species richness, is ranked second in terms of the number of aphid species it hosts (Table 1.1).

The lack of a correlation between ranking of the plant groups in terms of their species diversity and the numbers of aphids they host (*cf.* Table 5.4 in

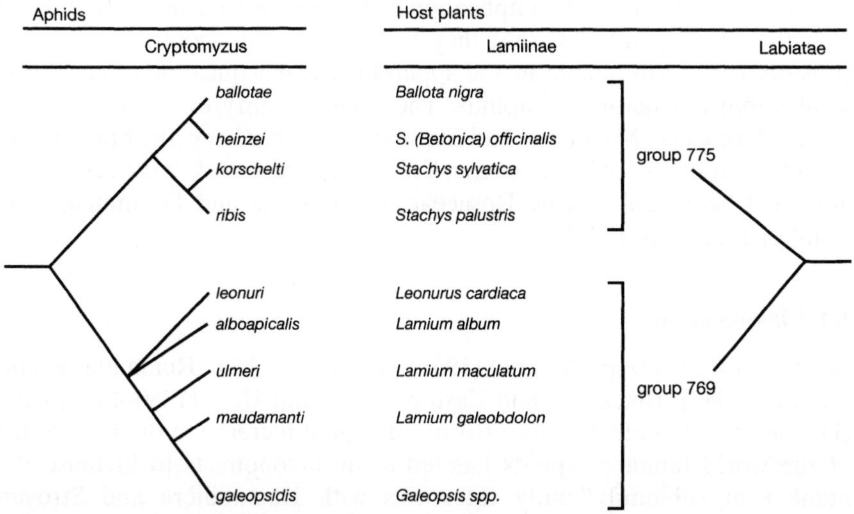

Figure 1.3 Association between the phylogeny of *Cryptomyzus* and that of its lamiaceous host plants. (After Guldemond, 1990a.)

Eastop, 1978, $r_s = 0.08$, $n = 90$, NS) is surprising, especially as several groups of aphids appear to have speciated in parallel with particular groups of plants (Börner, 1939). For example, the most primitive aphids living on the Rosaceae live only on the primitive species of this group of plants (Hille Ris Lambers, 1950). A phylogeny of the European species of *Cryptomyzus* based on allozyme and life-history characteristics indicates two groups of species, which mirror their two distinct taxonomic groups of host plants within the Lamiinae (Labiatae) (Figure 1.3; Guldemond, 1990a). However, within aphid genera there are often species that infest aberrant hosts, e.g. *Myzocallis asclepidis* lives on *Asclepias* whereas all the other species of the

Table 1.1 The species diversity and ranking based on this diversity of the top nine groups of plants ranked in terms of the numbers of species of aphids they host (after Eastop, 1978)

Plant Group	*No. of species*	*No. of aphids*	*Ranking* *Aphids*	*Plants*
Compositae	19 000	605	1	3
Coniferae	400	363	2	58
Rosaceae	2500	293	3	22
Gramineae	8000	242	4	8
Saliceae	530	216	5	56
Fagaceae	750	211	6	50
Betulaceae	150	108	7	68
Umbelliferae	2900	107	8	19
Leguminosae	12 500	106	9	4

genus live on *Quercus*. The implication is that aphids have not only evolved with their host plants but that they have frequently acquired new hosts (Eastop, 1972). This common and apparent casual acquisition of new hosts is an important feature of aphids. The relative rarity of aphids on Lauraceae, Moraceae, Myrtaceae and Euphorbiaceae has been attributed to the great speciation of aphids in the north temperate region after the Lauraceae flora had been replaced by Rosaceae, Compositae and Gramineae, etc. (Table 1.1; Eastop, 1978).

1.4 Classification

Börner (1952), Shaposhnikov (1964), Eastop (1977), Remaudière and Stroyan (1984), Blackman and Eastop (1994) and Heie (1980–1995) have classified the Aphididae. Since Börner the great increase in the knowledge of the world fauna of aphids has led some taxonomists to increase the number of subfamily/family categories with Remaudière and Stroyan (1984) favouring a classification incorporating 20 families. The status and relatedness of these taxa are debatable and await more definitive phylogenetic studies using molecular techniques. The phylogeny presented in Figure 1.2 is that of Heie (1967) whose recent keys to the aphids of Fennoscandia and Denmark (Heie, 1980–1995) are likely to remain the standard work for European ecologists for some time. The phylogenetic position of the Lachnidae, i.e. whether they should be regarded as one of the most primitive or one of the most advanced groups of aphids, is one of the more important phylogenetic issues that remain to be resolved.

Ease of identification largely determines the attractiveness of insects for study. Aphids are difficult to identify because even within clones individuals in most species can exist in one of several different forms, i.e. they are polyphenic. However, modern keys to aphids exist or are being developed for most of Northern Europe (Hille Ris Lambers, 1938, 1939, 1947, 1949, 1953; Shaposhnikov, 1964; Stroyan, 1977, 1984; Heie, 1980, 1982, 1986, 1992, 1994b, 1995), North America (Palmer, 1952; Richards, 1960, 1963, 1965, 1972), the Middle East (Bodenheimer and Swirski, 1957), East and West Africa (Eastop, 1958, 1961), India (Ghosh, 1980, 1982, 1984), Japan (Shinji, 1941; Miyazaki, 1971; Higuchi, 1972; Aoki, 1975; Moritsu, 1982; Akimoto, 1983), Korea (Paik, 1965, 1972), Australia (Eastop, 1966), New Zealand (Cottier, 1953), and for the world's crops and trees (Blackman and Eastop, 1984, 1994).

1.5 Distribution

Although predominant in temperate regions, aphids have a world-wide distribution. Within the Aphidoidae, the Aphididae and Drepanosiphidae, that make up 70% of modern aphids, are not restricted to a particular region, but the Greenideidae and Homaphididae, that make up 7%, are restricted mostly to South East Asia and Australia, although the presence of fossil Greenideidae in Yugoslavia indicates that this group once had a much wider distribution (Eastop, 1977). However, as the most important families of aphids are not restricted in their distribution we have to account for why there are so few species of aphids in the tropics and this is considered in Chapter 10.

Summarizing, aphids evolved 280 million years ago and from the beginning showed two of the more striking features of modern aphids: small size and parthenogenetic reproduction. The major diversification in aphids occurred in the Cretaceous in parallel with that of the Angiosperms. However, there is no correlation between the rank of a plant group in terms of its species diversity and the number of species of aphids it hosts. Aphids are very host specific, have a world-wide distribution and although certain species can become very abundant, there are surprisingly few species, especially in the tropics. Polyphenism and the telescoping of generations are also characteristics of aphids.

2 Feeding behaviour and food quality

Most, if not all, species of aphids feed on the phloem sap of plants, which they obtain by tapping the phloem elements with their stylets. Phloem elements are living cells and are located at some depth within a plant. The contents of these elements (phloem sap) are rich in sugars and relatively poor in amino acids, especially those that are essential for growth. This chapter describes the ways in which aphids locate and process their food.

2.1 Feeding behaviour

Aphid antennae bear many sensilla (Figure 2.1), amongst which there are some whose structure and electrophysical responses indicate that they are used in chemoreception or gustation and the perception of the leaf surface (Shambaugh *et al.*, 1978; Bromley *et al.*, 1979, 1980; Bromley and Anderson, 1982). In the laboratory aphids respond to plant odours both when walking (Pettersson, 1971; Visser and Taanman, 1987; Nottingham *et al.*, 1991a; Visser and Piron, 1995) and when flying (Nottingham and Hardie, 1993). Single-cell recording techniques have been used to assess the responsiveness of individual sensory neurones to odours (Bromley and Anderson, 1982) and recordings from the antennal nerve (electroantennograms) provide a measure of the total antennal responsiveness (Boeckh *et al.*, 1965). Although there is little doubt the antennal olfactory sensilla are receptive to both positive signals associated with host plant volatiles and negative signals associated with the odour blend of non-host plants their role in the location of plants from a distance is debatable (p. 28).

Aphids scan the surface of a plant accepted as a potential host with the tip of their proboscis. The tactile receptors on the tip of the proboscis (Figure 2.2) respond to contact and surface texture and enable aphids to detect the contours of veins, their preferred feeding site (Tjallingii, 1978b). They then probe into the plant with their mandibular and maxillary stylets, which together form a hollow needle-like structure (Figure 2.3). Aphids secrete a continuous stylet sheath as they penetrate plant tissue. The sheath material encases the stylets and is thought to be mainly lipoprotein, possibly containing some 10% phospholipid. The viscous precursor secretion

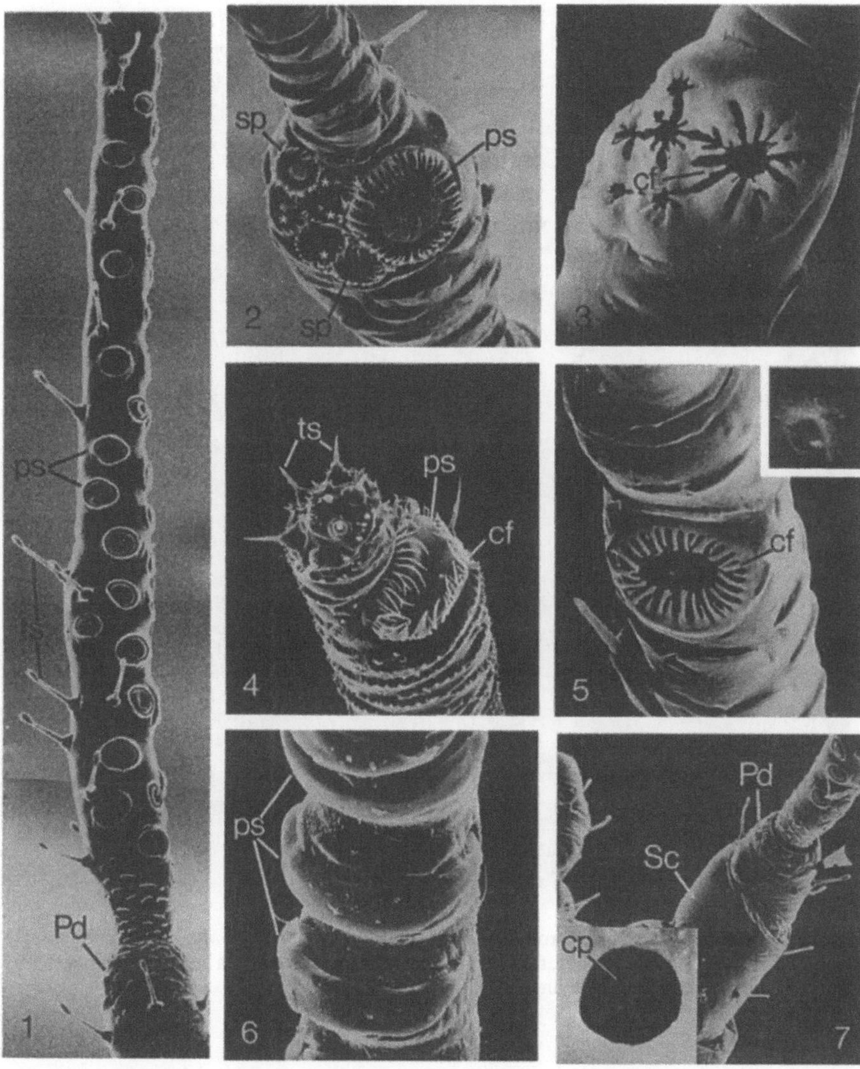

Figure 2.1 Scanning electron micrographs of aphid antennal sensilla. (1) Small placoid sensilla (secondary rhinaria) on third segment of alate *Nasonovia ribis-nigri*, ×200. (2) Primary rhinarium on sixth segment of alate *Aphis pomi* comprising one large placoid sensillum, two smaller ones and four coeloconic pegs, ×900. (3) Primary rhinarium on sixth segment of alate *Nasonovia ribis-nigri* almost totally enclosed by cuticular fringe, ×1000. (4) Primary rhinarium on sixth segment of alate *Pemphigus bursarius* (sexuparae) comprising a single enlarged placoid sensillum and four coeloconic pegs (not visible), ×680. (5) Primary rhinarium on fifth segment of *Macrosiphum euphorbiae*, ×1000. Inset: raised projection on outer cuticle of sensillum, ×9900. (6) Elongate placoid sensilla (secondary rhinaria) on third segment of alate *Pemphigus bursarius* (sexuparae), ×800. (7) Pedicel of alate *Nasonovia ribis-nigri* showing position of coeloconic sensillum (arrow); ×150. Inset: top of coeloconic sensillum at entrance to a pit, ×6750. (cf, cuticular fringe; cp, cuticular projections; Pd, pedicel; ps, placoid sensillum; Sc, scape; Sp, small placoid sensillum; ts, trichoid sensillum). (Reproduced with permission from Bromley *et al.*, 1979.)

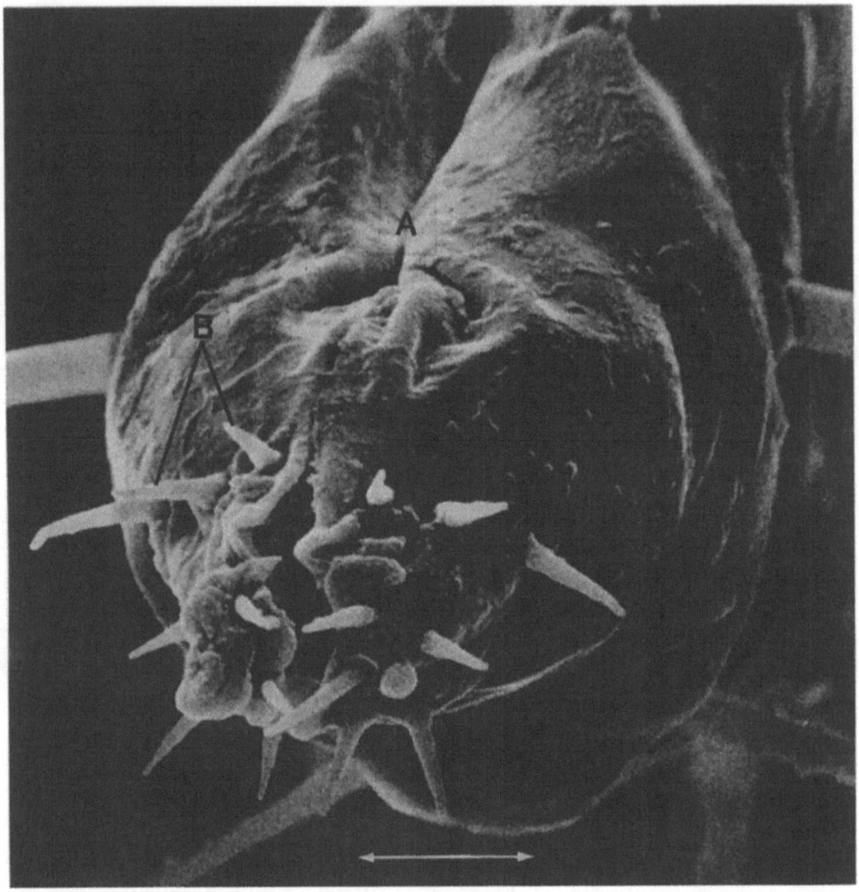

Figure 2.2 Scanning electron micrograph of the tip of the proboscis of *Myzus persicae* showing the labial groove (A) and 16 bilaterally symmetrically arranged sensilla (B). Scale bar is 15 μm. (Photographed by Dr W. F. Tjallingii.)

begins to gel immediately after it leaves the tips of the stylets, possibly by enzymic oxidation of sulphydryl groups to form disulphide bonds, accompanied by hydrogen bonding. Once formed the sheath is relatively impermeable (Miles, 1987). This salivary sheath gives rigidity to the very flexible stylets and enables aphids to control the direction of the probe by restricting bending except at the apex of the stylets (Pollard, 1973). The stylet sheath usually ends in the phloem, indicating that aphids feed on the contents of the sieve elements (Pollard, 1973).

The early studies using light microscopy indicated that the stylets followed both an intra- and intercellular route to the phloem (Pollard, 1973). A major technical advance pioneered by McLean and Kinsey (1964) and

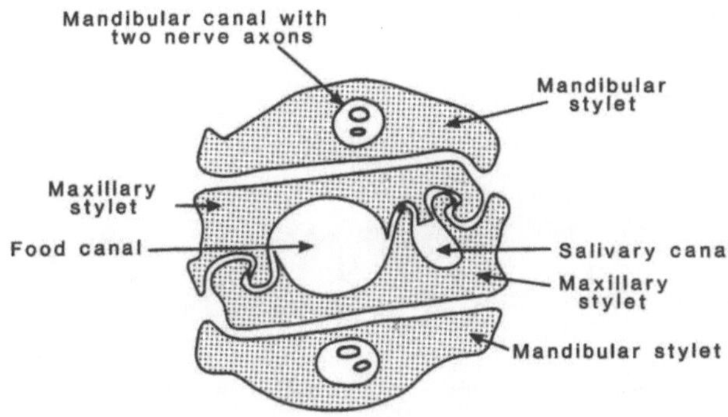

Figure 2.3 Diagram of a transverse section through the stylet bundle of an aphid.

perfected by Tjallingii (1978a, 1986), the electrical penetration graph (EPG), has greatly increased our understanding of aphid feeding behaviour. EPGs are obtained by making an aphid part of an electrical circuit (Figure 2.4). A very fine flexible gold wire from an amplifier is attached by means of conductive silver paint to the back of an aphid. The other electrode is put into the soil close to the roots of the plant or the excised leaf on which the aphid is feeding. As the aphid probes the plant changes in resistance are recorded in the form of changes in potential on an EPG (Figure 2.5). Several distinct wave forms have been recorded. Their significance has been established by cutting the aphid's stylets during the recording of a particular wave

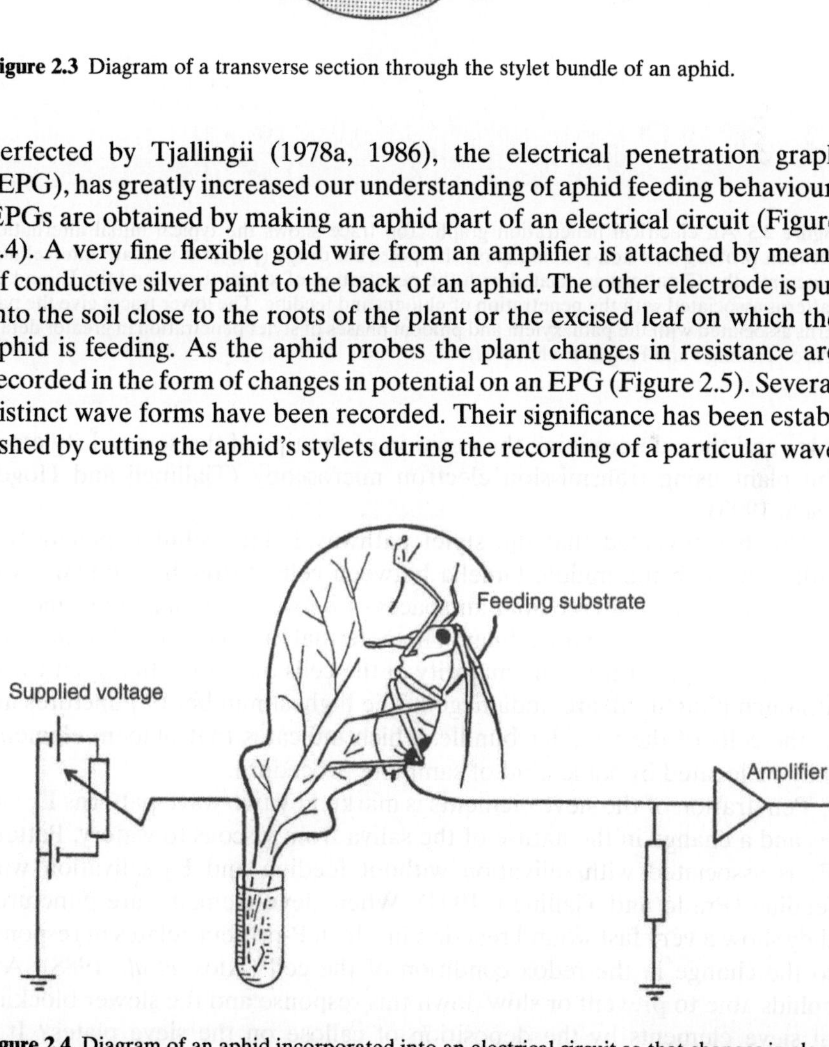

Figure 2.4 Diagram of an aphid incorporated into an electrical circuit so that changes in electrical potential can be recorded as the aphid's stylets penetrate the tissues of a plant. (After Tjallingii, 1978, 1985.)

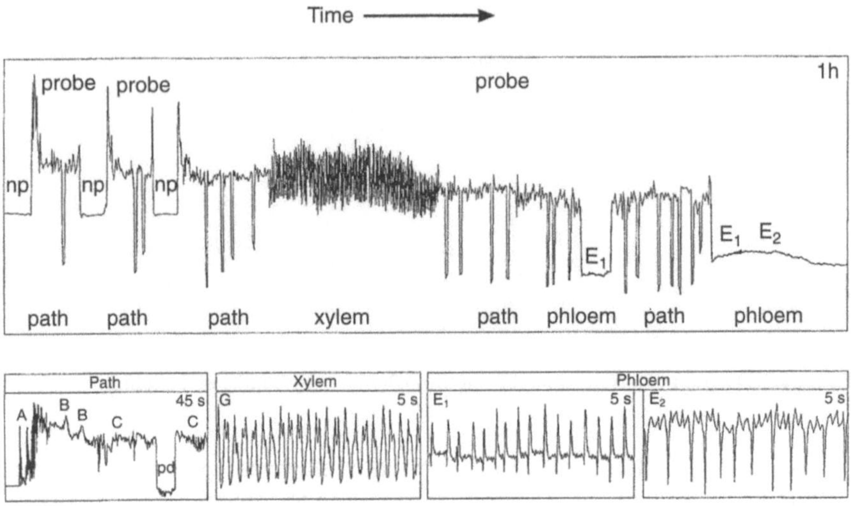

Figure 2.5 An electrical penetration graph. Top trace shows the typical initial alternation between probing and non-probing (np), the potential drops (pd) that indicate intercellular punctures, the G pattern associated with the penetration of xylem tissue, and the E_1 and E_2 patterns associated with the penetration of phloem and feeding. The lower traces give the patterns associated with the path, xylem and phloem phases of stylet penetration in greater detail. (After Tjallingii and Hogen Esch, 1993.)

form and then determining the position of the tips of the severed stylets in the plant using transmission electron microscopy (Tjallingii and Hogen Esch, 1993).

This has revealed that the stylet pathway is intercellular, penetrating either through the middle lamella between cells, through secondary wall material, through intercellular air spaces or between plasmalemma and cell wall, i.e. intramural/extracellular. The potential drops (Figure 2.5) indicate intracellular punctures. The majority of the cells bordering the stylet track, although punctured, are undamaged. The highest number of punctures are in the cells of the vascular bundle, which indicates that phloem elements may be located by some kind of sampling procedure.

Penetration of the sieve elements is marked by two wave patterns E_1 and E_2 and a change in the nature of the saliva from viscous to watery. Pattern E_1 is associated with salivation without feeding and E_2 salivation with feeding (Prado and Tjallingii, 1994). When sieve elements are punctured they show a very fast wound reaction in which P-protein gelates in response to the change in the redox condition of the cell (Alosi *et al.*, 1988). Are aphids able to prevent or slow down this response and the slower blocking of sieve elements by the deposition of callose on the sieve plates? It is possible that the watery saliva pumped into the sieve element (E_1) reduces P-protein gelation and callose deposition within the phloem element. The

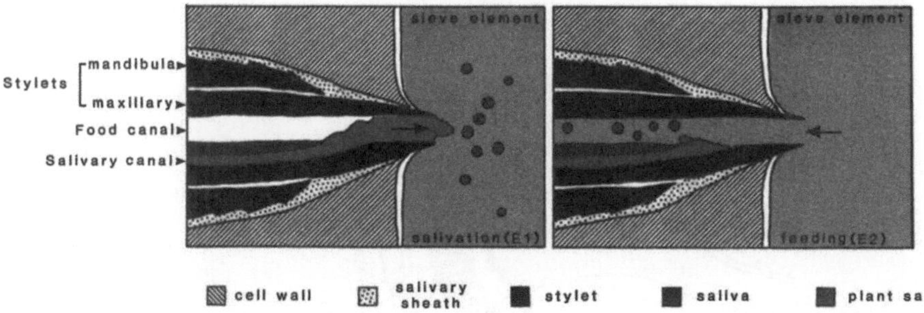

Figure 2.6 Diagram of the tips of the stylets and the flow of saliva after penetration of a phloem element. (After Prado and Tjallingii, 1994.)

saliva produced while feeding (E_2) is unlikely to be passed into the plant because the outlet of the salivary canal is inside the food canal, about 10 μm from the tips of the stylets (Figure 2.6). The saliva, therefore, is likely to be carried by the large volume of incoming sap back into the food canal. The function of this saliva might be to prevent gelation of the P-proteins inside the food canal and to detoxify secondary plant substances (p. 38).

Phloem sap is under 15–30 atmospheres of pressure, sufficient to force sap through the extremely fine food canal in the stylets and into the aphid's alimentary canal. This flow is controlled by a piston valve, opened by dilator muscles, in the aphid's pharynx (Figure 2.7). If the substrate pressure is negative or ambient aphids can pump fluid into their alimentary canal. The dilator muscles of the cibarial pump contract and increase the diameter of the pump lumen and decrease the pressure within the pump. This decrease in pressure causes the oesophageal valve to close and the collapse of the thin-walled oesophagus, which prevent regurgitation from the midgut. Opening the piston valve allows fluid to pass into the system. Closure of the piston valve followed by relaxation of the cibarial pump dilator muscles forces the fluid through the oesophagus into the stomach. Fluids may also be forced back through the stylets if the piston valve is not closed. In this way aphids are able to suck up sap, as when feeding on synthetic diets. The composition of sap is sampled in the pharynx by means of the gustatory epipharyngeal sensillae (Figure 2.7). Sap that fails to stimulate, or negatively stimulates the chemosensillae, initiates the relaxation of the cibarial pump muscles but not the closure of the piston valve. This results in the sap being extravasated through the stylets (McLean and Kinsey, 1984).

2.2 Food quality

The sap that passes into the gut of the aphids consists mainly of a concentrated solution of simple sugars and a weak solution of amino acids. Although there is little or no need to digest this food as it is already in a

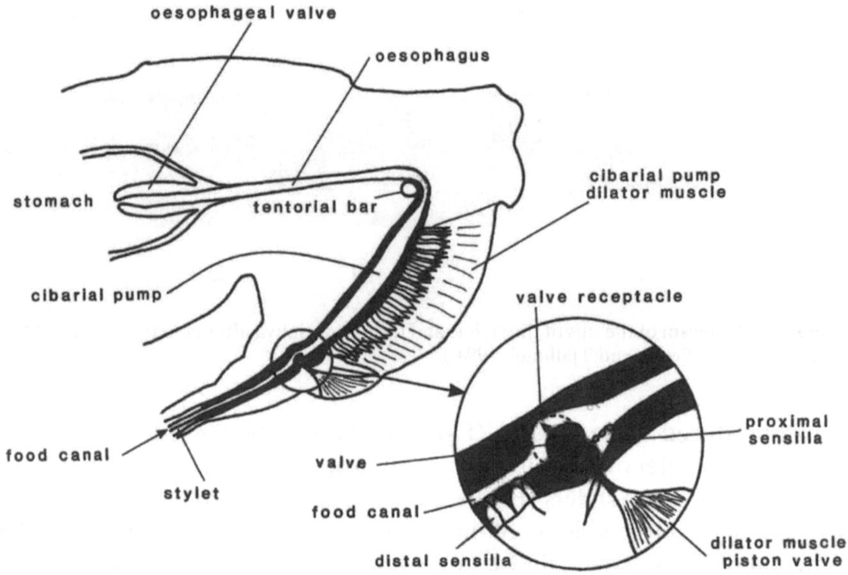

Figure 2.7 Diagram of a longitudinal section through the cibarial pump in the head of an aphid. (After McLean and Kinsey, 1984.)

soluble form (*cf.* Srivastava and Auclair, 1963; Rahbé *et al.*, 1995) the nature of the food does present problems. Its osmotic pressure can be much higher than that of an aphid's haemolymph and therefore they are in danger of desiccating. The low concentration of amino acids would also appear to put a severe constraint on the rate of growth they can achieve.

2.2.1 Osmoregulation

Aphids can substantially reduce the osmotic concentration of the ingested fluid, as honeydew is similar in osmotic concentration to the haemolymph (Downing, 1978; Figure 2.8). The presence of a so-called filter chamber(s) in the alimentary canal of some but not all aphids tends to suggest that a major problem facing aphids is the need to concentrate their food (Figure 2.9). Although this may appear desirable when considering the amino nitrogen component any concentration of the sugars would aggravate the osmotic problem. By converting mono- and disaccharides into trisaccharides like melezitose (Michel, 1942; Bacon and Dickinson, 1957) and oligosaccharides, aphids reduce the number of molecules in solution and thus the osmotic concentration of their food. If the osmotic concentration is close to that of their haemolymph there is little, if any, polymerization of the simple sugars into oligosaccharides and *vice versa* (Figure 2.10). There is a dramatic switch in the relative proportion of oligosaccharides to simpler

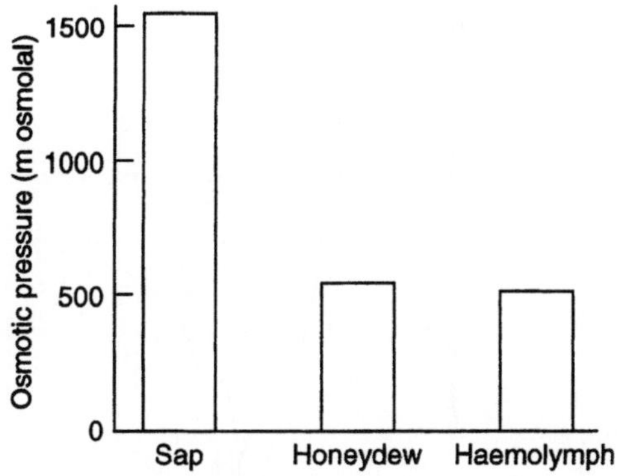

Figure 2.8 The osmotic pressure of plant sap compared with that of honeydew and haemolymph of *Myzus persicae*. (After Downing, 1978.)

sugars in the honeydew when the osmotic concentration of the incoming diet exceeds that of the haemolymph (Fisher *et al.*, 1984; Walters and Mullin, 1988; Rhodes *et al.*, 1996).

Another way of avoiding desiccation would be to use the filter chamber to dilute rather than concentrate the incoming food (Rhodes *et al.*, 1996). In their passage along the alimentary canal the simple sugars are converted into complex sugars and as a consequence the liquid in the hind gut is likely to have a lower osmotic pressure than that in the fore gut. Therefore, the filter chamber could function to dilute rather than concentrate the food as it is usually formed by the hind gut enclosing the stomach (Figure 2.9a and b). In those species that do not have a filter chamber the hind part of the gut lies alongside and is attached closely to one side of the stomach (Figure 2.9c). This would also allow the aphid to dilute the incoming fluid by drawing water from the hind gut rather than the haemolymph. The presence of a capacious stomach in an organism like aphids that feed continuously is surprising but its function may also be to dilute rather than store the incoming food.

2.3 Nitrogen metabolism

In contrast to that of sugar, the concentration of nitrogen in the food is relatively very low. In addition the ratio of essential to unessential amino acids is unbalanced in favour of the latter, with most of the dietary nitrogen consisting of asparagine and glutamine (Girousse *et al.*, 1990; Sasaki *et al.*, 1990; Douglas, 1993). As the symbionts of aphids do not fix atmospheric nitrogen (Smith, 1948) aphids are entirely dependent on their dietary nitrogen (Mittler, 1958b). In order to fuel their very high rates of growth (p. 83), therefore, aphids need to process large quantities of food and use the nitrogen it contains effectively.

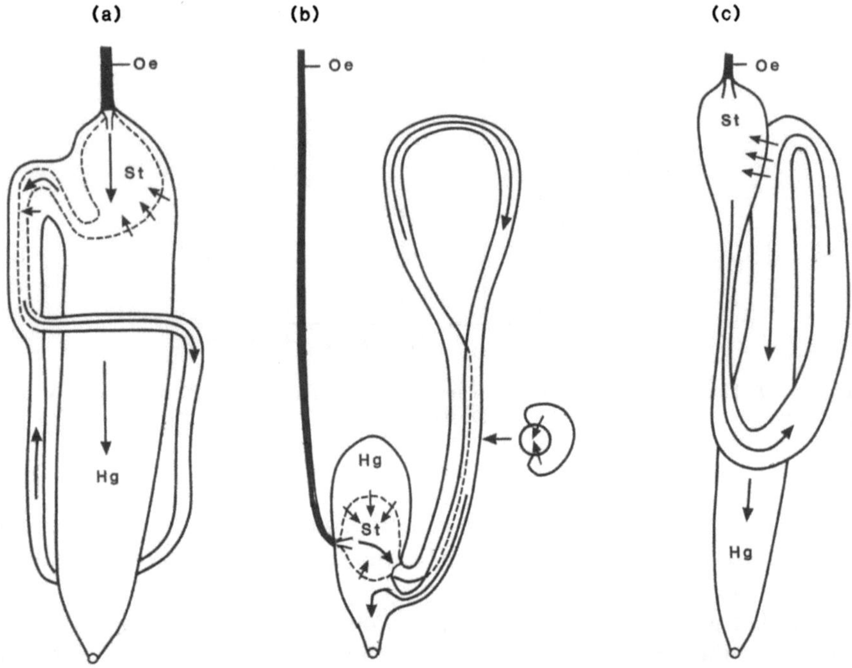

Figure 2.9 Dorsal view of the alimentary canal of (a) *Lachnus roboris*, (b) *Subsaltusaphis ornata* and (c) *Acyrthosiphon pisum*. In (a) and (b) the stomach (st) and part of the fore gut are enclosed completely or partially within the hind gut (hg) to form the so-called 'filter chamber'. In (c) part of the hind gut is attached closely to one side of the stomach (oe, oesophagus; the arrow indicates the flow through the system). (After Ponsen, 1979, 1990, 1991.)

2.3.1 Feeding rate and assimilation

Adult sycamore aphids process at least their own weight of phloem sap per day and the immatures several times their weight (Dixon and Logan, 1973). The quantity depends on the nitrogen content. For example, nymphs of the large willow aphid, *Tuberolachnus salignus*, process four times more sap when feeding on mature plant tissue, where the dietary concentration of nitrogen is 0.032 (w/v), than when feeding on senescent tissue where the concentration is 0.128 (w/v) (Mittler, 1958b). Compared to Lepidoptera (30–40%; Slansky and Feeny, 1977) aphids are very efficient at assimilating nitrogen from their food, which they achieve with an efficiency of 60% or greater even when feeding on wild species of plants (Mittler, 1958b). This is achieved in spite of having to process a relatively large volume of food and apparently in the absence of a mechanism for concentrating the dietary nitrogen prior to assimilation. It is surprising that this aspect of aphid physiology has received so little attention.

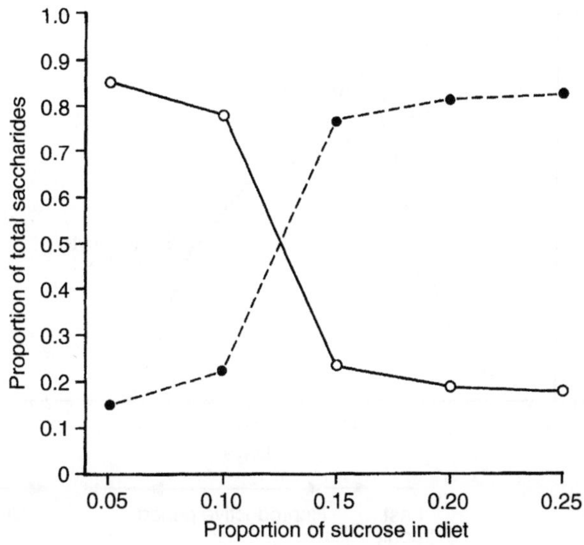

Figure 2.10 The proportion of the total saccharide content of honeydew in the form of oligosaccharides (filled circles), and mono- and disaccharides (open circles) in relation to the concentration of sucrose in the diet. (After Rhodes *et al.*, 1996.)

2.4 Symbionts

Most insects that live on nutritionally unbalanced diets possess symbiotic micro-organisms (Trager, 1970). In most aphids, the symbionts are confined to special groups of cells called bacteriocytes, which are most numerous at birth. During growth and about halfway into the reproductive period, the number of bacteriocytes gradually decreases and their volume increases, which suggests that they and their increasing symbiont population are essential for embryonic development. Towards the end of an aphid's reproductive period, bacteriocyte numbers decrease markedly and virtually none is present when an aphid ceases reproduction (Figure 2.11; Douglas and Dixon, 1987). Toth (1940) suggested that aphids utilize their symbionts to supplement their poor-quality diets. The slow growth and partial or complete sterility of experimentally produced aposymbiotic (i.e. symbiont-free) individuals indicates that the symbionts are important (Houk and Griffiths, 1980). In most aphids the symbionts are transmitted transovarially to the embryos (Buchner, 1965). Early in the development of an embryo a group of cells, which are destined to become the bacteriome, push into the blastoderm and a conical structure forms on top of the bacteriome rudiment. Symbionts pass from a neighbouring maternal bacteriocyte via a canal in the conical structure into the embryo's bacteriome. It is likely that this symbiosis has costs as well as benefits for aphids as they appear to control the size of the population of their endosymbionts. The highly fecund first generation individuals of

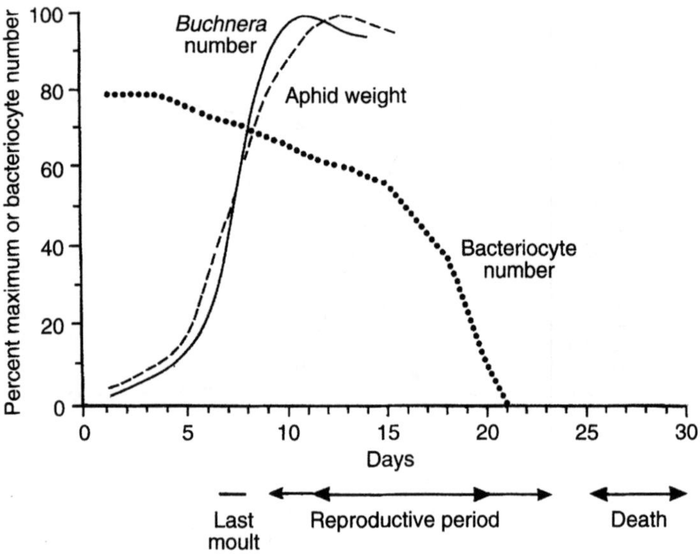

Figure 2.11 The trends in number of symbionts and aphid size, expressed in terms of percentage of the maximum, and number of bacteriocytes relative to the age of an aphid. (After Douglas and Dixon, 1987.)

Pemphigus have a large population of symbionts whereas the less fecund individuals of the latter generations host much small populations. The number of symbionts appears to be determined by the balance between division and degeneration. In the early generations of aphids the symbionts are frequently observed dividing and rarely degenerating whereas the reverse holds for late generation individuals. Degeneration of the symbionts is associated with aphid oenocytes invading the bacteriome (Tóth, 1933, 1937).

The primary symbiont of most aphids is a bacterium, *Buchnera aphidicola* (Munson *et al.*, 1991b), belonging to a subdivision of the Proteobacteria that includes *Escherichia coli* (Figure 2.12) and many other intestinal bacteria (Unterman *et al.*, 1989; Munson *et al.*, 1991a), including those of modern aphids (Harada and Ishikawa, 1993). Thus it is likely that the ancestor of *Buchnera aphidicola* was a gut microbe. Molecular data and fossil evidence indicate that this symbiont was acquired some 160–280 million years ago (Moran *et al.*, 1993; Moran and Baumann, 1994; Figure 2.12). In addition to the primary symbiont some aphids also harbour other prokaryotic intercellular symbionts – secondary symbionts. It is likely they have been acquired independently many times as they are not confined to a particular group of aphids. Their function is unknown but as monosymbiotic and disymbiotic species of aphid occur in the same clade, secondary symbionts are not likely to have an essential role (Figure 2.13; Fukatsu *et al.*, 1994; Moran and Baumann, 1994).

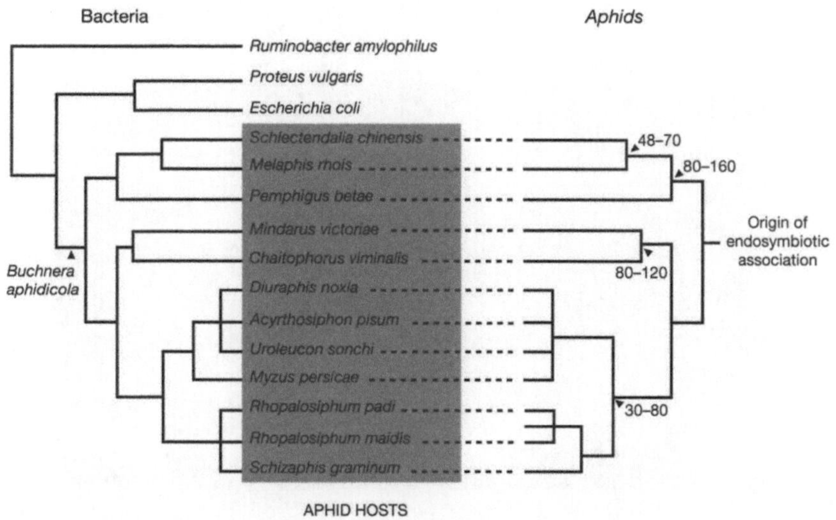

Figure 2.12 Phylogeny of the primary symbionts of aphids (*Buchneria aphidicola* complex) and that of their aphid hosts. The dates, in million of years ago, associated with events in the aphid phylogeny, were estimated from the age of aphid fossils and/or a knowledge of biogeography. (After Moran and Baumann, 1994.)

One group of the Cerataphidini do not have intercellular symbionts but harbour yeast-like cells in their haemocoel and fat body (Figure 2.14). These symbionts appear to belong to the Ascomycota Pyrenomycetes, a group to which the parasitic fungi of insects belong. Thus it is likely that these symbionts evolved from the fungal parasites of aphids. This novel relationship and the presence of secondary symbionts in some species of aphids tends to indicate that the primary symbiotic relationship has been challenged by various micro-organisms. Occasionally the challenger has succeeded to coexist with or displace the original symbiont (Fukatsu, 1994).

2.4.1 Role in upgrading food quality

The study of the role of symbionts in the nutrition of aphids has been facilitated by the development of chemically defined synthetic diets and the use of the antibiotic chlortetracycline and heat shock to produce symbiont-free (aposymbiotic) aphids. Comparative performance of aposymbiotic and symbiotic aphids on synthetic diets and the incorporation of radioactive components in the diet into aphid proteins has been used to determine the role of the symbionts.

Buchnera aphidicola is believed to complement an aphid's diet by synthesizing vitamins, sterols and certain amino acids (Douglas, 1988a,b). In particular, in *Myzus persicae*, the symbionts incorporate inorganic sulphate

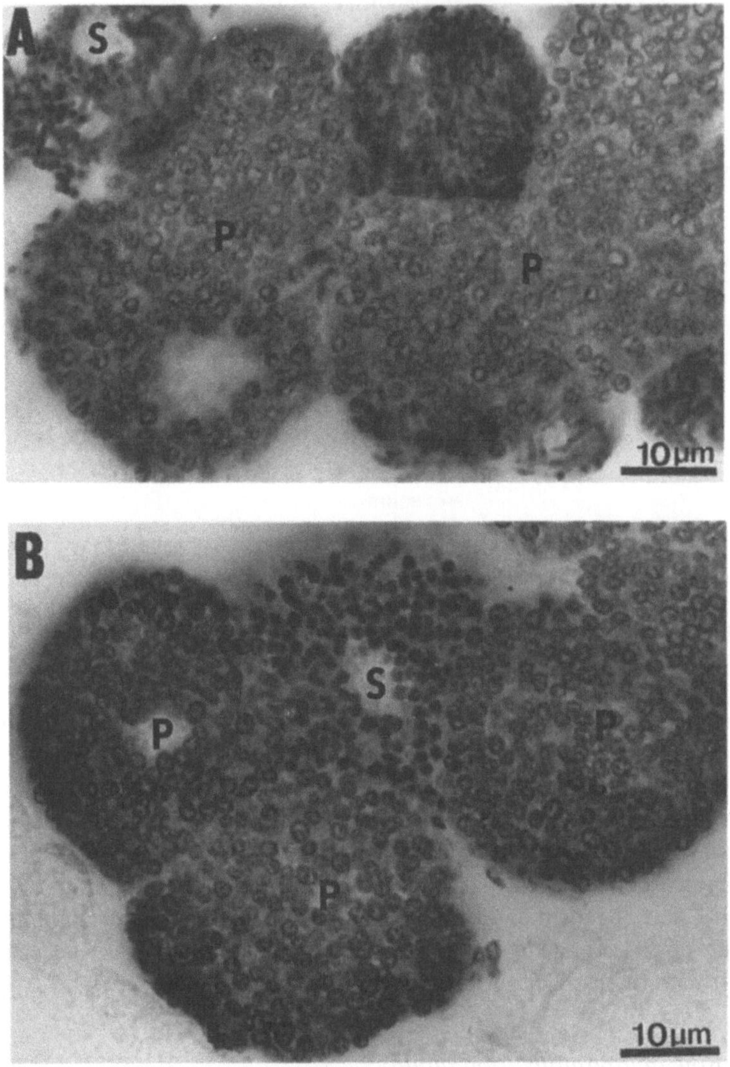

Figure 2.13 Photographs taken using light microscopy of the primary (P) and secondary (S) intracellular symbionts of two disymbiotic aphids: (a) *Stomaphis yanonis* and (b) *Sitobion ibarae*. (Photographed by Dr T. Fukatsu.)

into the sulphur amino acids methionine and cysteine (Douglas, 1988b) and synthesize tryptophan in *Acyrthosiphon pisum* (Douglas and Prosser, 1992), *Schizaphis graminum* (Munson and Baumann, 1993; Lai *et al.*, 1994, 1995). The gene (*trpEG*) responsible for tryptophan biosynthesis in *Schizaphis graminum* is present in the symbiont as four tandem repeats on a circular plasmid. This gene amplification is seen as an adaptation to an

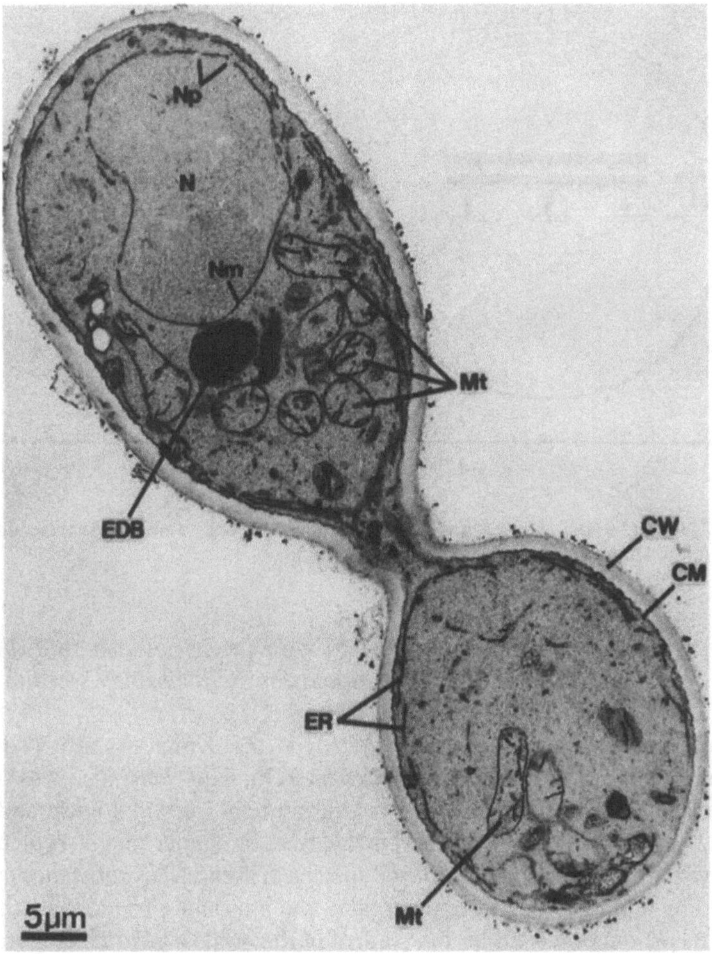

Figure 2.14 Electron micrograph of a yeast-like extracellular symbiont of *Hamiltonaphis styraci*. The nucleus (N), mitochondria (Mt) and thick cell wall (CW) are typical of fungi. (Photographed by Dr T. Fukatsu.)

endosymbiotic association in which the symbiont overproduces tryptophan for its rapidly developing host (Lai *et al.*, 1994). In the slower developing *Schizaphis chinensis* the *trpEG* of the symbiont occurs as a single copy on the bacterial chromosome (Lai *et al.*, 1995). The free amino acid composition of *Acyrthosiphon pisum* remains remarkably balanced and stable whatever symbiotic aphids feed on but not so in aposymbiotic aphids. In aposymbiotic aphids the unessential amino acids like asparagine, aspartic acid, glutamine and proline are present in higher concentrations and essential amino acids like isoleucine, leucine, phenylalanine and threonine in

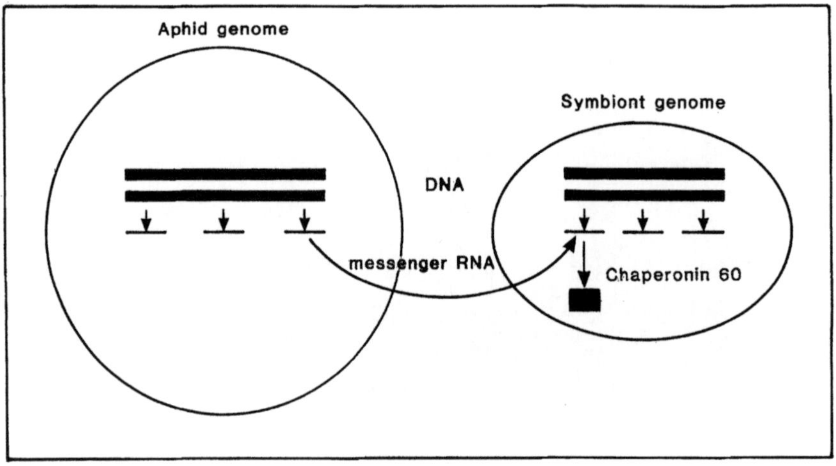

Figure 2.15 Symbionin or chaperonin 60 is produced by the symbiont genome under the control of the aphid genome. (After Ishikawa, 1984.)

lower concentrations. This is consistent with the hypothesis that the symbionts contribute to the nutrition of aphids by synthesizing essential amino acids (Liadouze *et al.*, 1995).

In addition to upgrading dietary nitrogen the symbionts also produce a protein that has a molecular weight of 63 kDa. This substance was initially called symbionin (Ishikawa, 1984) but subsequent immunological and histological studies have revealed that it is the same as chaperonin 60 (cpn 60) and has now been recorded from some 60 species of aphid. This substance controls the folding and assembly of polypeptides and although produced by the symbionts its production is under the control of the aphid's genome (Figure 2.15). Initially symbionin was thought to accumulate in the symbionts of an aphid's embryos and be used in their post-embryonic development (Ishikawa, 1984). However, it is now known that symbionin is mainly found in the symbionts, which suggests it is not used by the host but is important for the functioning of the symbionts (Fukatsu and Ishikawa, 1992b). That present in aphid haemolymph is thought to be from degenerating symbionts extruded by the bacteriocytes (Ponsen, 1972; Verbeek and van den Heuvel, 1994).

In addition to cpn 60 the symbionts also produce a smaller protein chaperonin 10 (cpn 10). The production of this chaperonin, under the control of the host's genome, is characteristic of organelles, like mitochondria and chloroplasts, which have also evolved from endosymbiotic bacteria (Margulis, 1970, 1981). However, cpn 10 is not produced by organelles. The secondary symbionts are thought to be phylogenetically younger than the primary symbionts because they produce proportionally more cpn 10 and

Table 2.1 Chaperonin 60 (symbionin) and chaperonin 10 production of bacteria, symbionts and organelles in relation to the period of time they have lived intracellularly in eukaryotes (after Fukatsu and Ishikawa, 1993)

Time	*Chaperonin*	
	60	*10*
Free-living bacteria	+++	+++
Secondary symbionts	+++	+++
Primary symbionts	+++	+
Organelles	+++	–

show variability in shape and incidence and in the molecular characteristics of their 16S rRNA (Untermann *et al.*, 1989; Fukatsu and Ishikawa, 1993). The decline in the production of cpn 10 from secondary through primary symbionts to organelles indicates that the prokaryote ancestors of the symbionts/organelles have all been subject to similar selection pressures and that the primary symbionts are more organelle-like than the secondary symbionts (Table 2.1). However, *Buchneria* still retains most of the genes that underlie the basic biosynthetic pathways present in free-living bacteria (Moran and Baumann, 1994).

2.4.2 Recycling of nitrogen

Aphids are unusual among insects in lacking Malpighian tubules, which eliminate nitrogenous waste and maintain ionic balance. The nitrogenous waste voided by aphids consists of very small amounts of ammonia and amino acids (Lamb, 1959; Hussain *et al.*, 1974; Whitehead *et al.*, 1992; Sasaki and Ishikawa, 1993; Sasaki *et al.*, 1993). As the amino acid composition of the honeydew (excreta) of aphids and plant sap is similar it has been assumed that most of the amino acids are not of aphid origin but constituents of the diet. Honeydew of aposymbiotic aphids, however, contains more asparagine and glutamine than that of symbiotic aphids feeding on the same diet. This has led to the suggestion that ammonia produced in nitrogen metabolism reacts with glutamic acid, catalysed by glutamic synthetase, to produce glutamine, some of which is converted into asparagine. These two amides are converted into various nitrogenous substances by the symbionts and reutilized by the host. Although most of the waste nitrogen is recycled in this way some is thought to be excreted in the form of arginine (Figure 2.16; Sasaki and Ishikawa, 1993; Sasaki *et al.*, 1993).

2.4.3 Nitrogen economy

As there is physiological evidence that indicates the symbionts of aphids both recycle and upgrade the nitrogen in their food, then symbiotic aphids

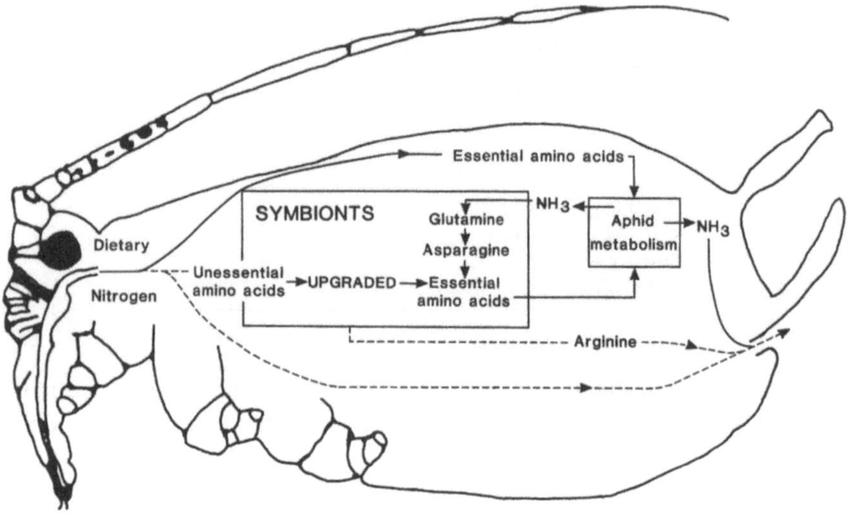

Figure 2.16 Diagrammatic representation of the role of symbionts in upgrading and recycling the nitrogen in the diet of aphids.

should out perform aposymbiotic aphids, especially on unbalanced diets and when the amino acid concentration is low. A good measure of performance in this context is the efficiency of conversion (EC) of food into aphid biomass. In the only case where EC has been measured, symbiotic aphids were more efficient (1.3–1.7 times) than aposymbiotic aphids, particularly on diets that contained very little amino nitrogen (Prosser *et al.*, 1992). Although this agrees with the above predictions the measurements were made over a relatively short period of time (24 h).

Aposymbiotic aphids produce relatively few or no offspring on plants and diets (Mittler, 1971; Sasaki *et al.*, 1991; Douglas, 1992), even on those that have a balanced amino acid composition (Sasaki *et al.*, 1991). Interestingly the offspring production of both symbiotic and aposymbiotic aphids reared on diets is a simple function of the weight they achieve, with the large more fecund than small individuals. The feeding rates of the aphids on diets is considerably less than on plants, especially that of aposymbiotic aphids. Although the developmental time on diets is more prolonged (1.5 times) it is insufficient to compensate for the lowered feeding rate (0.25 times) and as a consequence the aphids are considerably smaller at maturity. Thus diets are a poor substitute for plants. This has led Sasaki *et al.* (1991) to suggest there is a need to develop a diet specifically for aposymbiotic aphids. However, as the performance of both types of aphid is so poor the need may be for a more suitable diet as its acceptance by aphids appears to be confounding the interpretation of the results.

The poor reproductive performance of aposymbiotic aphids, particularly on balanced diets, has led to the proposal that the synthesis of essential amino acids is not the sole function of the symbionts. There is experimental evidence that they provide lipids, other than sterols and triglycerides (see Campbell and Nes, 1983; Rahbé *et al.*, 1993, 1994) essential for embryogenesis, which are absent from diets and in short supply in phloem sap (Douglas, 1989, 1992). It has also been suggested that the symbionts are involved in the detoxification of pesticides and, by implication, secondary plant substances (Amiressami, 1980; Amiressami and Petzold, 1976, 1977; Ball and Bailey, 1978; Jenkins, 1991) and in the production of aphid pigments. However, attempts to confirm that symbionts are involved in these functions have been unsuccessful. As aphids generally perform better on virus infected plants it is interesting that symbionin in the haemolymph of aphids appears to be a key protein in preventing the proteolysis of circulative viruses and in maintaining the infectivity of aphids (van den Heuvel *et al.*, 1994), i.e. the symbionts may have been important in the development of a mutualistic relationship between aphids and plant viruses. If so, it is puzzling why it is symbionin rather than the aphid's cpn 60 that is essential for the protection of the virus.

2.4.4 Presence in male and soldier aphids

Interestingly the males and soldiers of some species of aphids lack symbionts (Tóth, 1933; Fukatsu and Ishikawa, 1992a). As the symbionts are only transmitted through the female line (Buchner, 1965) it is tempting to suggest that symbionts selectively infest the embryos that will become reproductive females and avoid those that are destined to become males or sterile soldiers – the symbiont selection hypothesis. Alternatively, the host aphid controls the infection by symbionts – the host selection hypothesis (Fukatsu and Ishikawa, 1992a).

The little evidence there is tends to favour the host selection hypothesis. During the blastoderm stage of embryogenesis in aphids a well defined group of cells, the presumptive bacteriome, first became apparent. Just prior to gastrulation, cells of the embryos bacteriome are invaded by symbionts from the mother's bacteriome (Tóth, 1937). In the case of those species that have non-feeding (arostrate) males, like *Pemphigus* and *Stomaphis*, the male embryos do not develop a presumptive bacteriome. That is, the embryo appears to control whether it is infected by symbionts – it cannot become infected without first developing the cells to receive the symbionts. The presence of symbionts in males and soldiers that feed and grow (Uichanco, 1924; Ponsen, 1991; Ito, 1994) also tends to support the host selection hypothesis.

Summarizing, aphids feed by inserting tubular stylets into phloem elements located deep within plant tissues. Their food is phloem sap, which is a concentrated solution of simple sugars and a weak solution of mainly unessential amino acids. They have overcome the osmotic problem associated with the high concentration of simple sugars by converting them into more complex sugars, and possibly by diluting the incoming sap by first passing it into a large stomach and drawing water into the stomach from the more dilute content of the adjacent hind gut. They are entirely dependent on the nitrogen in their food to fuel their very rapid growth and reproductive rates. They do this (1) by efficiently processing very large quantities of sap, and (2) by utilizing their symbionts to upgrade the amino acids and recycle the nitrogenous waste. How aphids manage to process such large quantities of sap and extract the nitrogen so efficiently is unknown. However, they apparently do not utilize their 'filter chamber', which contrary to expectation is likely to function to dilute rather than concentrate the food.

3 Host specificity and speciation

Most aphids are autoecious (Figure 3.1), living on one or a few species of a particular genus of plants (Eastop, 1973). About 10% are heteroecious (Figure 3.1), spending autumn, winter and spring on a primary host plant, and the summer on a secondary host plant(s), which is rarely closely related to the primary host. For example the primary host of the rosy apple aphid, *Dysaphis plataginea*, is apple (Rosaceae) and its secondary host is plantain (*Plantago lanceolata*, Plantaginaceae); for the currant-yellow rattle aphid, *Hyperomyzus rhinanthi*, it is red currant (*Ribes rubrum*, Grossulariaceae) and greater yellow rattle (*Rhinanthus major*, Scrophulariaceae), respectively. Although heteroecious species are classified as polyphagous it is noteworthy that most of them live on one species of plant at a time, i.e. they are sequentially monophagous. The type of heteroecious life cycle shown by the black bean aphid, *Aphis fabae*, and the peach-potato aphid, *Myzus persicae* (Figure 3.1), in which there is a wide range of secondary host plants, is very rare. Even in these highly polyphagous species it is possible that particular genotypes do best on certain species of plant (Takada, 1979; Weber, 1985a–c), i.e. most aphids show a very high degree of host specificity, even the heteroecious species.

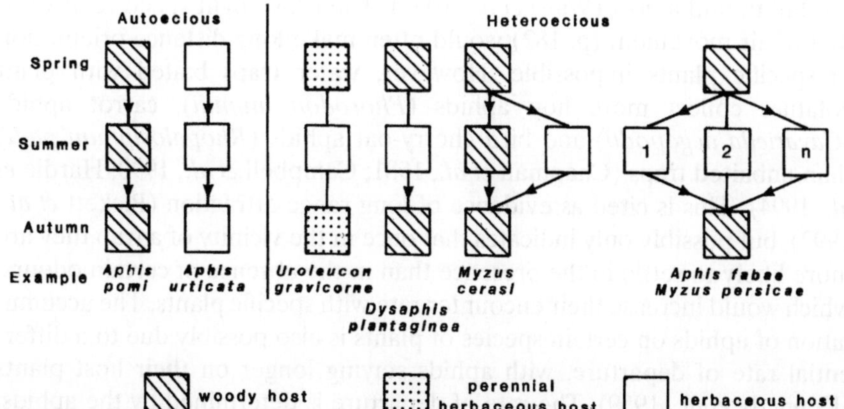

Figure 3.1 Life cycles of aphids (n, number of secondary hosts).

Why are aphids so host specific? This question can be answered on one of two levels: (1) by describing the behavioural responses, and chemical and morphological cues that enable an aphid to discriminate host from non-host and between nutritionally good and poor parts of their host plants, i.e. the proximate factors that determine host plant selection, or (2) by addressing the ultimate or evolutionary reason, i.e. the selection pressures that have shaped host plant usage and speciation in aphids.

3.1 Proximate factors in host selection

Plants are colonized primarily by flying aphids that, because of their low speed, have little control over the direction of their flight. However, once within the layer of relatively still air around vegetation, called the 'boundary layer' (p. 183), aphids can control their landing on plants and respond to either olfactory or visual cues, or both. After settling, an aphid recognizes a potential host by the structure and chemistry of its surface and internal tissues.

3.1.1 Host location

Aphids respond to visual cues, especially yellow. Colour is a good indicator of the nutritive status of a plant, as both highly nutritious young and senescent foliage tends to be yellower than the nutritionally poorer mature leaves, but a poor indicator of a plant's taxonomic status (Kennedy *et al.*, 1961). In spite of the distinctive odour of many plants and strong physiological evidence that aphids can perceive plant volatiles (Nottingham *et al.*, 1991b; Pickett *et al.*, 1992; Visser and Piron, 1994, 1995; Hardie *et al.*, 1995) there is no ecological evidence that aphids locate plants from a distance using odour. In fact host location is very hazardous for aphids as 99 out of 100 fail to find a host (Ward *et al.*, 1998). Their low flight speed relative to that of air movement (p. 182) would often make long distance orientation to specific plants impossible. However, water traps baited with plant volatiles collect more hop aphids (*Phorodon humuli*), carrot aphids (*Cavariella aegopodii*) and bird cherry-oat aphids (*Rhopalosiphum padi*) than unbaited traps (Chapman *et al.*, 1981; Campbell *et al.*, 1990; Hardie *et al.*, 1994). This is cited as evidence of long range attraction (Pickett *et al.*, 1992), but possibly only indicates that once in the vicinity of a trap they are more likely to settle in the presence than in the absence of certain odours, which would increase their encounter rate with specific plants. The accumulation of aphids on certain species of plants is also possibly due to a differential rate of departure, with aphids staying longer on their host plants (Kennedy *et al.*, 1959). The rate of departure is determined by the aphids' response to external and internal features of the plant.

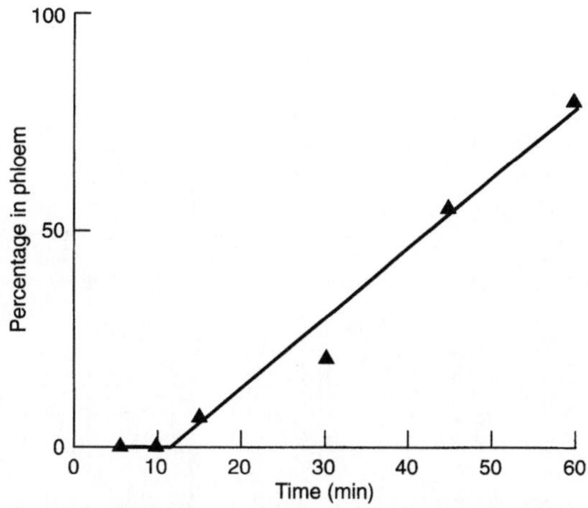

Figure 3.2 Percentage of *Aphis fabae* with stylets inserted in a phloem element in relation to the time from start of penetration. (After Henning, 1966.)

3.1.2 Host recognition

After alighting on a plant aphids walk over the surface testing it with their antennae and by probing it with their mouth parts. The antennae have many sensilla, amongst which there are some whose structure and electrophysiological responses indicate that they are used in close-range chemoreception or gustation and the perception of the leaf surface (p. 8). Sensilla at the tip of the proboscis are sensitive to tactile and chemical stimuli. The chemoreceptors described by Tarn and Adams (1982) probably detect chemicals on the surface of plants. Although the mandibular stylets contain nerves (Forbes, 1966; Parrish, 1967) there is neither morphological nor unequivocal electrophysiological evidence for chemoreceptors at the tips of the stylets. However, in the dorsal wall of the cibarial pump, there is a group of sense endings called the epipharyngeal organ, that has the typical structure of a contact chemoreceptor (Wensler and Filshie, 1969; Mclean and Kinsey, 1984). This organ is in direct contact with fluid in the food canal and possibly detects chemicals in the ingested sap (Figure 2.7).

While walking over and probing the surface of a plant, an aphid obtains information about the physical properties and chemistry of the surface, and possibly the internal chemistry of the plant. This initial investigation of the surface involves little or no stylet penetration, but often enables an aphid to sense the suitability of a plant within 60 s. This is the time it takes an aphid to penetrate the epidermis and exude saliva on the surface (Harris, 1977; Pollard, 1977). Once an aphid settles it probes deep into the plant.

It takes approximately 40 min for the black bean aphid (*Aphis fabae*) to reach the phloem of its host plant (Henning, 1966; Figure 3.2), and possibly considerably longer for aphids that feed on the phloem elements situated

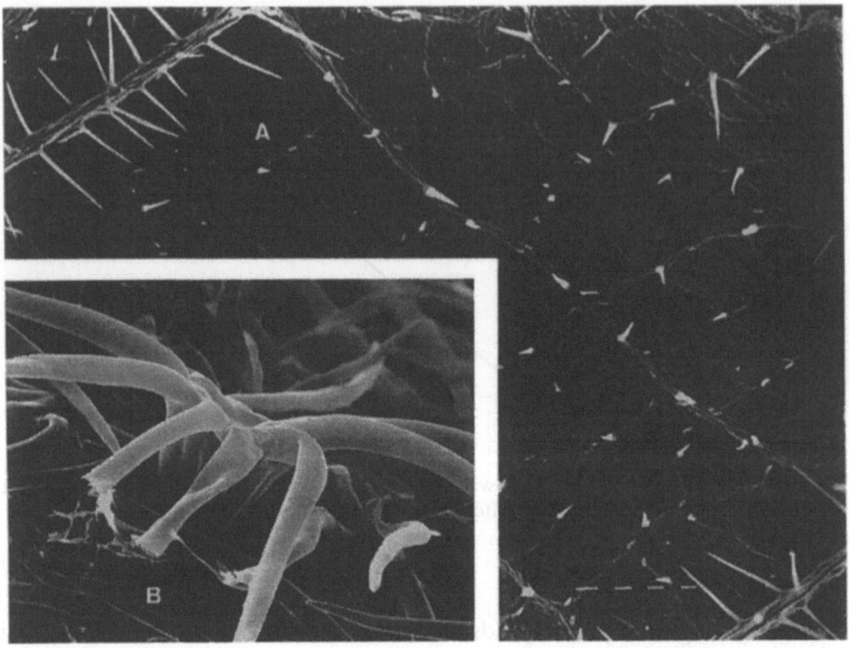

Figure 3.3 Scanning electron micrographs showing the simple hairs positioned along the leaf veins of (a) *Tilia platyphylla* and a stellate hair on a leaf of (b) *Tilia petiolaris*. (Photographed by M. Jukes.)

deep within the woody tissues of plants. As reaching the phloem takes so long aphids would benefit from being able to recognize a suitable plant from the nature of its surface. Volatile substances and features on the surface of a plant, such as the nature of the waxes (Klingauf, 1972; Avé *et al.*, 1987; Birch *et al.*, 1990; Visser and Piron, 1994, 1995) or hairiness, could be very important in the initial stages of selection.

3.1.3 Morphological characteristics

The leaves and stems of many species of plants are covered with small epidermal hairs and/or hooks. The function of this pubescence has rarely been investigated and it is unknown whether host-specific aphids respond to the pubescence in selecting their host plants. The undersides of the leaves of limes like *Tilia petiolaris* appear white because of the dense covering of stellate hairs (Figure 3.3). Adult lime aphids are able to feed on the leaves of this lime because there are few hairs on their preferred feeding site, the main veins. However, newly born nymphs starve because the dense barrier of interdigitating stellate hairs prevents them from reaching the smaller veins. When the pubescence is removed young aphids do as well on this

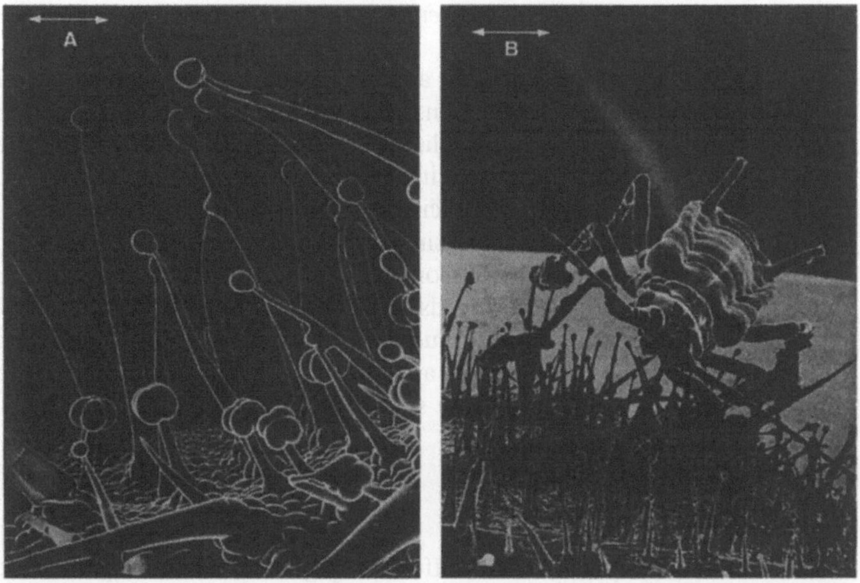

Figure 3.4 Scanning electron micrograph of the two types of glandular hairs on the leaves of *Solanum berthaultii* (a) and of an aphid entangled in the secretion of the glandular hairs (b). Scale bar is 97 μm in (a) and 290 μm in (b). (Photographed by Dr R. W. Gibson.)

species of lime as on their native host, *Tilia platyphylla*. The underside of the leaves of the silver lime (*Tilia tomentosa*) is covered with stellate hairs 25–35 μm long. In Turkey the leaves of this species provide food for an undescribed species of lime aphid whose proboscis is 20–40 μm longer than that of the European lime aphid: the specialist aphid has overcome the barrier by virtue of its longer reach (Carter, 1982; Blackman and Eastop, 1994). In the case of alder the high trichome density on the underside of the leaves of *Alnus incana*, in reducing aphid survival, render this species of alder resistant to colonization by *Pterocallis alni* (Gange, 1995).

The foliage of the wild potato (*Solanum berthaultii*) is covered with two types of glandular hairs. Type A has a short stalk and a four-lobed head that ruptures on contact and releases a quick-setting fluid, and type B has a longer stalk and exudes a sticky droplet at the tip (Figure 3.4a). The exudate of the type B hairs supposedly entangles aphids that then struggle and rupture type A glands, whose contents physically trap the aphids (Figure 3.4b) (Tingey and Laubengayer, 1981; Gibson and Pickett, 1983). It is the high density of these glandular hairs on the foliage, and not the presence of high levels of glycoalkaloids in the tissues of species of wild potatoes, which accounts for their resistance to aphids (Tingey and Sinden, 1982). However, many plants that have glandular hairs have specific aphids, e.g. the undersides of the leaves of *Geranium macrorrhizum* have a dense covering of such hairs that *Amphorophora tuberculata* avoids triggering by moving very slowly.

Although latex canals (lacticifers) and resin ducts are commonly located around the phloem (Esau, 1965; Fahn, 1979) and seem ideally positioned to intercept probing aphids, whose delicate stylets would be clogged by the latex, aphids can bypass these lacticifers (Botha *et al.*, 1975a,b). However, *Uroleucon pseudoambrosiae* can elicit latex release onto the surface of its host plant, lettuce (*Lactuca sativa*), simply by walking on the surface of the plant. This is particularly obvious on the reproductive structures of the plant, such as bracts and flower heads, which appear to be better defended with lacticifers than the non-reproductive structures like stalk and leaves. Although relatively host specific the aphid frequently becomes mired in the latex and a high proportion of those so trapped die (Figure 3.5; Dussourd, 1995).

3.1.4 Chemical characteristics

In addition to the physical defence afforded by glandular hairs of plants, the hairs may also produce and release insecticides that are highly effective against aphids. For example, nicotine, nornicotine, and anabasine are produced by the glandular hairs of species of tobacco (*Nicotiana*) and *Petunia* (Thurston *et al.*, 1966; Thurston, 1970) and 2-trideconone by those of the wild tomato (*Lycopersicon hirsutum f. glabratum*) (Williams *et al.*, 1980). The type B glandular hairs of potato even release the aphid alarm pheromone, (*E*)-β-farnesene, and maintain a concentration around the foliage sufficient to result in an alarm response in aphids and make them avoid the foliage (Gibson and Pickett, 1983) and the type IV trichomes of tomato exude acylsugars that deter *Macrosiphum euphorbiae* and *Myzus persicae* from feeding (Rodriguez *et al.*, 1993).

During superficial probing an aphid imbibes small quantities of plant sap (McLean and Kinsey, 1968) specific chemical components of which could stimulate the gustatory epipharyngeal organ. Many aphids are known to respond to specific secondary plant metabolites. This has been established by making non-host plants (Figure 3.6) and synthetic diets acceptable by the addition of secondary plant substances. The glucoside phlorizin, present in the leaves of apple and other Pomoideae, promotes colonization by the apple aphids *Aphis pomi* and *Rhopalosiphum insertum* (Klingauf, 1971). Similarly, sinigrin, characteristic of Cruciferae, induces the cabbage aphid (*Brevicoryne brassicae*) to settle and feed (Wensler, 1962; Nault and Styer, 1972); and rutin and quercetin have the same effect on the dock aphid (*Aphis rumicis*) (Herger, 1975).

Aphids that reach the phloem of resistant cultivars of their host plant tend to cease feeding shortly after the phloem is penetrated (Nielson and Don, 1974; Harrewijn, 1990). Similarly, aphids feeding on non-host plants initially

Figure 3.5 Photographs of nymphs (a) and a winged adult (b) of *Uroleucon pseudoambrosiae* entangled in the latex of their host plant, *Lactuca sativa* (Photographed by D. Dussourd.)

ingest phloem sap at normal rates, but then suddenly cease feeding, withdraw their stylets and leave the plant (Kloft, 1977). Such observations have led to the suggestion that the phloem sap of some resistant and non-host species is nutritionally unsuitable. However, grafting does not influence the

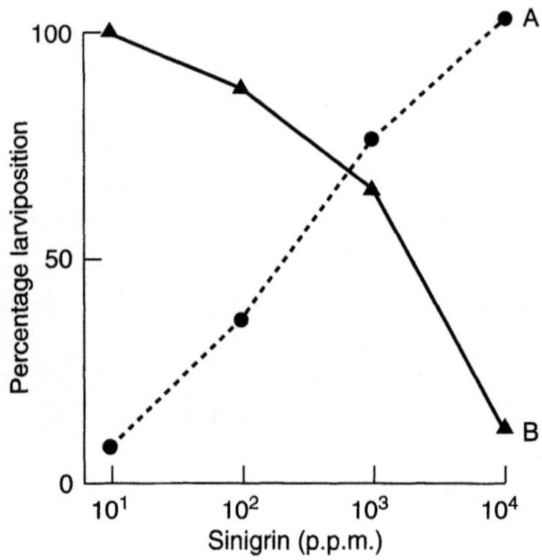

Figure 3.6 Percentage of *Brevicoryne brassicae* (A) and *Acyrthosiphon pisum* (B) producing offspring on cuttings of broad bean in relation to the concentration of the sinigrin solution supplied to the cuttings. Broad bean is a good host for *Acyrthosiphon pisum*. (After Nault and Styer, 1972.)

growth and reproduction of *Aphis gossypii* on either resistant or susceptible muskmelon plants. The aphid still does 10 times better on the 'susceptible' part of a graft, be it the scion or the stock, than on the 'resistant' part. Since neither resistance nor susceptibility is translocated across a graft union and the aphid finds no difficulty in locating the phloem of resistant plants, it is thought that other parts of the phloem such as the companion cells or cell walls produce a phytoalexin on contact with aphid stylets or saliva (Nielson and Don, 1974; Kennedy and Kishaba, 1977). Recently small proteins called lectins have been implicated in this form of resistance, e.g. wild brassica species contain a chitin binding lectin, which at high concentrations is an effective defence against the brassica aphid specialist, *Brevicoryne brassicae*. This lectin possibly acts by binding onto the chitin surrounding the food canal in the stylets and the fore gut of the aphid (Cole, 1994a,b).

Secondary plant substances have been viewed as 'flags' by which host-specific aphids recognize their host plants (van Emden, 1978), e.g. the cabbage aphid (*Brevicoryne brassicae*) will not settle and feed on plants lacking sinigrin (Figure 3.7) but can be reared on leaves of broad bean provided they are impregnated with sinigrin (Wensler, 1962). However, although it does not deter colonization, sinigrin has a negative effect on the growth of the polyphagous aphid *Myzus persicae* (van Emden, 1972) (Figure 3.8). High concentrations of the flavenoids catechin and epicatechin in the late buds and flowers of rose are also thought to deter the rose aphid (*Macrosiphum rosae*) from feeding on these structures (Miles, 1978, 1985; Peng and Miles, 1988). The broom (*Aphis cytisorum*) and ragwort

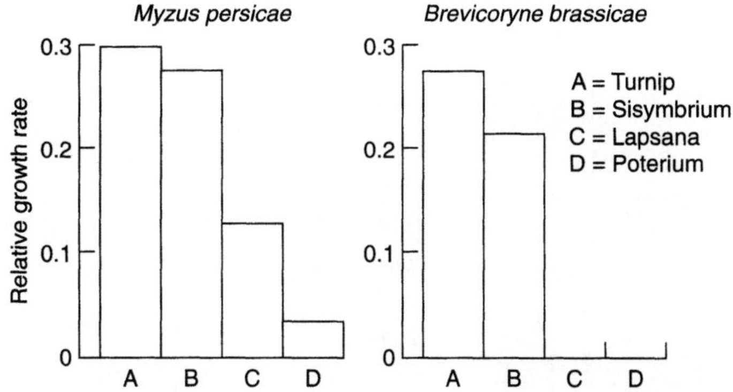

Figure 3.7 The relative growth rate of *Myzus persicae*, a generalist, and *Brevicoryne brassicae*, a specialist aphid, on plants that have (A + B) and lack (C + D) the secondary plant substance sinigrin. (After van Emden, 1972.)

aphids (*Aphis jacobaeae*) prefer to feed on the flowering stem of their host plants, which contain more alkaloids than other parts of the plants (Figure 3.9), but tend to infest only those plants that have a below average alkaloid content (Wink *et al.*, 1982; Vrieling *et al.*, 1991). Similarly, the hydrogen cyanide present in cyanogenic plants of white clover (*Trifolium repens*) is thought to explain why fewer *Aphis craccivora* are on these plants than on acyanogenic phenotypes of clover (Dritschilo *et al.*, 1979); high levels of phenol glucosides could account for the low suitability of some poplar leaves for galling by *Pemphigus betae* (Zucker, 1982); the high alkaloid content of 'bitter' lupins probably makes them resistant to aphids (Figure 3.10) (Wegorek and Krzymanska, 1971; Wink *et al.*, 1982); and the high levels of hydroxamic acids in the early development stages of certain grasses account for the poor population development on them of *Metopolophium dirhodum* (Figure 3.11), *Rhopalosiphum maidis* and *Schizaphis graminum* (Long *et al.*, 1977; Argandona *et al.*, 1980, 1983; Givovich *et al.*, 1994).

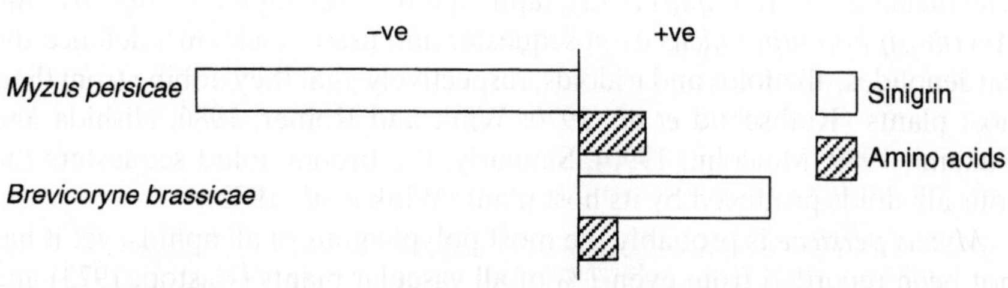

Figure 3.8 The relative effects of sinigrin and amino acid concentration on the growth of a generalist aphid, *Myzus persicae*, and the specialist, *Brevicoryne brassicae*. (After van Emden, 1972.)

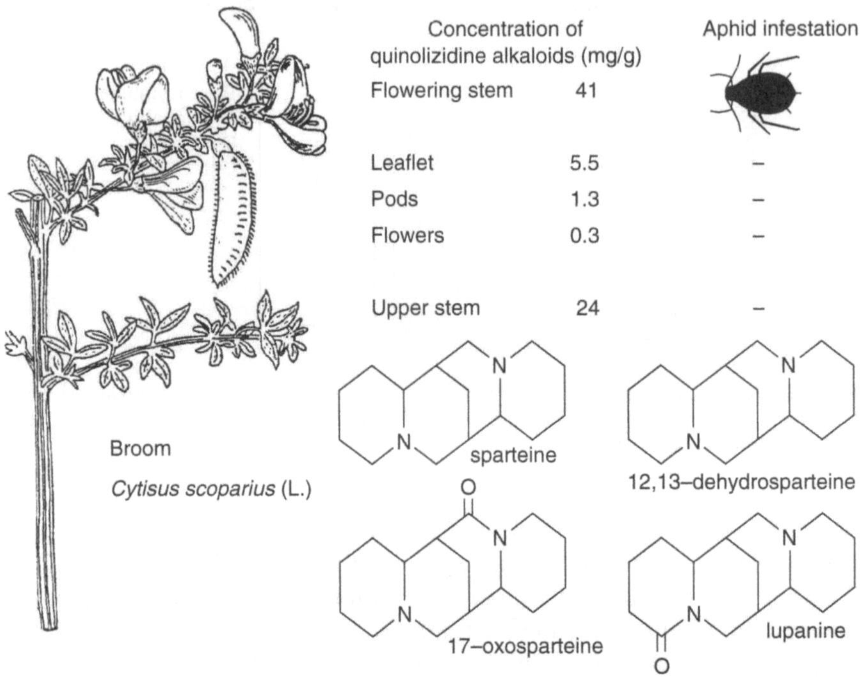

Figure 3.9 The total concentration and distribution of the quinolizidine alkaloids, sparteine, 12,13-dehydrosparteine, 17-oxosparteine and lupanine in broom in relation to aphid infestation. (After Wink *et al.*, 1982.)

Although the presence of secondary plant substances in tissues other than the phloem is well documented, there is growing evidence to indicate that these substances or their glycosides are to be found in phloem sap (Eschrich, 1970; Wink *et al.*, 1982; Dreyer *et al.*, 1985; Hartmann *et al.*, 1989; Molyneux *et al.*, 1990; Montllor, 1991; Wink and Witte, 1991; Givovich *et al.*, 1994). Certain defensive chemicals like the non-protein amino acid canavanine and phenols have been recorded from phloem sap (Ziegler, 1975; Dixon, 1975a). The oleander aphid (*Aphis nerii*), lupin aphid (*Macrosiphum albifrons*) and *Acyrthosiphon nipponicus* even sequester and use in their own defence the cardenolides, alkaloids and iridoids, respectively, that they imbibe from their host plants (Rothschild *et al.*, 1970; Wink and Romer, 1986; Nishida and Fukami, 1989; Malcolm, 1990). Similarly, the broom aphid sequesters the four alkaloids produced by its host plant (Wink *et al.*, 1982).

Myzus persicae is probably the most polyphagous of all aphids, yet it has not been recorded from even 1% of all vascular plants (Eastop, 1973) and particular genotypes of this aphid are associated with particular species of plants (Takada, 1979). This aphid has a marked preference for highly nutritious plants (van Emden and Bashford, 1971) perhaps because smaller

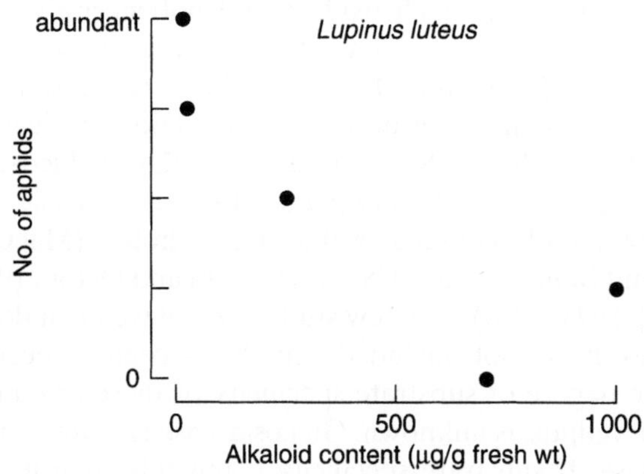

Figure 3.10 Abundance of aphids on lupin in relation to its alkaloid content. (After Wink *et al.*, 1982.)

quantities of toxicant are ingested, which are quickly detoxified, whereas the larger quantities ingested when feeding on poor-quality hosts are not all quickly detoxified, as has been suggested for *Eriosoma* feeding on various resistant varieties of apple (Sen Gupta and Miles, 1975). In selecting for high-quality food, polyphagous aphids like *Myzus persicae* possibly compensate for the cost of detoxifying secondary plant metabolites, which they do in several stages by means of enzymes. Their saliva contains peroxidase

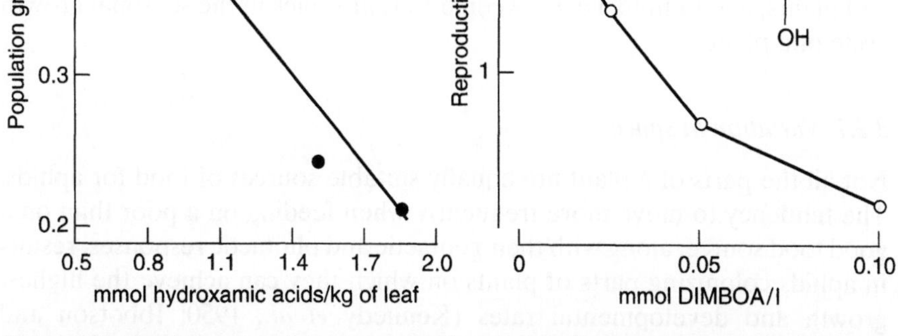

Figure 3.11 Population growth rate on cereals and reproductive rate on synthetic diets of *Metopolophium dirhodum* in relation to (a) the concentration of hydroxamic acids in the leaves of its cereal hosts and (b) the concentration of the hydroxamic acid DIMBOA in the diet. (After Argandona *et al.*, 1980.)

and polyphenol oxidase, which oxidase and polymerize a wide range of allelochemicals. This first step in detoxification occurs to some extent before the food is ingested (Peng and Miles, 1988; Leszczynski *et al.*, 1993). Once ingested further detoxification occurs via the action of glucosinolase, and isoenzymes of glutathione S-transferase and UDP-glucose transferase, which convert lipophilic secondary plant substances via a semi-hydrophilic stage to water soluble conjugates that are excreted (MacGibbon, 1975; MacGibbon and Benzenberg, 1978; Leszczynski and Dixon, 1990; Leszczynski *et al.*, 1992, 1993, 1994). The few studies that have been done on detoxifying enzymes have not included any host-specific species of aphid. Therefore, the degree of substrate specificity of these enzymes, especially in host-specific aphids, is unknown. Glucosinolase is more substrate specific than glucosidase. If substrate specificity is the rule then it is yet another barrier to host shifts, except possibly to plants in the same genus or family, i.e. paucity of genetic variation in detoxifying enzymes may constrain or bias the evolution of host plant associations. Detoxification would also appear to pose a problem for the seasonal host transfer seen in host-alternating species of aphids, e.g. the carrot-willow aphid, *Cavariella aegopodii*, which moves seasonally between plants belonging to the Salicaceae and Umbelliferae that differ greatly in the nature of their secondary plant substances. This is an area worthy of further study.

3.2 Food quality

The growth and reproduction of aphids are dependent upon the state of growth, or level of soluble nitrogen in their host plants (Kennedy *et al.*, 1950; Mittler, 1958a; Dixon, 1970a). There is more nitrogen in the phloem sap of plants that are growing, flowering or senescent, because nutrients are then being translocated. The sap is not as nutritious when plants cease growing and their leaves are mature (Figure 3.12). Thus the food of aphids shows marked variations in quality in space, i.e. between different parts of a plant and plants, and in time, i.e. between different stages in the seasonal growth cycle of a plant.

3.2.1 Variation in space

Not all the parts of a plant are equally suitable sources of food for aphids. The tendency to move more frequently when feeding on a poor than on a good food source, along with their geotactic and photactic responses, results in aphids colonizing parts of plants on which they can achieve the highest growth and developmental rates (Kennedy *et al.*, 1950; Ibbotson and Kennedy, 1950). For example, early in the season *Aphis spiraecola* preferentially infest the most rapidly growing shoots of *Spiraea* but then density-dependent dispersal results in the shoots all becoming equally infested per

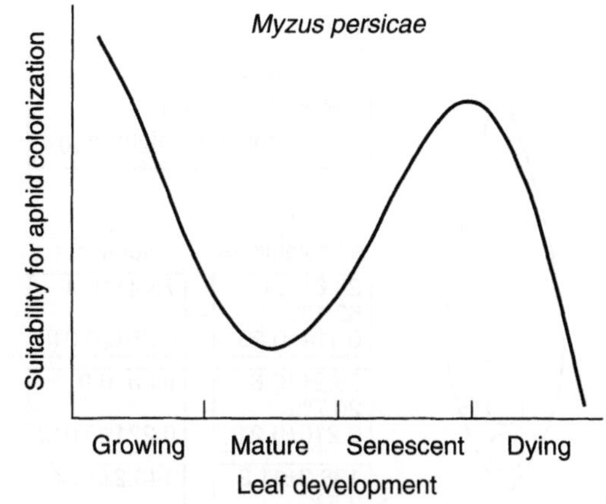

Figure 3.12 Relative suitability for colonization by *Myzus persicae* of leaves of successive stages of development. (After Kennedy *et al.*, 1950.)

unit length (Osawa, 1992), i.e. the distribution of aphids on a plant depends both on spatial variation in the quality of plant sap and the intensity of competition for this resource.

That the best feeding sites are often in short supply has been elegantly shown by Whitham (1979) who studied leaf selection in galling aphids. *Pempigus betae* produces the largest and most fecund galls on the basal part of the largest leaves of cottonwood (*Populus angustifolia*). There are fewer large leaves than small leaves and when two stem mothers (p. 129) attempt to settle on the same leaf, a kicking and shoving contest follows that can last for 2 days. The larger aphid usually wins, displacing the loser to a more distal position along the midrib. However, the loser's chances of initiating a gall in this position are less and, even if successful, it will have fewer descendants than the aphid in the prime basal position. The potential genetic fitness of a third aphid attempting to colonize a leaf is even lower (Figure 3.13), and so by means of territorial behaviour, large aphids can deny other individuals access to the best leaves (Witham, 1979).

In *Epipemphigus niisimae* there is also evidence of a shortage of highly suitable leaves for galling. When there is a large number of galls on a twig, there are more likely to be one or more dead stem mothers (intruders) in some of the galls than where there are few galls per twig (Figure 3.14). Like *Pempigus betae*, *Epipemphigus niisimae* also fights, this time for possession of a gall, and a large individual is capable of using its stylets to stab and kill a small intruder (Aoki and Makino, 1982).

For *Pempigus betae* the size of a leaf reflects its quality. However, each additional aphid founding a gall on a leaf, apart from having a lower fitness, also reduces the fitness of the aphid(s) already present. As a consequence, the average fitness of aphids galling a leaf declines with an increase in the

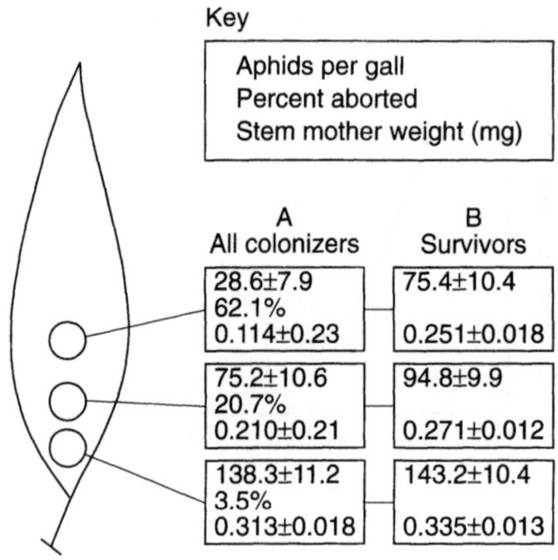

Figure 3.13 Effect of gall position on relative fitness when three *Pemphigus betae* stem mothers colonize the same leaf. (After Whitham, 1978.)

number of competitors. A large leaf, 14.6 cm long, with three galls, will sustain the equivalent number of aphids per gall as a medium leaf of 12.3 cm with two galls or a small 10.2 cm leaf with one gall (Figure 3.15). The observed relationship between number of galls per leaf and leaf size follows this closely, showing that the average fitness of aphids on a leaf is independent of leaf size (Whitham, 1980). This appears to agree with the ideal free distribution hypothesis put forward by Fretwell and Lucas (1969), which postulates that as population density increases, the habitat patches should be occupied in a sequence that reflects the ranking of their value as feeding sites, i.e. an aphid should gall a small leaf, or form the

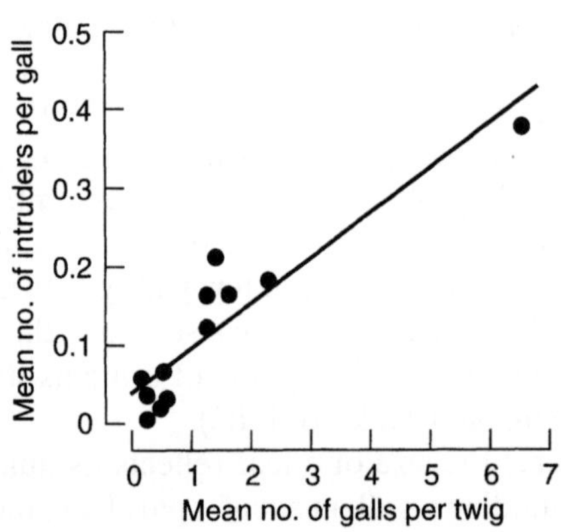

Figure 3.14 Mean number of additional (intruder) stem mothers per gall of *Epipemphigus niisimae* in relation to the number of galls per twig. (After Aoki and Makino, 1982.)

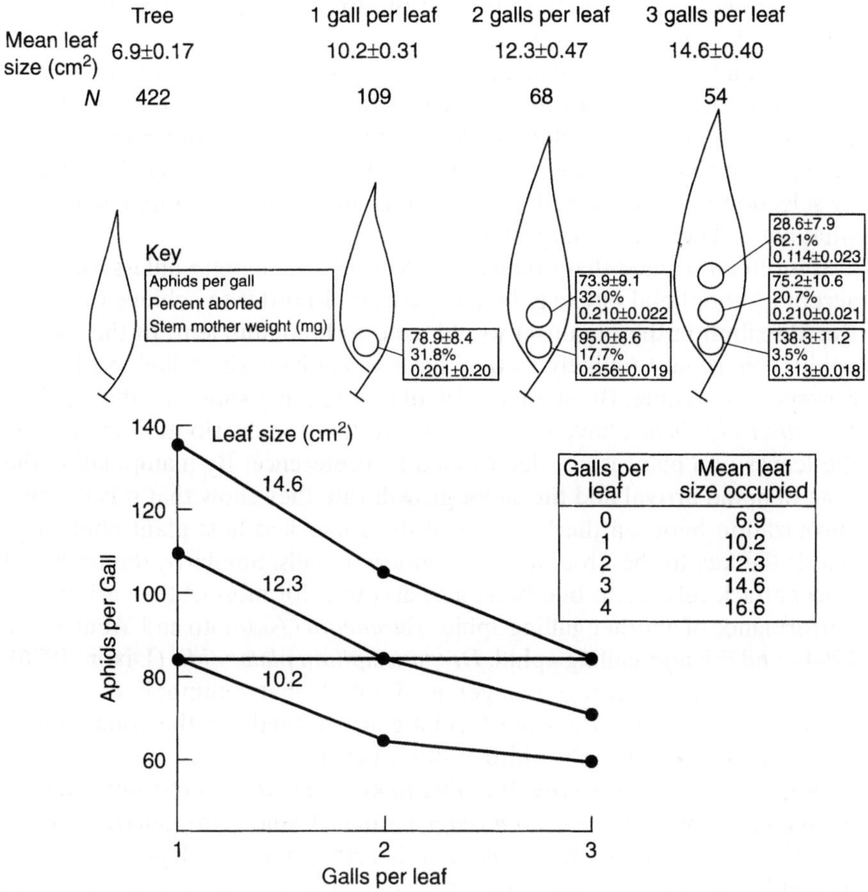

Figure 3.15 Distribution of galls and estimated fitness of *Pemphigus betae* in relation to leaf size and number of galls per leaf. *N*, sample size. See text for details. (After Whitham, 1980.)

second gall on a medium-sized leaf, rather than the fourth gall on a large leaf. It is individual fitness that is important, however, not average fitness. If efficient at maximizing their fitness, then it is puzzling why aphids that initiate the third gall on a large leaf do not colonize unifested small leaves, which are just as abundant, and on which they could give rise to 3 times as many descendants.

Dunn's (1960) detailed observations on the rapid gall initiation in other species of *Pemphigus* should also be borne in mind when interpreting the distribution of galls. Aphids initiate galls in the early stages of leaf enlargement, when the leaves are growing very rapidly. When selecting a leaf at this stage an aphid is possibly not responding to its size or even nutritive status, but the rate at which it is growing. The faster the tissue is growing, the more effective an aphid is at galling it (Forrest and Dixon, 1975). Generally, as leaves grow faster at the base than at the tip, the best position to initiate galls is at the base of the large leaves. Then, too, the leaves do not all emerge from the bud together. The largest leaves of poplar are located at the middle of each shoot and are not the first leaves to emerge from the buds. Poplar aphids hatch just *after* bud burst (Whitham, 1978). As in other species, the timing of egg hatch of the poplar aphid (p. 43) has possibly been determined by selection to coincide with the unfurling and initial very rapid growth of what are to become the largest leaves.

It is often assumed that organisms select the best habitat unless forced to accept a suboptimal one, e.g. by competition. In order to achieve this ideal free distribution the organism would need to be omniscient. In the case of aphids this is very unlikely as they have a relatively short life and limited powers of searching. Burstein and Wool (1993) using another galling aphid, *Smynthurodes betae*, have elegantly shown that the distribution of galls on the leaves of a plant is not determined by preference. By manipulating the time of aphid arrival and the shoot growth rate they show that it is the synchronization between the hatching of the galler and host plant phenology that is the key to the observed distribution of galls. Similarly, the timing of insect attack relative to bud burst also accounts for most of the variation in performance of another galling aphid, *Tetraneura* (Akimoto and Yamaguchi, 1994), and the non-galling aphid, *Drepanosiphum platanoidis* (Dixon, 1976).

Eriosoma yangi belongs to a genus of aphids that commonly form galls. However, this species does not form a gall, but feeds on the young leaves of elm until it reaches the third instar when it then seeks out the gall of another species of *Eriosoma*. It is able to kill the owner of the gall and take it over for its own use. *Eriosoma yangi* has been found in the galls of as many as four other species of *Eriosoma* and therefore does not depend upon the availability of only one other aphid species (Akimoto, 1981). It appears that, in certain circumstances, to usurp a gall of another species has advantages over inducing one's own gall. This would be favoured by an abundance of galls and intraspecific gall usurpation of the type reported for *Epipemphigus niisimae* (p. 39) could have been the first step in the evolution of gall parasitism.

Aphids do not merely imbibe phloem sap. They simultaneously secrete substances into plants that rapidly affect the plant, in many cases to the aphids' advantage, i.e. they can manipulate host quality. Both the black bean aphid (*Aphis fabae*) and the cabbage aphid (*Brevicoryne brassicae*) can change the metabolism of their host plants, and although there are no outward signs, there are local changes in the plant around the feeding area that result in the development of larger and more fecund aphids (Way and Cammell, 1970; Dixon and Wratten, 1971). A few aphids, like *Myzus ligustri* and *Periphyllus acericola*, induce a localized yellowing of the tissues of mature leaves, which also makes the tissues attractive to other species of aphids, which feed on the same hosts but cannot induce senescence.

Some aphids are able to induce plant tissues to grow around and enclose them in complex and often colourful galls. As well as protecting aphids from the weather and general predators, galls offer a richer food supply than ungalled leaves and, very significantly, the production of galls also extends the usual period for which high-quality food is available (Forrest, 1971). The shapes of the very different galls induced by different species of *Pemphigus* on poplar are determined by the probing patterns of the aphids that initiate the galls (Dunn, 1960). Even the position of the ostiole, through which the aphids subsequently leave, is determined by the aphid making 10 times as many probes in the area immediately around the future ostiole than in the area of the ostiole itself (Schwarzbach, 1962). The ostiole opens when the gall begins to dry out (de Geer, 1773) and will close when dampened (Frank, 1896), as a consequence of changes in the water content of the intercellular pectic material (Rohfritsch, 1966).

Miles (1968a,b) proposed that the form of the gall can be attributed to a single chemical substance, indole acetic acid, present in the saliva, and the behaviour of the insect and the nature of the plant tissue attacked are also important. Certainly the behaviour of the gall-forming aphid is important in aphids such as *Pemphigus*. *Dysaphis devecta* or *Dysaphis plataginea*, if they are confined to the stem of an apple seedling, can induce leaves several centimetres away to develop leaf galls characteristic of each species, but it is unlikely that the marked differences in form and colour of these leaf galls on the same host are attributable to a single substance. More probably the saliva differs in each species and therefore in the way it affects leaf development (Forrest and Dixon, 1975) and improves the quality of the phloem sap for the aphids.

3.2.2 *Variation in time*

Plant growth varies during the course of a year, especially in temperate regions where there is a pronounced effect of the seasons. Bud burst, leaf growth, flowering/fruiting and leaf senescence are reflected in changes in the quality of phloem sap (Figure 3.16), which affects both the size and fecundity of aphids (Dixon, 1970a). Thus aphids should synchronize their development with that of their host plants.

As indicated above, the success of a galling aphid in synchronizing its egg hatch with bud burst determines its fitness. In the non-galling sycamore aphid, *Drepanosiphum platanoidis*, average egg hatch times appears to be well synchronized with average bud burst time of sycamore, *Acer pseudoplatanus*. However, egg hatch on individual trees, especially early or late bud burst trees, is unlikely to be synchronized with bud burst. Close synchronization results in the development of large, quickly maturing and highly fecund adults, with those hatching earlier or later being less fit (Dixon, 1976; Figure 3.17). A high incidence of inter tree movement

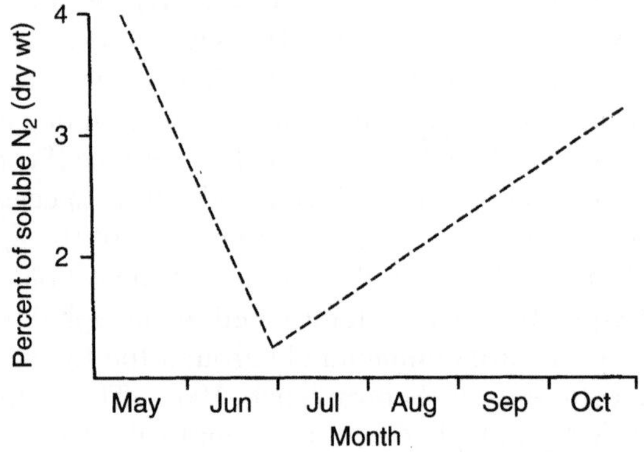

Figure 3.16 Seasonal changes in the percentage soluble nitrogen in the leaves of sycamore.

(Dixon, 1969) would reduce adaptation to particular trees. In other species like *Kaltenbachiella japonica*, where the winged forms, unlike those of the sycamore aphid, are relatively sedentary, the incidence of inbreeding is high, which leads to a very rapid increase in homozygosity, i.e. subpopulations on individual trees become genetically differentiated through adaptation to the phenological heterogeneity between trees (Komatsu and Akimoto, 1995). Thus providing the incidence of interplant movement is low, selection will synchronize the development of aphids with that of particular plants. Where the incidence of interplant movement is high selection synchronizes average egg hatch to the average bud burst, i.e. the scale

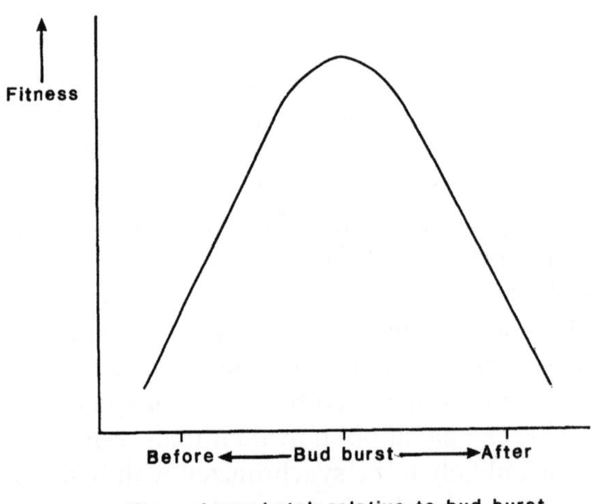

Figure 3.17 Fitness of first generation sycamore aphids is determined by the degree to which their hatching from overwintering eggs is synchronized with bud burst. (After Dixon, 1976.)

of adaptation, fine or coarse, depends on the degree of interplant movement. However, why some species of, for example, tree-dwelling aphids show a high and others a low incidence of dispersal between host trees is puzzling.

In autumn tree-dwelling aphids are faced with another event, leaf fall. As with bud burst, selection favours those individuals that can anticipate and complete egg laying just before leaf fall. Their success in tracking individual trees will depend on the gene flow between subpopulations. However, as with bud burst, there is a very close synchronization between egg laying and leaf fall in many aphids (Dixon, 1971a). Between the deadlines of bud burst and leaf fall, each generation of aphids experiences particular ranges of food quality and temperature. Phenotypic plasticity enables them to track the variation in conditions experienced within a generation and generation-specific strategies (p. 161) the bigger but specific differences in habitat quality experienced by each generation, i.e. phenological adaptation not only involves synchronization with deadlines but generation-specific tracking of the specific changes in habitat quality.

3.3 Evolution of host specificity

Most phytophagous insects are highly host specific. Work on the adaptive significance of this specificity has focused on factors determining the performance of individuals colonizing, or ovipositing on, preferred and non-preferred hosts. Four groups of factors can be distinguished:

1. Features of the host – its chemistry, architecture, phenology, etc.
2. Competition with other species of phytophaga on the same host.
3. Natural enemies foraging preferentially on certain hosts.
4. Within species interactions – direct density dependence resulting from intraspecific competition, or inverse density-dependence if aphids on the 'wrong' host fail to mate.

Thirty years ago, Ehrlich and Raven (1964) postulated that 'the evolution of secondary plant substances and the stepwise evolutionary responses to these by phytophagous insects have clearly been the dominant factors in the evolution of butterflies and other phytophagous insects'. Although this idea (host adaptation) was not new (*cf.* Börner, 1939) it gained immediate and broad acceptance, and still dominates the literature. Recently, however, it has come under attack from two quarters. First, one of its central assumptions – that specialist species generally outperform generalist species – is weakened by the fact that most species studies do not show the expected strong trade-offs between performance on different hosts: adaptation to one (group of) host(s) does not lead to poor performance on others (Gould, 1979; Rausher, 1984; Via, 1984; Hare and Kennedy, 1986; Futuyma and Phillippi, 1987; James *et al.*, 1988). Indeed it is likely that many other unsuccessful attempts were not published. Second, several authors have argued strongly that predation has played a major role in the evolution of host range; that is, phytophagous insects have evolved to prefer 'enemy-free space' (Lawton, 1978; Bernays and Graham, 1988). Evidence in support of this hypothesis too is equivocal (*cf.* special feature: *Ecology* **69**, 1988). The three hypotheses put forward to account for host specificity in aphids are presented below.

3.3.1 Host plant utilization hypothesis

Is there any evidence from studies on aphids for negative trade-offs in host plant utilization? Via (1991) noted a negative genetic correlation between performance on alfalfa and clover in the pea aphid (*Acyrthosiphon pisum*) and Mackenzie (1996) also presents evidence of a genetically based trade-off in the black bean aphid (*Aphis fabae*) between fitness (performance) on broad bean and garden nasturtium (Figure 3.18). In addition sympatric host races in several species of aphids have been reported to do better on their preferred host (Blackley, 1982; Guldemond, 1991). However, studies on other host races of the pea aphid have not revealed trade-offs in performance on different hosts and some host races even perform worse on their preferred host than on other hosts. In addition, there are several 'generalist' species like *Metopolophium dirhodum* that can outperform specialist species like *Hyalopteroides humilis* and *Metopolophium albidum* on their respective host plants.

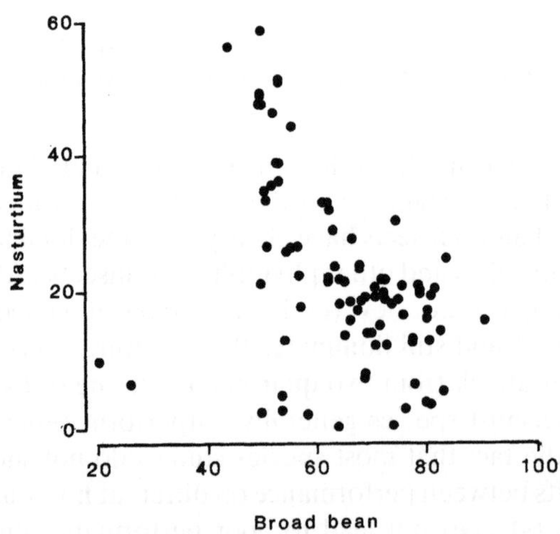

Figure 3.18 A trade-off in host utilization in *Aphis fabae* between fitness on broad bean and nasturtium. Each point is a clone's mean fecundities on both hosts. (After Mackenzie, 1996.)

It is true that plant species vary markedly in their suitability as hosts for particular aphid species (e.g. Weber, 1985a–c) and the fact that clones may remain for several generations on one host means that small differences in nutritional quality may have large effects on the long-term fitness of colonists (Kindlmann and Dixon, 1994). There are several reasons, however, for doubting that trade-offs in performance on hosts of different quality is sufficient to account for aphids' host specificity. Some 10% of aphids host alternate (p. 128) seasonally between primary and secondary host plants, which are taxonomically unrelated, e.g. Salicaceae/Umbelliferae; Rosaceae/Gramineae; Grossulariaceae/Labiatae. This seasonal host transfer tends to argue against plant chemistry or other plant features being all-important in determining host plant specificity. Indeed, the *individual* migrants must be able to feed on both hosts. For some of these aphids it is also known that they can complete their life cycle on either the primary or the secondary host plant, which argues against host specificity being a consequence of adaptation to the chemistry/morphology of a specific host plant.

3.3.2 Optimal host range hypothesis

Dispersal is particularly hazardous for aphids (p. 148) and the level of risk is important when assessing whether certain selection pressures can account for the degree of host specificity observed. Dispersal in aphids can be viewed as a series of 'trials' in which they land on plants at random and after probing decide whether to fly or stay on the plant. Due to their short life and/or limited fuel reserves the number of 'trials' is finite and small. Assuming that (1) the aphid can only live on two host plants, A and B, and the population growth rates on these plants are r_{mA} and r_{mB}, where $r_{mA} > r_{mB}$, (2) the proportion of the ground covered by these plants is C_A and C_B, (3) the length of time each season these plants remain suitable for the aphids is t, and (4) there is exponential population growth, then the numbers of aphids on these plants at the end of the season will be $\exp(tr_{mA})$ and $\exp(tr_{mB})$, respectively. Further, assuming that an aphid has only fuel and/or time for one more trial and it is on B, then if it decides not to fly but to stay and reproduce the relative increase in population size, its benefit, R_S, will be:

$$R_S = \exp(t \cdot r_{mB}) \tag{3.1}$$

If it decides to fly and try to find A then there is a cost as it may not find either A or B and dies, the probability of which is $(1 - C_A - C_B)$. The probability of finding B is C_B and the benefit $\exp(t \cdot r_{mB})$, the same as if it did not fly. The probability of finding A is C_A and the benefit $\exp(t \cdot r_{mA})$. Therefore, the benefit associated with flying is:

$$R_F = C_B \cdot \exp(t \cdot r_{mB}) + C_A \cdot \exp(t \cdot r_{mA}) \tag{3.2}$$

Figure 3.19 gives the minimum difference in population growth rates on A

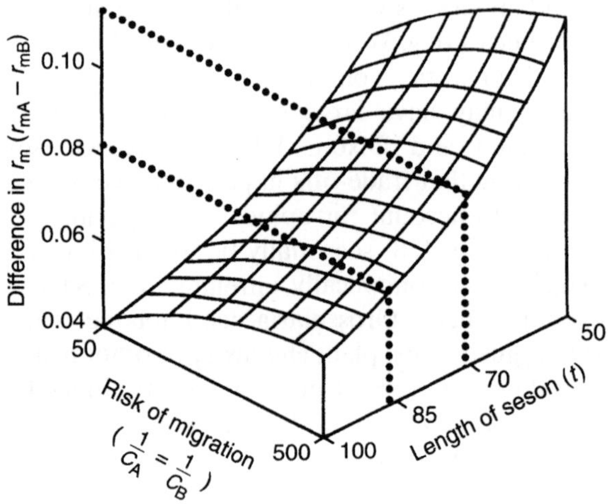

Figure 3.19 The difference in population growth rates on the two plants, A and B, needed for it to be advantageous to change host if not on A, relative to the risk of migration and length of the season. (After Kindlmann and Dixon, 1994.)

and B, $r_{mA} - r_{mB}$, required if $R_F > R_S$. The dependent variables are the length of the season (50–100 days) and the risk associated with dispersal, expressed as the inverse of the relative cover of A, which is for convenience assumed to be the same as that of B. The risk associated with dispersal is expressed in terms of the number of aphids necessary if one is to survive and find a host plant. Empirical data suggests that this number is in the order of 100 (Taylor, 1977). Moreover, the relative covers of the host plants of aphids seem to fall within the range indicated by the risk of dispersal in Figure 3.19 (Dixon *et al.*, 1987).

Aphid population growth rates, r_m, may be as large as 0.5, if time is measured in days (Kindlmann *et al.*, 1992). The difference in the population growth rates on the two host plants needed to make it advantageous to seek out host plant A (less than 0.1 in most cases, Figure 3.19) is small and such differences are frequently observed on plants of different species. Therefore, the best strategy is 'fly' until host A is located. If aphids have fuel and/or time for several flights and are on B then the preference for 'fly' as opposed to 'stay' should be even more pronounced.

If the relative cover of the host plants differ but the population growth rates on both hosts are the same, i.e. $r_{mA} = r_{mB}$, then the numbers of aphids on A and B at the end of the season depend on the relative covers of A and B and the number of flights (D) (Figure 3.20). The benefits can be calculated assuming that host specificity on A gives:

$$R_A = [1 - (1 - C_A)^D] \cdot \exp(t \cdot r_{mA}) \tag{3.3}$$

and on B gives:

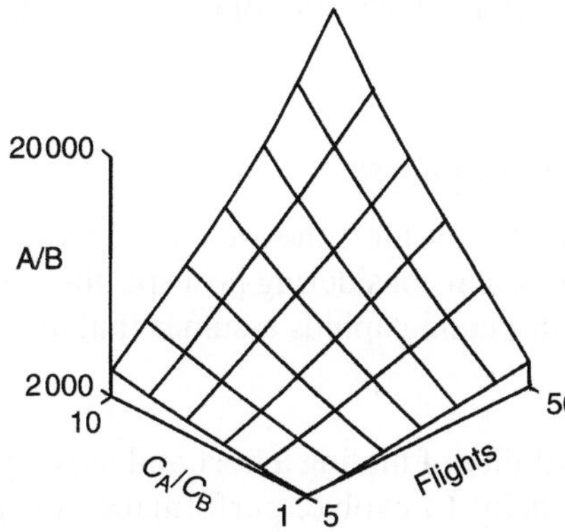

Figure 3.20 The number of aphids on A relative to B at the end of a season, relative to the proportional cover of A and B, and the number of flights available to locate A. (After Kindlmann and Dixon, 1994.)

$$R_{\mathrm{B}} = [1 - (1 - C_{\mathrm{B}})^{D}]\cdot\exp(t\cdot r_{\mathrm{mB}}) \tag{3.4}$$

The extremely large values of the ratio of benefits (A/B) indicates that there is likely to be a strong selection pressure for preference for the slightly more common host plants.

The above model assumes that the aphid populations grow exponentially, which would appear to be at variance with what happens in nature where within a season populations tend to increase and then crash in numbers. The increase phase is clearly exponential. There is evidence that aphids scramble for resources (Dixon, 1994) and that the collapse in abundance is mainly driven by competition for resources resulting in many aphids dispersing to seek resources elsewhere. Those that remain continue to grow exponentially within the constraints imposed by host quality and natural enemy activity.

The above model indicates that very small differences in the rates of population increase achieved on two host plants amplified by clonal parthenogenesis over several generations makes it advantageous for aphids to seek out the species of plant on which they perform best in spite of the huge losses incurred in finding this plant, providing it is not rare. This would lead to most aphids being on the favoured host and most genetic changes occuring there, on which selection could act resulting in more host-specific adaptations, such as changes in morphology, physiology and phenology that render the aphid less able to exploit other hosts. Reinforced by such feedbacks, monophagy represents an evolutionary stable state (Kindlmann and Dixon, 1994).

It is clear that the very high rates of population increase achieved by aphids could have been a major contributing factor enabling them to become host specific. However, it is unlikely to have been the factor because other groups, which lack the high rates of population increase of aphids, are equally host specific.

3.3.3 Rendezvous host hypothesis

Ward (1992a,b) similarly considers the great risk associated with dispersal to be an important issue when considering host specificity in aphids. His game-theoretic model for dispersing aphids assumes that an aphid's fitness is:

$$F = \Sigma p_i w_i \tag{3.5}$$

where p_i is the probability of finding a host and w_i the performance on that host. For host specificity to evolve, performance on poor hosts must be below a threshold that depends simply on how many dispersers encounter the more suitable host before dying of starvation, desiccation, etc. For instance, a host-specific aphid whose dispersers have a 5% chance of finding a host could increase their expected fitness by settling on any novel host on which they can produce about 5% as many offspring as on the normal host. Equation (3.5) is most relevant to dispersers whose decision to settle commits them to remaining on the chosen host for life – as is the case with many migrating aphids, as they autolyse their wing muscles after settling.

These arguments led Ward (1987a, 1991a) to propose that selection favours host specificity because host plants serve not only as habitats and source of nutrients but also as rendezvous for the sexes. If mating occurs on the host plant and the sexes find each other by first finding a host, the individuals that settle on uninfested hosts have little chance of mating. This means that selection favours genes for extreme specificity on whichever host is most commonly colonized, even if other hosts are more suitable in other respects. A similar explanation has been proposed for host site specificity in monogenean trematodes (Rhode, 1979) and host specificity in flower mites (Colwell, 1991).

Ward (1991a) offers three examples from aphid biology in support of the rendezvous host hypothesis:

1. Of the aphids that have winged males and live on grasses in central Europe the non-host-alternating species have a significantly narrower summer host range than the host-alternating species.
2. Some of the most polyphagous non-host-alternating aphids (e.g. *Aulacorthum solani, Macrosiphum euphorbiae*) produce unwinged males. That is, the ability to self-fertilize may uncouple the selection on host range from that for mate-finding.
3. Asexual clones are often more polyphagous than their sexual relatives, i.e. the loss of the sexual generation uncouples host range from mating success.

Basic to the rendezvous host hypothesis is that long range mate location is primarily dependent on locating a particular host plant. However, van

Emden *et al.* (1969) proposed that sex pheromones might prove to be a more powerful attractant for male aphids than the host plant. This idea has received some support mainly from the catches of males in water traps baited with sex pheromone (Campbell *et al.*, 1990; Hardie, 1991; Pickett *et al.*, 1992, 1994). The catches were particularly high when both a sex pheromone and a plant extract were used as a 'lure' (Campbell *et al.*, 1990; Hardie *et al.*, 1994). However, at least one species (*Sitobion fragariae*) only appears to produce and mainly respond to nepetalactone, which is one component of the sex pheromone of at least five species of aphids from two tribes. Thus the sex pheromone in this case does not appear to be very specific (Hardie, 1991). Similarly, Steffan (1983, 1987, 1990) also presents evidence that the sexual pheromone is not very species specific, is usually only perceived over a relatively short distance of 2–10 cm and that in host-alternating aphids the primary host (p. 130) acts as a rendezvous host for sexual reproduction. In addition it is not clear whether males are *attracted* to or are just more likely to *settle* on and be caught by baited traps. Laboratory studies on males using electroantennograms and olfactometers have also revealed that they respond to the odour of conspecific females, but, unlike field studies, not to the odour of their host plant on its own or in combinations with the sex pheromone. However, in cages males will colonize the 'right' host plant even in the absence of mating females (Guldemond, 1990; Pickett *et al.*, 1992; Guldemond *et al.*, 1993). In addition, in those species of host-alternating aphids in which the return migrants are monomorphic (p. 132) it is the migrant that locates the primary host on which the sexes are then born and subsequently mate.

Thus the laboratory studies indicate that mate recognition appears to be triggered by the sex pheromone produced by the mating female. However, it is uncertain what role, if any, this pheromone plays in mate location, particularly long distance location, in spite of claims to the contrary (Pickett *et al.*, 1992). In host-alternating aphids the remigrants, which give birth to the mating females, are the first to arrive on the primary host in autumn. These remigrants must respond to some feature of the host plant. The males arrive later and could use the sex pheromone produced by the mating female for long range orientation. If they do, as is strongly implied in the discussions in some of the published work, then the rendezvous host hypothesis cannot account for host specificity in aphids.

The mechanism by which aphids locate their mates in the field is an ecological problem and as such can only be resolved by field experiments. The presence of a sophisticated sense organ, however apparently relevant, is not proof that it is used for a particular purpose. For example, adult ladybirds have well-developed eyes, which are known to be sensitive to a wide range of wavelengths of light and can perceive structures; however, there is no evidence that they aid ladybirds in any way when they are hunting for prey.

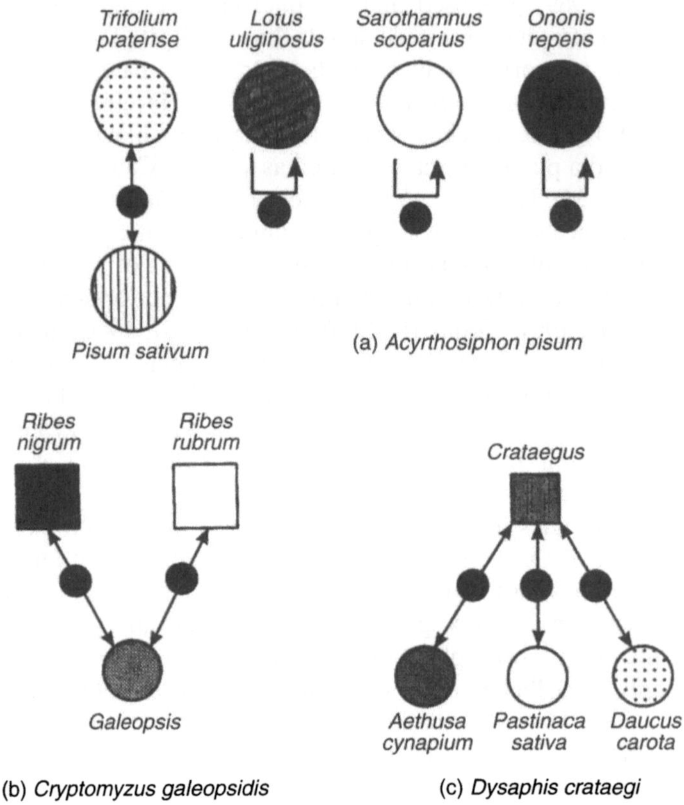

Figure 3.21 (a) Host race complex of the pea aphid, *Acyrthosiphon pisum* (one taxon host alternates between *Trifolium pratense* and *Pisum sativum*, but others complete their life cycles on specific host plants). (b) Host races of the host-alternating *Cryptomyzus galeopsidis* in which a new primary host has been colonized. (c) Host races of *Dysaphis crataegi* in which new secondary hosts have been colonized. (After Guldemond and Mackenzie, 1994.)

3.4 Speciation

In addition to allopatric speciation in which populations of an ancestral species diverge in geographical isolation living on the same host plant, speciation in phytophagous insects may be initiated by individuals shifting to a new, previously unused, host plant. If this occurs in the area of the ancestral population, sympatric speciation may occur (Bush, 1975; Brooks and McLennan, 1991). The common sympatric occurrence of morphologically similar subspecies, races and biotypes of aphids, which differ in their use of host plants, has led to the suggestion that sympatric speciation is possible in aphids (Müller, 1971a, 1985; Guldemond, 1990a,b; Ward, 1991b; Guldemond and Mackenzie, 1994; Mackenzie and Guldemond, 1994). Indeed, this sort of information is often the strongest evidence there is for speciation via host switching (Brooks and McLennan, 1991).

There are two situations in which speciation by host shift may occur in aphids. In the first the cycle of sexual reproduction is shifted to another host plant. This happens if a host shift occurs in a monoecious (non-host-alternating) species, or if a new primary host, where sexual reproduction takes place, is colonized by a heteroecious (host-alternating) species (p. 128). Examples are the monoecious host races of *Acyrthosiphon pisum* (Figure 3.21a) (Müller, 1971a, 1980), *Acyrthosiphon pelargonii* (Müller, 1983), *Acyrthosiphon solani* (Müller, 1976), and *Uroleucon jaceae* and *Uroleucon cichorii* (Mosbacher, 1963). Host races in which a shift to a new primary host has occurred are *Cryptomyzus galeopsidis* (Figure 3.21b) (Guldemond, 1990a,b, 1991) and *Myzus cerasi* (Dahl, 1968; Gruppe, 1988). The second situation involves a shift to a new secondary host whilst retaining the primary host, on which sexual reproduction of both the ancestral and new population occurs. In this case the possibility of gene flow between the two populations is much greater. Examples are found in *Dysaphis crataegi* (Figure 3.21c) (Stroyan, 1958), *Aphis fabae* (Figure 7.5) (Müller, 1982) and *Aphis frangulae* (Thomas, 1968).

Two processes of divergence can be distinguished: an abrupt shift to a new host plant because of a mutation in host recognition and/or reproductive performance, which makes the ancestral host plant less recognizable/suitable, or a more gradual process of selection leading to adaptation to a new host, where conditioning to a host plant and the presence of trade-offs are the driving forces for divergence. The abrupt shift model is essentially the sympatric host race formation model of Bush (1975). There is no rigid theoretical boundary between these two processes and both may act simultaneously. When the primary host is retained (Figure 3.21c), however, the process is necessarily more gradual (Guldemond and Mackenzie, 1994).

Aphids are possibly particularly good candidates for sympatric speciation because:

1. **Host specificity**. They show a very high degree of host specificity with 99% of all species confined to one or a few closely related species of plants (Eastop, 1973), which is associated with a genetically determined highly developed host plant preference and reproductive performance (Guldemond, 1990a,b; Via, 1991; Mackenzie, 1996). The fact that each clone is usually made up of many individuals, each with a prodigious rate of increase makes it advantageous to attempt to find a better host (p. 48). For example, if a poorer host supports a tenth of the fecundity that a preferred host does, it must be much more than 10 times as abundant to compensate. Therefore, in spite of huge losses incurred in finding host plants being a specialist is not maladaptive but the optimal strategy for aphids (Kindlmann and Dixon, 1994; Mackenzie and Guldemond, 1994; Dixon, 1994).
2. **Cyclical parthenogenesis**. Because of this mechanism a mutant aphid that has colonized a new host plant can rapidly give rise to a population

of genetically identical females. Later in the season genetically identical males, except for the loss of one X chromosome (p. 90), are produced and in monoecious species inbreeding is likely to result. The offspring homozygous for the mutant are likely to have a higher fitness on the new host and the small size of this newly emerged population is likely to enhance the effect of drift, which leads to a greater chance of fixation of the well-developed genotypes.

3. **Induction**. That a change in habitat may initiate a profound evolutionary change was first suggested by Lamarck (1809) and later developed as a theory with specific reference to parasitic insects by Walsh (1864). Work on both insect herbivores (Singer, 1983) and parasitoids (Collins and Dixon, 1986) indicate that the physiological state of the female is important in host acceptance. In aphids, for example, large *Myzus persicae*, which have been reared on a high-quality plant, accept a poor-quality host less readily than a small individual. This could contribute towards the maintenance of a host based polymorphism.

   More importantly aphids are reported to show inherited induced performance changes, which endow 'native' phenotypes with a competitive advantage. Such changes have been reported in *Dysaphis anthrisci*, which after a period of eight generations on a non-host, *Chaerophyllum bulbosum*, could successfully be transferred to *Chaerophyllum maculatum*, which was previously unacceptable. Interestingly the change in the aphid also involved morphology (Shaposhnikov, 1985). Similarly, culturing *Acyrthosiphon pisum* and *Myzus persicae* on non-hosts for several generations have also resulted in these aphids eventually doing better on the non-host (Markkula and Roukka, 1970; Lowe, 1973). Both *Myzus persicae* and *Aphis fabae* reared on poor-quality hosts/diets increased their fecundity over three generations (Mackenzie, 1992).

   If induction can result in the establishment of phenotypic races by changing the host preference and by both changing the host preference and putting 'foreign' phenotypes at a competitive disadvantage by induced trade-offs due to induced physiological and/or morphological changes then the establishment of genetically differentiated races will occur as the host plant forms accrue locally beneficial genes, i.e. induction may be a 'critical component' of the early stages of host race formation.

4. **Host alternation**. Some 10% of species show seasonal transfer between host plants – host alternation (p. 128). The loss of this way of life is thought to have been an important mode of speciation (Hille Ris Lambers, 1950). In *Cryptomyzus galeopsidis* the tendency to host-alternate is determined by a single gene (complex). If this is general then it would greatly facilitate the switch to a monoecious form by segregation during inbreeding on a new secondary host (Guldemond, 1990a,b), which may result in instantaneous speciation.

5. **Sexual females**. They are born on the host plant and are usually wingless. Consequently mating is very likely to be on the host plant.
6. **Males**. Many species of aphids have wingless males. When a host shift occurs in species with apterous males the possibility of gene flow by migration of males is severely limited. The fact that about half of the non-host-alternating species have apterous males suggest that this may be an important, though not essential, feature facilitating host shift in aphids.

The most contentious scenario for sympatric speciation is that in which the new population does not occupy a habitat (host plant) that immediately isolates it from gene flow from the parental population. This involves two steps: (1) a resource utilization polymorphism must be established with the two populations thriving on different host plants and individuals having low fitness on the wrong hosts, and (2) selection against intermediates must result in assortative mating to avoid the cost of hybridization. However, the conditions under which this selection in the second step will generate a response have usually been considered restrictive. Strong selection against hybrids is required and this reinforcement is opposed by two key processes: recombination and gene flow.

A model developed by Mackenzie and Guldemond (1994) indicates that when an empty niche is invaded by a rare genotype and there is heterozygote dysfunction, establishment of a polymorphism may depend on the rate of increase achievable by an organism. Cyclical parthenogens, by virtue of avoiding both recombination and the cost of producing males, possess the highest between mixis rate of increase of all sexual metazoans. Aphids, by telescoping of generations, have further enhanced this power of increase and hence may be able to sustain polymorphism under conditions which obligately sexual species cannot.

A primary condition required to establish a resource-based polymorphism is that hybrid dysfunction should occur. If the hybrids between the two forms are not at a selective disadvantage then one of the parental forms will predominate. Aphids do indeed appear to show genetically based trade-offs in fitness (p. 46). The existence of intraspecific trade-offs and the widespread occurrence of host-related hybrid dysfunction within aphids indicate that the primary requirement, selection against intermediates, is in place.

The mechanism of evolution of reproductive isolation, termed reinforcement by Blair (1955), was proposed by Dobzhansky (1940, 1951). The production of inferior hybrids is reduced by genes for mating within the incipient species. Assortative mating genes spread 'until the possibility of gene exchange is ... severely limited or stopped' (Dobzhansky, 1951). Although its significance is still hotly debated, reinforcement remains a potentially important evolutionary process which awaits a fully convincing example (Butlin, 1995). The power of recombination to shuffle genes affecting dysfunction and mate choice when hybridization occurs between two populations is probably the strongest limitation to the reinforcement process in aphids sharing a host plant on which they mate.

Stam (1983) and Butlin (1990) have suggested an alternative model, which involves 'gene swapping'. Early mating individuals in a population producing sexuales late are more likely to mate with members of a population producing sexuales early and *vice versa.* Thus genes for earliness accumulate in the early population and for lateness in the late population. This differs from reinforcement in that no selection is occurring for differentiation. There is no reason why gene swapping and reinforcement should not occur simultaneously if hybrid dysfunction is strong.

One consequence of reinforcement and/or gene swapping could be allochronic isolation. This has been recorded in aphids. The sexuales of *Acyrthosiphon pisum destructor* are produced in November and later than those of *Acyrthosiphon pisum* ss, which occurs on the same primary host (Müller, 1980). *Uroleucon cichorii* ss produces sexuales a month later than the other host races, which live on other host plants in the same habitat (Mosbacher, 1963), and two forms (species?) of *Pemphigus populitransversus* have sex in autumn and spring, respectively (Faith, 1979; Sokal *et al.*, 1991).

Opportunities for reinforcement exist in the many host-alternating sympatric sibling species that share the same primary host, i.e. mating occurs on the same plant species, though there may be disruptive selection for utilization of secondary hosts. These species differ in both the composition of their sexual pheromones and in the time of release of the pheromone (Pettersson, 1971; Guldemond and Dixon, 1994; Thieme and Dixon, 1996). For example, the sister species *Cryptomyzus galeopsidis* and *Cryptomyzus maudamanti*, which share the same primary host, red currant, and between which gene flow is possible, though F_2 performance is poor (Guldemond, 1990c), differ in their daily patterns of pheromonal release and male activity (Figure 3.22). This does not prove that reinforcement has occurred as it may represent character displacement after the taxa speciated allopatrically. However, the distribution of the secondary host plant of *Cryptomyzus maudamanti* (*Lamium galeobdolon*) completely overlaps with that of *Galeopsis*; the secondary host of *Cryptomyzus galeopsidis*. Moreover, the distribution of *Cryptomyzus maudamanti* falls within that of *Cryptomyzus galeopsidis* (Guldemond, 1991b), which tends to exclude allopatric speciation. It is not possible to be sure of the past distributions of the host plants of *Cryptomyzus*; however, what is known does not favour allopatric speciation. Thus, it is likely that the difference in mate recognition shown by these sister species are most likely the result of reinforcement.

Although very rigorous requirements need to be met if sympatric speciation in the face of gene flow is to occur aphids are likely to have speciated by this route because they: (1) exhibit trade-offs in host utilization and considerable dysfunction; (2) have very high rates of increase, which may give

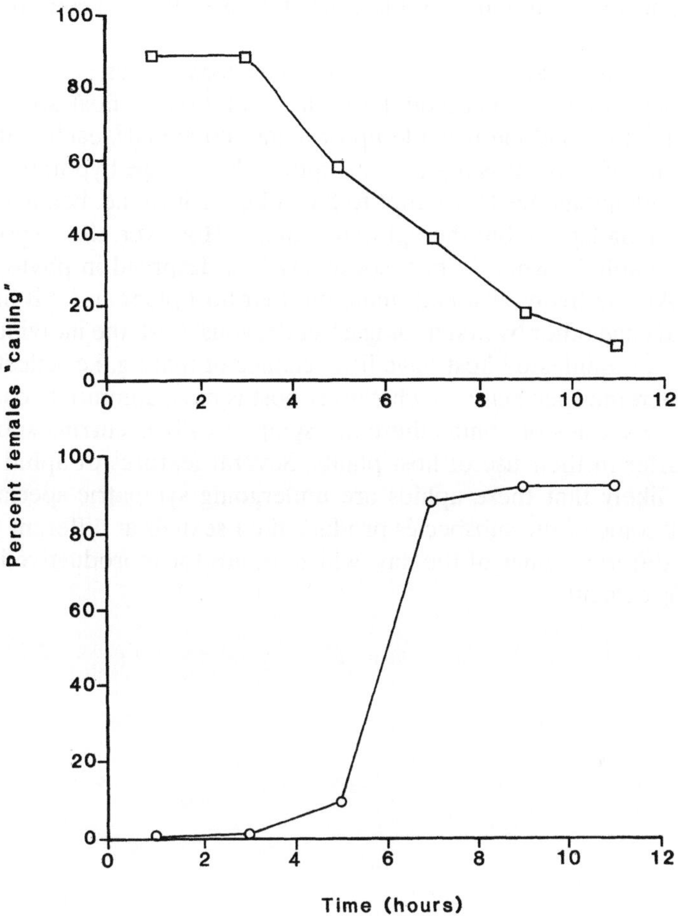

Figure 3.22 The percentage of mating females 'calling' during the course of a 12 h day in the sister species *Cryptomyzus galeopsidis* (top) and *Cryptomyzus maudamanti* (bottom). (After Guldemond and Dixon, 1994.)

rise to intense selection pressures; (3) are able to exploit empty niches and hence achieve habitat-based polymorphisms; (4) are able to phenotypically 'adjust' to host plants and to pass on these adjustments, predisposing them to host race formation; and (5) may overcome the restrictions to reinforcement or bypass them by gene swapping (Mackenzie and Guldemond, 1994).

Summarizing, dispersal is hazardous for aphids. Therefore, their marked host specificity would appear to be maladaptive. Although it is likely that plant odours affect settling rate-specific host plant identification most likely occurs after settling and in response to chemical and/or morphological cues. However, like the morphological, physiological and phenological adaptations associated with living on particular plants, aphids' responses to specific 'flags' in host selection are not the cause but most likely a consequence of host specificity.

The great risk associated with dispersal has been addressed by two of the hypotheses proposed to account for the evolution of host specificity in aphids. That each clone is made up of many individuals, each with a prodigious rate of increase is basic to the optimal host range hypothesis. In this case it is advantageous to attempt to find a better host, i.e. being a specialist is not maladaptive but the optimal strategy. However, this hypothesis is specific to aphids, whereas host specificity is widespread in phytophagous insects. Alternatively, as aphids mate on their host plant and it is likely the sexes find each other by first finding a 'rendezvous' host, the individuals that settle on an uninfested host have little chance of mating, i.e. selection will favour extreme specificity on whichever host is most commonly colonized.

In a few species of aphids there are sympatrically occurring subspecies, which differ in their use of host plants. Several features of aphid biology make it likely that these aphids are undergoing sympatric speciation. In addition, some of the subspecies produce their sexuals at different times or mate at different times of the day, which argues for reproductive isolation by reinforcement.

4 Size

A central problem of evolutionary biology is to provide an explanation for the design of organisms (Stearns, 1982) and an important feature of design is size. Small size has been a feature of aphids ever since their evolution in the Permian. Even so, there is a range of size within and between species. In general, it is thought that bigger organisms have a better chance of reaching maturity and producing more gametes (Calow, 1978). One component of this is the ability of individuals to survive when environmental quality changes. It has been suggested that as the size of the fat body (F) increases superproportionally, and energy consumption (E) subproportionally with increase in body size, then large species and individuals are at an advantage when starved (Calder, 1984). This also appears to apply to aphids as both inter- and intraspecifically body fat increases proportionally ($F = M^1$; Figure 4.1) and energy consumption scales subproportionally ($E = M^{0.75}$) relative

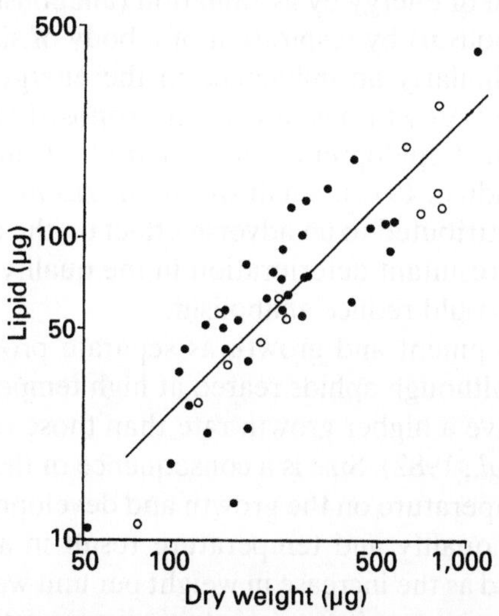

Figure 4.1 The relation between the logarithm of the lipid content and the logarithm of the dry weight for the winged individuals (filled circles) of 35 species and the unwinged individuals (open circles) of 12 species. The relationship for the total data set is $y = -0.57 + 0.99x$, $r = 0.86$, $n = 47$.

to body size (M), which indicates that large species and individuals are more resistant to starvation than small species and individuals. However, as aphids spend most of their time feeding, it is more likely they are going to be killed by natural enemies or, because of their small size, physical hazards, than die of starvation. Therefore, it is of interest to consider what other factors might determine size in aphids.

4.1 Size within a species

Adults of several species show as much as a 10-fold range in size (Way and Banks, 1967; Murdie, 1969a,b; Dixon and Dharma, 1980a). Small individuals develop when they are reared in crowded conditions, on mature plants or at high temperatures; large individuals result when nymphs are reared in isolation, on young or senescent plants or leaves, or at low temperatures. Well-fed animals generally grow larger than poorly fed ones; therefore, the effect of food quality on size is not surprising. The inverse relationship observed between size and temperature is more puzzling, but is a widespread phenomenon in insects.

Müller (1966a) postulated that size in aphids is a consequence of the balance between catabolism and anabolism. At high temperatures catabolism consumes most of the available energy and little is left for growth (anabolism). This can be expressed in the form of the production function, $P(m)$:

$$P(m) = am^{\alpha} - bm^{\beta} \tag{4.1}$$

where am^{α} is the gain of energy by assimilation (anabolism) and bm^{β} is the loss of energy (catabolism) by respiration of a body of size m (Ziolko and Kozlowski, 1983). Similarly an imbalance in the energy requirements of basic metabolism and growth has also been proposed to account for the smallness of certain Lepidoptera when reared at high temperatures (Marthavan and Pandian, 1975). Small size in *Myzus persicae* at high temperatures has been attributed to an adverse effect on the aphid's symbionts (Lawson, 1941) and resultant deterioration in the quality of food available to the aphid, which would reduce anabolism.

By treating development and growth as separate processes, Chambers (1979) showed that although aphids reared at high temperatures are small, they nevertheless have a higher growth rate than those reared at low temperatures (Dixon *et al.*, 1982). Size is a consequence of the relative effect of food quality and temperature on the growth and developmental rates. Both an increase in food quality and temperature result in an increase in the growth rate measured as the increase in weight per unit weight per unit time ($\overline{RGR}$). However, increases in temperature disproportionately decrease the time it takes to reach maturity (Figure 4.2, I). Similarly the time to maturity is affected by an increase in food quality, but the effect is

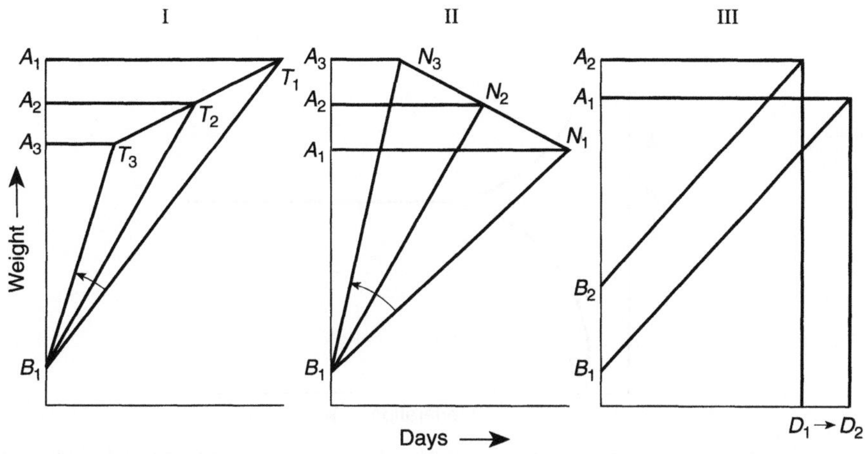

Figure 4.2 The relationships between adult weight (A_1, A_2, A_3) and developmental times for: aphids reared on food of the same quality at a range of temperatures (T), where $T_1 < T_2 < T_3$, (I); at one temperature and a range of food qualities (N), where $N_1 < N_2 < N_3$ (II); and at one temperature and a particular food quality but starting with nymphs of different birth weights (B_1, B_2), where $B_1 < B_2$ and D_1 and D_2 are the developmental times (III). Weight is plotted on a logarithmic scale. The relative growth rate (RGR) in each case is the slope of the relationship between weight and developmental time.

subproportional (Figure 4.2, II). As a consequence aphids are small when reared at high temperature and large when reared on high-quality food.

Time to maturity is also affected by birth weight so that the generally small nymphs born to small mothers take longer to mature than the large nymphs born to large mothers (Figure 4.2, III). However, the increase in developmental time does not compensate completely for the difference in weight at birth and produce a generation of adults uniform in size. Usually for a particular set of conditions a clone of initially large aphids will decrease in size and a clone of small aphids will increase in size over a number of generations, converging on a particular size (Figure 4.3). The final size is a consequence of the effect of temperature and food quality on the developmental and growth rates. A lower temperature or higher food quality increases the final size and a higher temperature or lower food quality decreases the final size.

The adaptive significance of size in aphids can be studied in terms of energy partitioning. For this one has to know that aphids differ from other insects in that even at birth they have well-developed gonads containing developing embryos, which also have embryos developing in their gonads (p. 83). In addition, the offspring must be a particular minimum size if they are to be able to feed (p. 66). In order to take these features into account the pattern of energy partitioning between growth and reproduction in aphids proposed by Kindlmann and Dixon (1989a) differs from that in other models

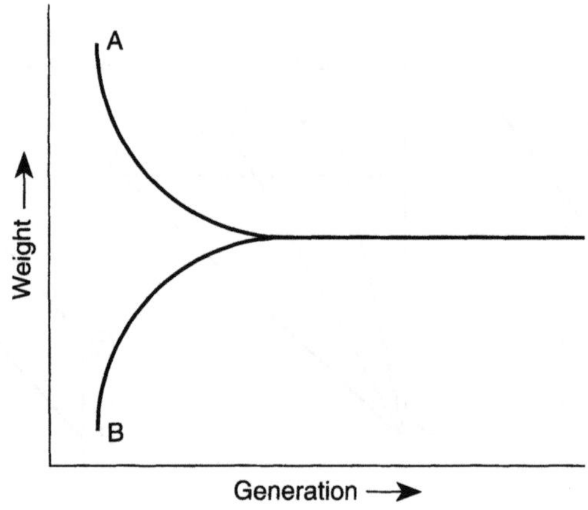

Figure 4.3 The trend in adult weight from generation to generation of initially large (A) and small (B) aphids reared at the same temperature and on food of the same quality.

in the incorporation of a constraint on the gonadal growth rate and the minimum size of the soma. What limits growth rate in organisms is still an open question (Maynard-Smith, 1969). Ziolko and Kozlowski (1983) also include a constraint on the gonadal growth rate, but their fitness criterion is lifetime fecundity, which is appropriate for organisms with non-overlapping generations, but not for aphids. The models of Sibly and Calow (1986) and Stearns and Koella (1986) assume that adult and juvenile mortality is dependent on the growth rate. In aphids there appears to be no good empirical or theoretical evidence for this claim. In rapidly changing environments the evolutionary response of short-lived organisms is likely to be more to immediate factors like food quality and temperature, than to infrequently acting mortality agents. Therefore, mortality in the aphid model is assumed to be constant. In addition the fecundity function is assumed to be triangular with the peak at the instant of maturity (Kindlmann and Dixon, 1992).

Given that a particular size of soma and gonads at birth ($s_0 + g_0$) and at the instant of maturity ($s_A + g_A$) are optimum it is pertinent to ask: what is the minimum time (D) required for development to reproductive maturity. Equation (4.1) can be modified to make it more appropriate for aphids:

$$P(s,g) = as^{\alpha} - b(s + g)^{\beta} \tag{4.2}$$

where a is the assimilation rate, b is the respiration rate, s and g are the size of soma and gonads, respectively, and α and β are exponents defining the relationships with weight. In this case D is minimized if all the energy is used initially for somatic growth, after which the energy is devoted to gonadal growth, i.e. the best strategy is to first grow and then switch to reproduction, a bang/bang strategy. This is the reproductive strategy adopted by many insects.

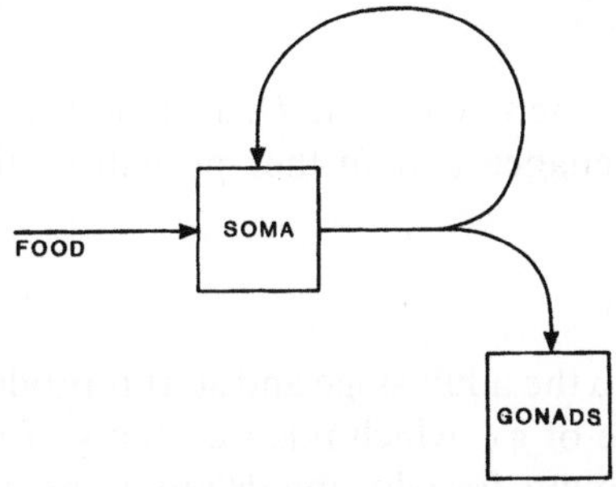

Figure 4.4 Assimilation of energy by the soma and subsequent partitioning between soma and gonads in unwinged aphids.

However, aphids from birth show a simultaneous commitment to somatic and gonadal growth (Figure 4.4). A possible explanation for this is that there is a constraint on the relative growth rate of either the soma or gonads. They simply cannot grow (develop) any quicker (*cf.* Maynard-Smith, 1969). If the constraint were only on the growth rate of the soma, one would not expect the growth rate of the soma to decline during development. If the somatic growth rate is maximum and constant until the soma reaches the optimum size, then for the same reasons as given above the developmental time (D) is shortest. This contradicts the observation that the soma grows logistically throughout development (Brough *et al.*, 1990). However, if the constraint is on the gonadal growth rate and the upper bound is denoted by R then the shortest time (D) for the development of gonads of size g_A from the initial size g_0 under particular conditions is achieved when gonadal growth rate is equal to R during nymphal development. If, after subtraction of energy needed for the maximum growth rate of the gonads, there remains surplus energy great enough for the soma to achieve s_A during D, then the strategy of assuring maximal gonadal growth rate and allocation of the surplus energy to somatic growth leads to the shortest possible time D for development:

$$D = ln\ (g_A/g_0)\ /R \tag{4.3}$$

This indicates that developmental time is dependent on the relative size of the gonads at birth, i.e. degree of telescoping of generations (g_A/g_0) and the maximum growth rate of the gonads (R). A constraint on somatic growth would result in some exponential somatic growth, which is not what is observed. If there is a constraint on somatic growth rate, it is always larger than the achieved growth rate. If there are no constraints on either soma or gonads, the optimal strategy is bang/bang.

The ratio s_0/g_0 at birth ensures somatic growth plus maximum gonadal growth, i.e. $s_0 + g_0$, satisfies:

$$as_0{}^{\alpha} - b(s_0 + g_0)^{\beta} - Rg_0 > 0 \quad (4.4)$$

but soma growth will decline to 0 at time D. The energy assimilated by the soma is used for maintenance and further growth of the soma and the gonads. When

$$as_A{}^{\alpha} - b(s_A + g_A)^{\beta} = Rg_A \quad (4.5)$$

the aphid should moult to the adult stage and start reproducing. This results in a reduction in the size of g_A, which releases energy for further gonadal growth and possibly accounts for why aphids are iteroparous.

As this model is derived from others that are framed in terms of growth rather than development, Kindlmann and Dixon retained this approach. However, the constraint on gonadal growth they introduced should be framed in terms of time required to complete development, rather than in terms of growth. Parthenogenesis has enabled aphids to start developing at ovulation, and, more importantly, inside immature or even embryonic mothers. Then given there is a constraint on the rate of development there are great advantages in terms of fitness in telescoping generations than in indulging in a bang–bang reproductive strategy. If telescoping of generations is taken into account and the developmental time of aphids is measured from **ovulation** to maturity then the developmental time is comparable to that in other similar sized insects (p. 83; Kindlmann and Dixon, 1989a).

4.2 Optimum body size

Because of the way aphids feed it has so far not proved possible to quantify food quality and therefore to relate changes in assimilation rate and/or gonadal growth rate directly to changes in environmental conditions. However, the aphid model predicts similar qualitative trends to those observed in empirical data. Results for the pea aphid reared at a range of temperatures and two food qualities are given in Table 4.1. As predicted, an increase in temperature and/or food quality results in an increase in embryonic growth rate (R) and relative growth rate ($\overline{RGR}$) and a decline in developmental time (D). Improving food quality and a decline in temperature both result in an increase in adult (W_A) and birth (W_i) weights. Similar trends have been recorded in the weights of *Aphis fabae*, *Cavariella aegopodii* and *Sitobion avenae* (Kindlmann and Dixon, 1992).

This model also indicates that at a particular temperature and food combination there is only one size at which r_m is maximized (Figure 4.5). As the observed trends in size with change in environmental conditions are predicted by the model it is likely that the phenotypic plasticity in size is adaptive, i.e. in insects, like aphids, which are food limited, variation in adult size within a species is not a means of coping with a variable environment but a consequence of maximizing r_m.

Table 4.1 The effect of temperature and food quality on the embryonic growth rate (*R*), birth weight (W_i), adult weight (W_A), developmental time (*D*) and relative growth rate ($\overline{RGR}$) of apterous pea aphids, *Acyrthosiphon pisum*

Food quality	*Temperature (°C)*	*R*	W_i	W_A	*D*	$\overline{RGR}$
Good	10	0.24	119	4713	28	0.13
	15	0.40	119	4273	16	0.22
	20	0.66	132	4026	11	0.31
	25	0.85	89	2901	8	0.43
Poor	20	0.50	121	1890	12	0.23

4.3 Species size diversity

Different species vary in length from 0.7 to 7 mm, and those that feed on the deeply located phloem elements in the trunks of trees are larger than those that feed on the more accessible phloem elements of leaves. To reach phloem elements deep within a trunk of a tree an aphid needs long stylets. They are not coiled up within the body, as in scale insects, but are characteristically enclosed within a proboscis. The stylets are exposed by telescoping the distal segments within the basal segments of the proboscis (Figure 1.1). *Stomaphis quercus* feeds on the trunks of mature oak trees and has a proboscis that is nearly twice the length of its body. The telescoping of the proboscis, necessary to expose the stylets, results in the invaginated proboscis running internally, the full length of the body (Figure 4.6). Thus the structure of the proboscis that evolved as long ago as the Jurassic (Shaposhnikov, 1977) makes it physically impossible for small aphids to feed on tissues very deep within a plant.

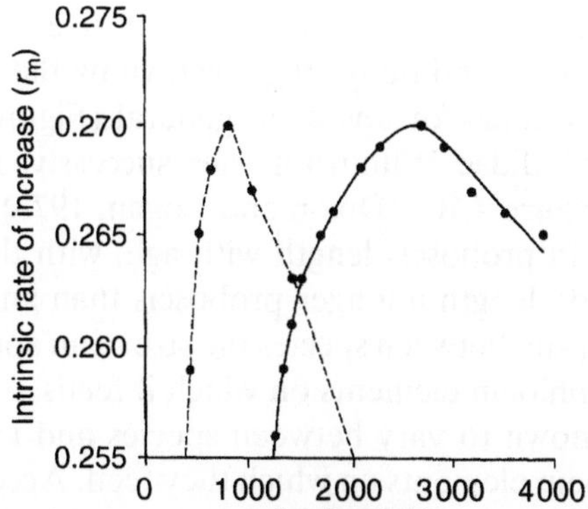

Figure 4.5 The relationship between the population growth rate, r_m, and adult size for two sets of conditions of food quality and temperature. (After Kindlmann and Dixon, 1992.)

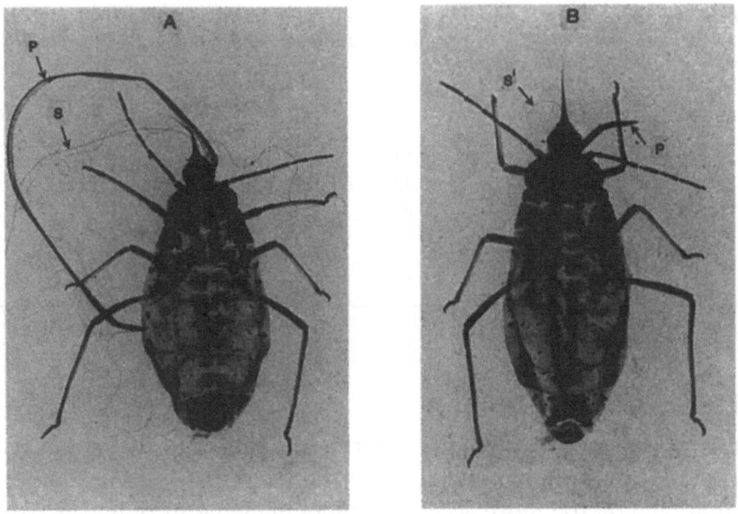

Figure 4.6 Photographs of slides of *Stomaphis quercus* showing its long proboscis and fine stylets (A); and the proboscis in the feeding position, invaginated on itself as far back as it can go within the abdominal cavity (B). P, proboscis, S, stylet, S^1, base of cut stylet.

The range in size, with large species of aphids feeding on the trunks of trees and small species of aphid feeding on leaves, is well illustrated by the aphids feeding on oak (Figure 4.7) and trees generally (Figure 4.8). Similarly, the generations of *Periphyllus* that feed on the twigs of sycamore are larger and have a proportionally longer proboscis than the generations of the same species that feed on the leaves of sycamore (Figure 7.24). On birch leaves the smallest aphid, *Callipterinella minutissima*, feeds on the smallest veins; the intermediate-sized *Betulaphis quadrituberculata* feeds on medium-sized veins; and *Euceraphis punctipennis*, the largest aphid, feeds on the largest and deepest veins. The positive relationship between stylet/proboscis length and body length shown by the aphids living on oak (Figure 4.7b) also holds for aphids in general (Figure 4.9), including the scale like Hormaphididae. Within a species, successive instars feed on larger veins of a leaf (Figure 4.10) (Dixon and Logan, 1973), which is associated with an increase in proboscis length with age, with the first instar having relative to its body length a longer proboscis than an adult (Figure 4.11). Thus, both within and between species the size of an aphid is correlated with the depth of the phloem elements on which it feeds.

Birth size is known to vary between species and is associated with the depth of the phloem elements on which they feed. Accepting this constraint on minimum birth size, the optimum energy partitioning model developed to account for telescoping of generations and phenotypic plasticity (Kindlmann and Dixon, 1989a, 1992) can be used to predict the interspecific range

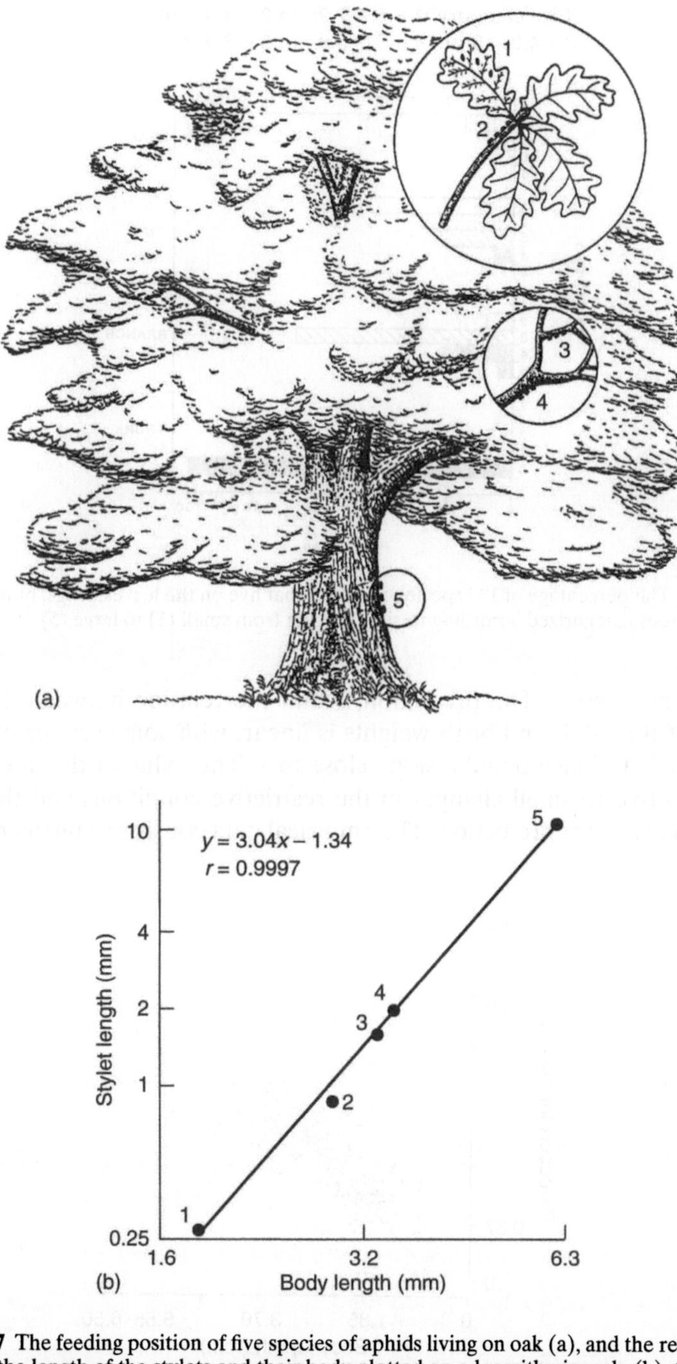

Figure 4.7 The feeding position of five species of aphids living on oak (a), and the relationship between the length of the stylets and their body plotted on a logarithmic scale (b). (1) *Tuberculoides annulatus*; (2) *Thelaxes dryophila*; (3) *Lachnus roboris*; (4) *Lachnus iliciphilus*; (5) *Stomaphis quercus*.

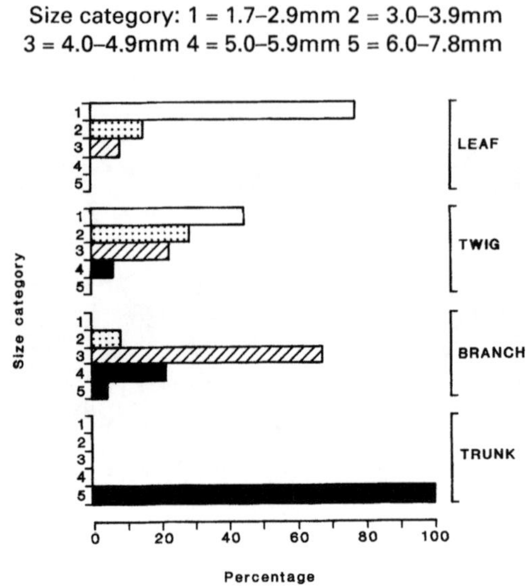

Figure 4.8 The percentage of 133 species of aphids that live on the leaves, twigs, branches and trunks of trees categorized according to size, ranging from small (1) to large (5).

in adult body sizes. The prediction is that the relation between the logarithms of the adult and birth weights is linear, with some scatter of points about the relationship and a slope close to 1. The value of the intercept is very sensitive to small changes in the restrictive conditions and therefore cannot be used for prediction. The empirical data consists of birth and adult

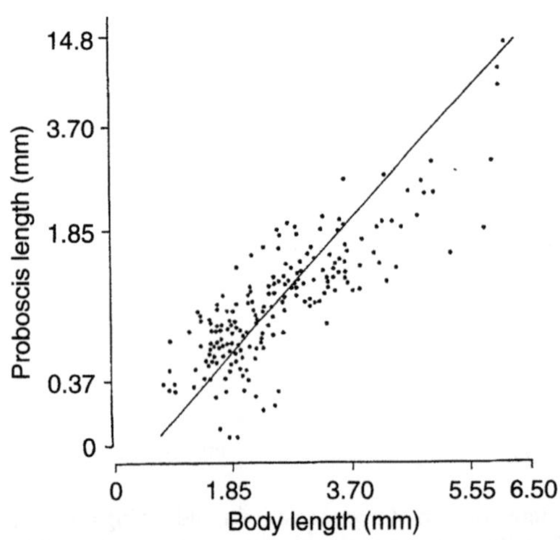

Figure 4.9 The lengths of the probosces of 205 species of aphid in relation to the lengths of their bodies.

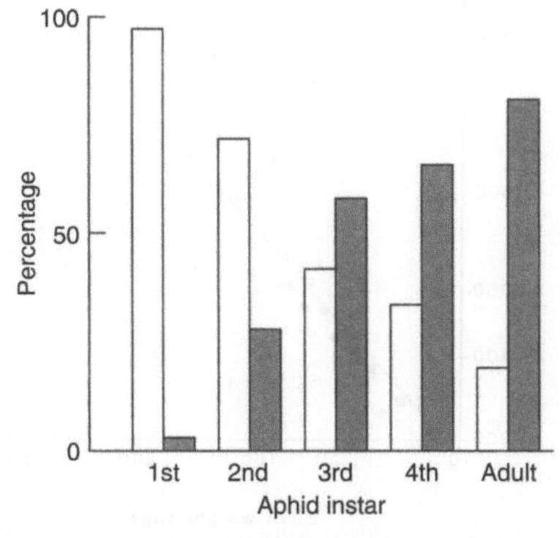

Figure 4.10 The percentage of first, second, third, fourth and adult instars of the sycamore aphid feeding on minor (open bars) and large (shaded bars) leaf veins.

weights of 18 species of aphids from the three largest subfamilies of the Aphididae reared on their respective host plants at 20°C. These species nearly represent the complete range of size shown by aphids and includes the largest species, *Stomaphis quercus*, and several of the smallest species on the British list. As predicted, the relationship is linear and has a slope

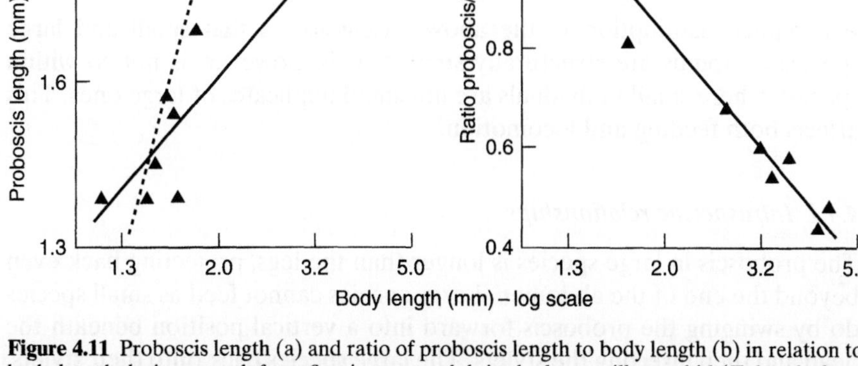

Figure 4.11 Proboscis length (a) and ratio of proboscis length to body length (b) in relation to body length during growth from first instar to adult in the large willow aphid (*Tuberolachnus salignus*). The dotted line indicates a slope were equal to 1.

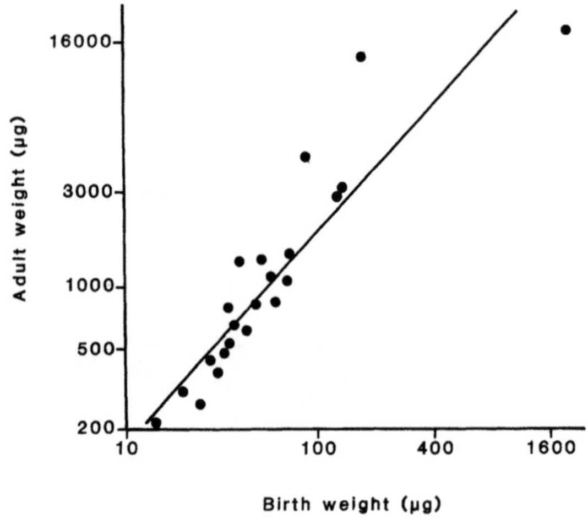

Figure 4.12 The relation between adult and birth size observed in 18 species of aphid ($y = 1.19 + 1.05x$, $r = 0.91$). (After Dixon and Kindlmann, 1994.)

that is not significantly different from 1 (Figure 4.12), i.e. a doubling of birth size results in a doubling of adult size. The intercept of this relationship indicates that the optimum adult weight for maximizing r_m is on average 15.5 times the birth weight (Dixon and Kindlmann, 1994).

It also follows from the above, and the marked affect of both food quality and temperature on r_m, that interspecifically there is unlikely to be a negative correlation between r_m and body weight in aphids. That is, it is not a question of 'failing to find' this relationship in aphids as suggested by Gaston (1988) because both theory and empirical evidence indicate that aphids differ in this respect from many other taxa (p. 79).

4.4 Geometrical similarity

An implicit assumption in the above discussion is that small and large species of aphids are structurally similar. This, however, is not so within species, where small individuals are not small duplicates of large ones. This affects both feeding and locomotion.

4.4.1 Intraspecific relationships

The proboscis in large species is longer than the legs, projecting back even beyond the end of the abdomen. Large species cannot feed as small species do by swinging the proboscis forward into a vertical position beneath the head and then inserting the stylets. The large species back onto their stylets, which are then exposed in the usual way by telescoping the segments of the

Table 4.2 The relationships between the logarithm of the wing area (y) and the logarithm of the weight (x) for five species of aphids

Species	*n*	*Relationship*	*r*	*b* < *0.66*
Drepanosiphum platanoidis	177	$y = -0.58 + 0.56x$	0.89	$P < 0.01$
Tuberolachnus salignus	43	$y = -0.52 + 0.54x$	0.81	$P = 0.05$
Acyrthosiphon pisum	41	$y = -0.49 + 0.44x$	0.81	$P < 0.01$
Clethrobius comes	34	$y = -0.98 + 0.61x$	0.88	NS
Cinara pinea	15	$y = -0.26 + 0.30x$	0.77	$P < 0.01$

proboscis. This is most strikingly seen in the offspring of large species, which at birth have a proportionally longer proboscis than the adults (Figure 4.11b). This change in the proportional length of the proboscis as on aphid grows also occurs in small species of aphids (Dixon, 1987b).

Features associated with locomotion, which do not scale proportionally with body size, are the area of the tibial pulvilli and the area of the wings. In *Megoura viciae* the area of the tibial pulvilli, which the aphid uses to adhere to the smooth surface of plants, increases with size of the aphid but not proportionally so (Dixon *et al.*, 1990). In the sycamore aphid, *Drepanosiphum platanoidis*, the large willow aphid, *Tuberolachnus salignus*, the pea aphid, *Acyrthosiphon pisum*, the birch aphid, *Clethrobius comes*, and the pine aphid, *Cinara pinea*, there are significant linear relationships between the logarithm of the wing area and logarithm of body weight. The relationships for these species are similar in indicating that wing area scales with less than the 0.66 power of body weight predicted if all individuals are structurally similar (Table 4.2). In four of the species the exponent is significantly less than the predicted 0.66, i.e. within species the large individuals have proportionally smaller wings and therefore a greater wing loading than the small individuals.

4.4.2 Interspecific relationships

Large species of aphids feed on phloem elements located deep within plant tissues, as in the case of those that feed on the trunk and branches of trees, and small species on the more shallowly located phloem elements in leaves. However, the relationship between proboscis length and body length (Figure 4.9) is not directly proportional. Large species have a proportionally much longer proboscis than small species.

Aphids differ in shape with conifer needle and some grass feeders (e.g. *Eulachnus* spp. and *Atheroides* spp.) more elongate than the species that feed on other plant structures. Nevertheless the winged forms of aphids do appear to be geometrically similar in structure. Measurements of slide material of 53 species, including representatives from seven of the 10 subfamilies of Aphididae, indicate a close relationship between the wing area

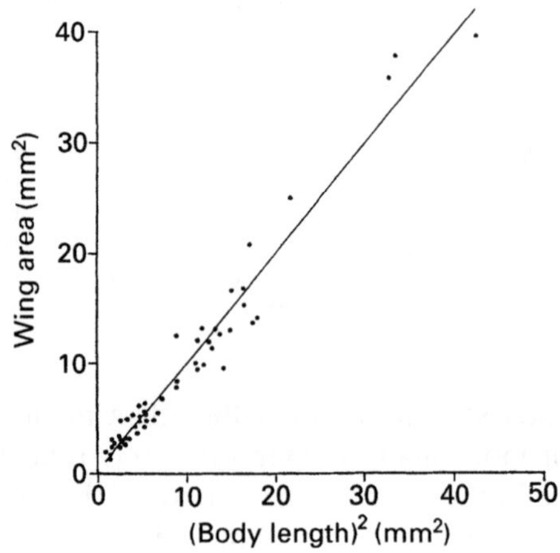

Figure 4.13 The relation between wing area and the square of the length of the body for 53 species of aphid.

and the square of the length of the body. This relationship passes through the origin, has a slope of 1 (Figure 4.13) and conforms to the expectation that if aphids species are structurally similar, then wing area should be directly proportional to the square of the body length. Unlike birds or butterflies the different groups of aphids do not appear to have evolved different flight strategies. Even aphids selected by a very experienced aphidologist, Jaroslav Holman, because he thought they had particularly long and narrow or short and broad wings, conformed to the above relationship. Further support comes from the relationship between wing area and body weight for the 10 species for which data is available. This indicates that wing area scales relative to the body weight with an exponent not significantly different from the theoretical prediction of 0.66 (Figure 4.14).

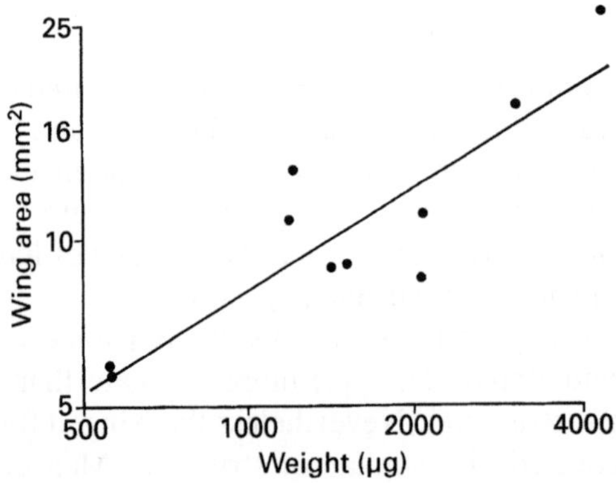

Figure 4.14 The relation between logarithm of wing area and logarithm of body weight for 10 species of aphids ($\log y = -0.97 + 0.63 \log x$, $r = 0.87$).

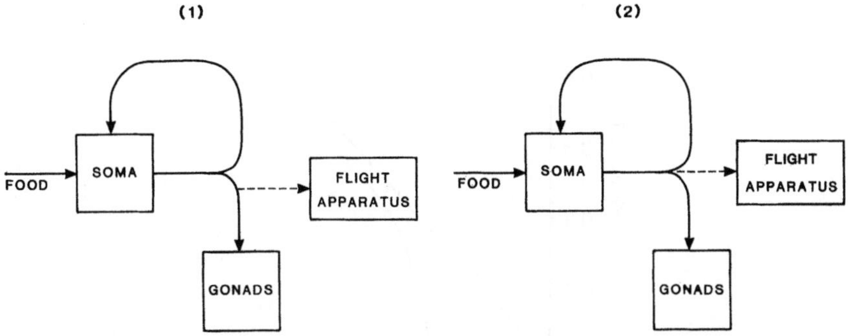

Figure 4.15 Assimilation of energy by the soma and subsequent partitioning between soma, gonads and flight apparatus in a winged aphid. In (1) the energy for the flight apparatus comes off the gonadal loop, and in (2) the energy is partitioned between gonads, flight apparatus and soma.

Although the range in adult size in aphids is three orders of magnitude, aphids are structurally very similar except for the adaptive superproportional relationship between the length of the mouth parts and body size. This uniformity supports the view that the basic biology of aphids is very similar, and it is the need to be able to reach the phloem elements in different parts of plants that has mainly determined their interspecific diversity in size.

4.5 Size of migrants

Migration is the process by which organisms redistribute themselves in space (Taylor and Taylor, 1983). In aphids this can be achieved by walking (Harrington and Taylor, 1990) or by flying. Although there is a continuum from short to long distance movements, there is a very clear distinction between those individuals that migrate by flying between habitats as opposed to walking within habitats. This is reflected in their morphology and the distance they are likely to travel, with the unwinged individuals relatively more sedentary than the winged individuals, which here are referred to as the migrants.

Whether cruising or hovering, aerodynamic theory predicts that large individuals need to invest proportionally more resources in flight than small individuals. In winged aphids the flight apparatus, including wing muscles, wings, etc., competes for resources (Figure 4.15). If, as concluded above, aphids are structurally similar then they need a flight apparatus, the size of which scales with the 7/6 power of their body weight (Dixon and Kindlmann, 1998). Therefore, a part of the energy assimilated, proportional to adult weight 7/6, needs to be devoted to the building of the flight apparatus during larval development, after which the rest of the energy assimilated is partitioned in exactly the same way as in unwinged aphids. There are two

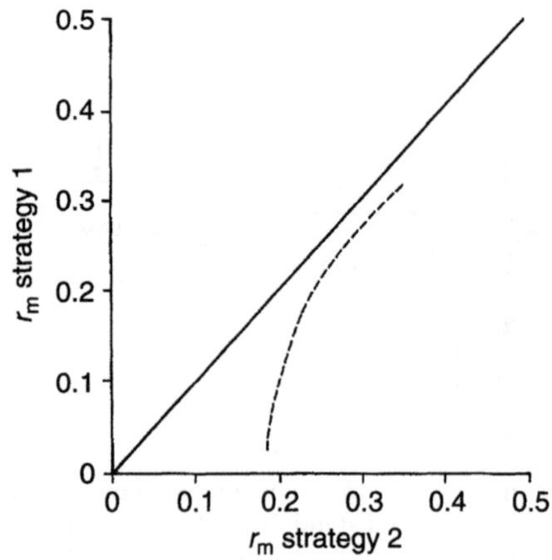

Figure 4.16 The instantaneous population growth rates predicted by the optimal partitioning of energy model when the flight apparatus competes (strategy 1) with gonads for resources relative to that when it competes (strategy 2) with the soma and gonads.

ways of achieving this: the wing apparatus can compete directly with the soma and gonads (Figure 4.15(2)) or only with the gonads (Figure 4.15(1)) for resources. If the former, then the winged aphid is smaller and takes longer to reach maturity than unwinged aphids developing under the same conditions, i.e. it faces similar problems to an unwinged aphid developing on poor-quality food: there is less energy available for partitioning between soma and gonads. If the latter strategy, the combined gonads and flight apparatus are equivalent to the gonads of unwinged aphids (Figure 4.15(1)), i.e. the flight apparatus competes directly with gonads for resources. The optimal partitioning of energy model (p. 61) predicts that this would result in winged and unwinged aphids taking the same time to develop and achieving the same size, but result in a smaller r_m (Figure 4.16).

The resource partitioning that results in the greater r_m predicts that migrants should be smaller and take longer to develop than the non-migrants developing under the same conditions. The prediction regarding the developmental time resolves the previous puzzling 6–26% prolongation in the duration of development recorded for winged individuals of several species of aphids (Dixon and Howard, 1986; Tsumuki *et al.*, 1990). However, what evidence is there that migrants are smaller than non-migrants?

The volumes of the bodies ($l \times b^2$) of the winged forms of 68 species are significantly smaller than those of the unwinged forms. In addition, the very close linear relationship between the logarithm of the volume of the winged and the unwinged forms of these 68 species has a slope that is less than 1, which indicates that large species have proportionally smaller migrants than the small species (Figure 4.17), i.e. the empirical data not only supports the

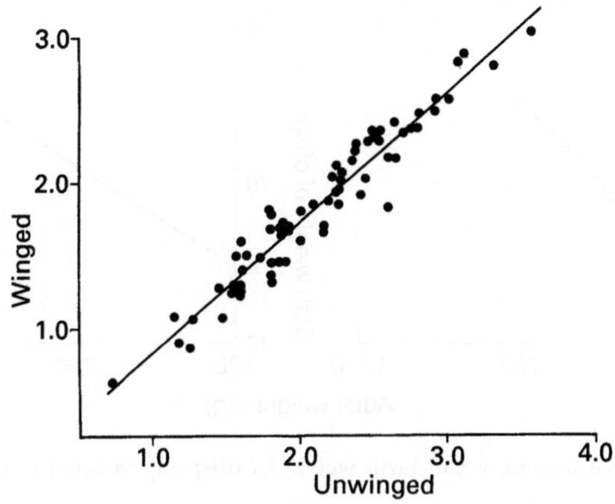

Figure 4.17 The relation between the logarithm of the volume ($l \times b^2$) of the winged and unwinged individuals of 68 species of aphid.

prediction that winged forms should be smaller than unwinged forms but also the additional prediction that large species incur proportionally greater costs than small species in developing a flight apparatus. The greater cost is because the minimum power necessary for take-off and flight scales relative to body weight with the exponent 7/6. This prediction is supported by the observation that the relative size of the head and thorax and associated appendages, the locomotory component of an aphid's body, scales relative to body size with an exponent of 1.13 (Figure 4.18), which is significantly greater than 1 and not significantly different from the 1.17 (7/6) predicted.

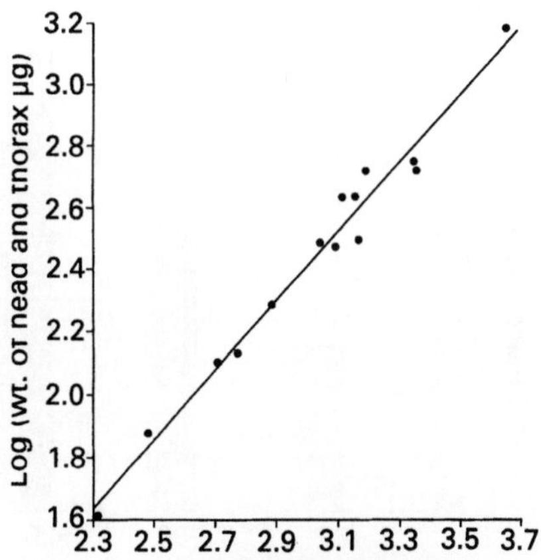

Figure 4.18 The relation between the logarithm of the weight of the head, legs, thorax and wings, and logarithm of the total weight for 14 species of aphid.

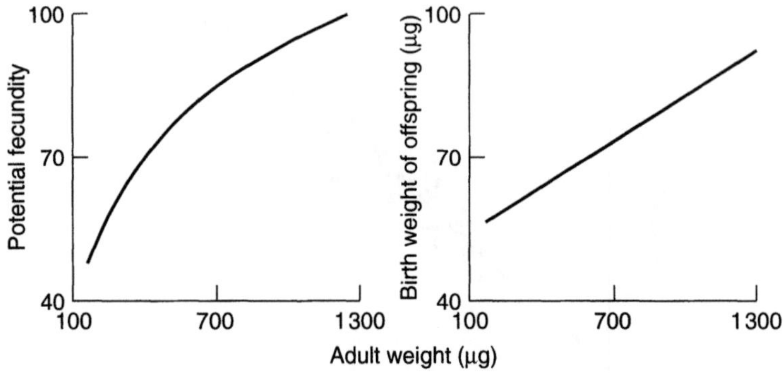

Figure 4.19 Potential fecundity and birth weight of offspring in relation to adult weight in *Aphis fabae*.

4.6 Reproduction

The largest individuals in many species produce the most, and the largest, offspring (Figure 4.19) which are fitter than small offspring because they are more likely to survive harsh conditions (Figure 4.20) and avoid capture by predators (Figure 4.21), or parasitization by hymenopterous parasites (Mackauer, 1973). Thus, in general within a species, large individuals have a greater reproductive potential than small individuals.

Although large species produce large offspring there is not a very close association between these parameters (*cf.* Figure 4.12). However, there is a close association between the logarithm of the birth weight of a species and

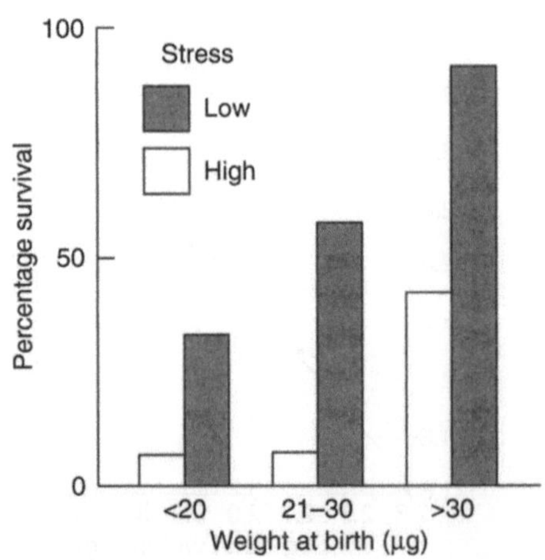

Figure 4.20 Percentage survival in relation to birth weight of *Aphis fabae* when reared on poor- and good-quality (stippled and unstippled) host plants, on which 88 and 44% died, respectively.

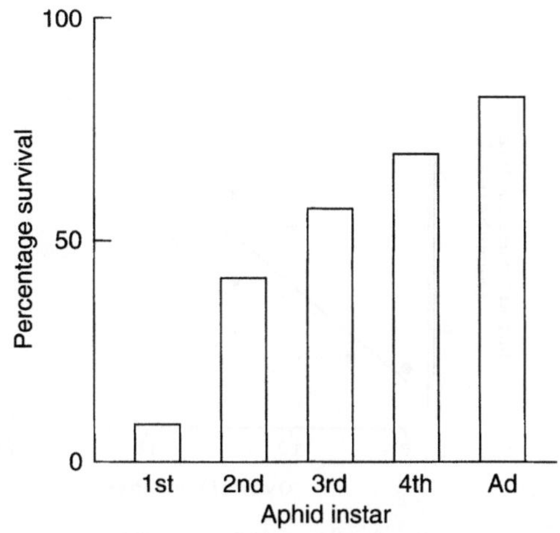

Figure 4.21 Percentage of the various instars of the nettle aphid, *Microlophium evansi*, that survive encounters with fourth instar larvae of the 10-spot ladybird beetle, *Adalia decempunctata*.

the logarithm of the mother's weight divided by the number of ovarioles in her gonads (Figure 4.22). This indicates that at the interspecific level birth weight is a constant proportion of adult weight if ovariole number is kept constant, and that a species with few ovarioles gives birth to relatively larger offspring than a similar-sized species with many ovarioles. Species with a high number of ovarioles have a high initial reproductive rate and *vice versa* (Figure 4.23). However, species that have a large number of ovarioles have

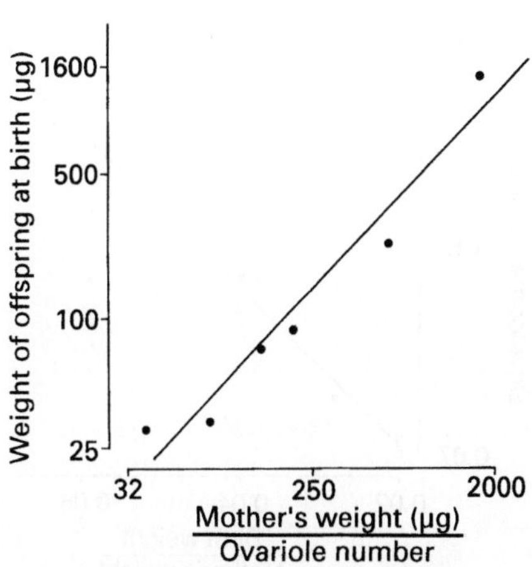

Figure 4.22 The birth weight in relation to the ratio of mother's weight to ovariole number for six species of aphids of a wide range of weights and including the largest British species, *Stomaphis quercus*. (After Dixon, 1987d.)

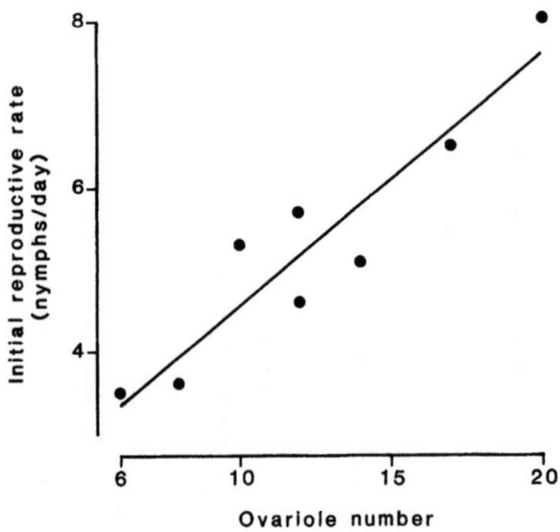

Figure 4.23 The initial reproductive rate at 20°C in relation to ovariole number for eight species of aphids. (After Dixon, 1987d.)

a higher initial reproductive rate, proportionally lower birth weights and markedly longer developmental times than those species that have few ovarioles (Figure 4.24). Accepting that there is a constraint on the minimum birth size in each species, and the optimum adult size is 15 times the birth size (p. 70) it is surprising that aphids do not all have the same number of ovarioles. That there is variation in ovariole number between species might indicate that for some species the advantage of a shorter developmental time more than offsets the reduced reproductive rate.

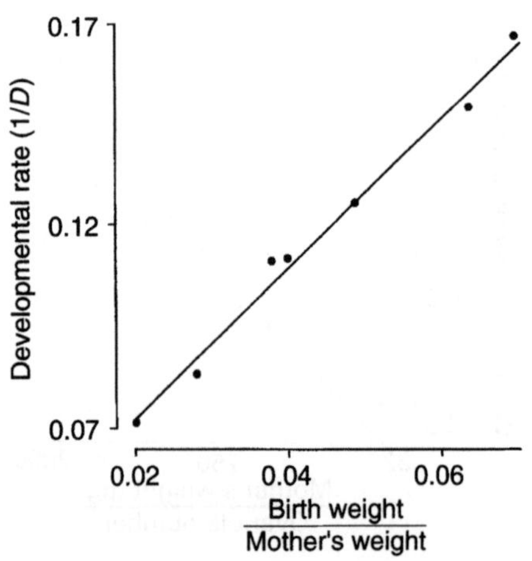

Figure 4.24 The developmental rate (1/*D*) at 20°C in relation to relative size at birth for seven species of aphids. (After Dixon, 1987d.)

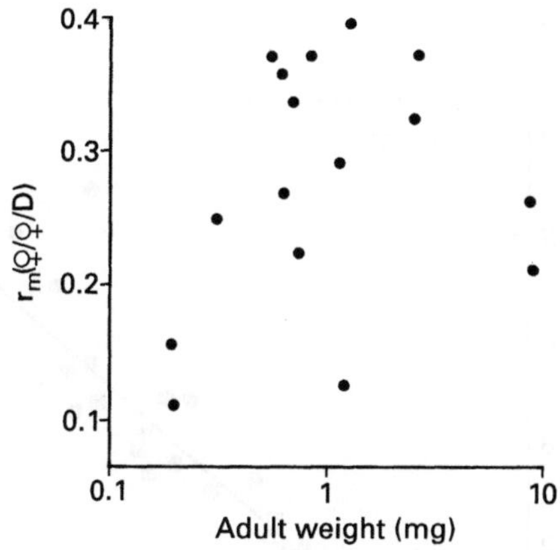

Figure 4.25 Intrinsic rate of natural increase (r_m) at 20°C in relation to body weight for the unwinged morph of 16 species of aphids. (After Dixon, 1990b.)

The intrinsic rate of increase (r_m) of a species is dependent on its developmental rate and age-specific fecundity and survival schedules, and generally the rate of development is more important than fecundity and the rate of reproduction in early adult life more important than the total number of nymphs born in determining r_m. The inverse relationship between r_m and body size described for heterothermic animals from a range of taxonomic groups by Fenchel (1974) does not hold for aphids, even assuming that temperature and weight can be dissociated (*cf.* p. 64; Figure 4.25). In addition, changes in food quality can affect r_m.

On reaching maturity an aphid's embryos make up a large proportion of its weight and its larger embryos have embryos developing within them, i.e. growth and reproduction occur simultaneously, with the young that are to be produced in early adult life being ovulated and achieving most of their embryonic growth during their parents' development (Dixon, 1985a). Therefore irrespective of its size as an adult, an aphid that has achieved a high growth rate is likely to be able to maintain a high reproductive rate and a high rate of increase (r_m). This has been shown for *Rhopalosiphum padi* reared on a range of hosts of different nutritional qualities and at several different temperatures (Leather and Dixon, 1984). It also holds for the six species for which there are measurements of the mean relative growth rate and r_m (Figure 4.26).

In those unicellular organisms that reproduce by division, all tissue is potentially immortal, and they are capable of achieving an r_m/RGR ratio close to 1. Multicellular organisms have a body made up of a soma, the assimilatory and mortal part, and gonads, the reproductive and potentially immortal part. Investment in soma results in an r_m/RGR ratio of less than 1. For realistic values of assimilation, gonadal growth and size of soma at birth, the optimal partitioning model predicts that the r_m/RGR ratio for

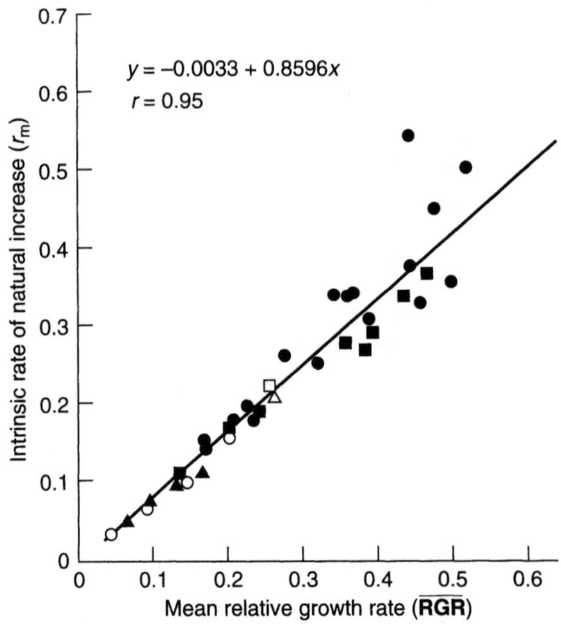

Figure 4.26 Intrinsic rate of natural increase (r_m) in relation to relative growth rate for six species of aphids reared under a wide range of conditions. Open circles, *Drepanosiphum acerinum*; filled triangles, *Drepanosiphum platanoidis*; open squares, *Eucallipterus tiliae*; filled circles, *Rhopalosiphum padi*; filled squares, *Sitobion avenae*; open triangles, *Tuberolachnus salignus*. (After Barlow and Dixon, 1980; Leather and Dixon, 1984; Llewellyn *et al.*, 1974; Watson, 1983; Wellings, 1981.)

aphids will be in the region of 0.9, which agrees with empirical data (Figure 4.26; Kindlmann *et al.*, 1992). That is, the cost to aphids of having a soma, is a reduction of approximately 0.1 in their potential r_m/RGR ratio.

Summarizing, aphids show a wide range of size both within and between species. Within species both food quality and temperature affect size. However, for any particular temperature and food quality there is only one size at which r_m is maximized, i.e. the phenotypic plasticity in size is adaptive. Accepting there is a constraint on the minimum size at birth, with species feeding on deeply located phloem elements being large at birth and *vice versa*, then interspecifically the optimum adult weight for maximizing r_m is approximately 15 times the birth weight.

Although within species small individuals are not small duplicates of large ones, nevertheless, different species, except for the relative sizes of their probosces, are structurally very similar. Accepting that aphids are geometrically similar then the resources that winged forms invest in a wing apparatus should scale with 7/6 power of their body weight. The consequences of this is that the winged form of a species should be smaller than the unwinged form and more markedly so in the large species.

Both empirical data and theory indicate that aphids should not show the interspecific inverse relationship between r_m and body size characteristic of several other taxa. However, aphids that achieve a high rate of growth (RGR) maintain a high reproductive rate and high rate of increase. The investment in soma necessary to support the high growth and reproductive rates reduces the maximum possible r_m/RGR ratio by approximately 10%.

5 Resource tracking: mechanism – cyclical parthenogenesis

The term 'parthenogenesis' was coined by Richard Owen in 1849, who defined it as the successive production of procreating individuals from a single ovum. He also viewed the increase in numbers by parthenogenesis as similar to growth, a view which was supported by Huxley in 1858 and much later by Janzen (1977), who stressed that in apomictic parthenogens natural selection acts at the level of a clone, which he designated the 'evolutionary individual' (p. 100), i.e. a clone produced by parthenogenesis is equivalent to the body of a sexually reproducing organism although split up into a number of physically separate units.

Most species of aphid reproduce both asexually and sexually, with several generations of apomictic parthenogenesis between each bout of sexual reproduction, i.e. they show cyclical parthenogenesis. Owen (1849) thought sexual reproduction occurred on the exhaustion of the spermatic force associated with fertilization. However, a hundred years earlier, Bonnet (1745) had correctly attributed the appearance of sexuals to the effect of long chilly autumnal nights. The production of sexual morphs is triggered by environmental cues (p. 114) and usually has an annual periodicity. Parthenogenesis is widespread in plants but in animals is commonly found only in rotifers, cladocera, cynipids and aphids. The earliest aphids are likely to have been oviparous, like the closely related Adelgidae and Phylloxeridae, with parthenogenesis and viviparity evolving as early as the Palaeozoic (p. 3). Although parthenogenesis is widespread in animals and plants it is relatively rare (Maynard-Smith, 1984) but predominant in climax situations that are maintained by disturbance, which tends to temporarily denude an area of its animals and plants. The high rate of increase of parthenogens and the ability of an individual to establish a colony gives them considerable advantage over sexual species when colonizing new habitats (Cuellar, 1977). That all species of aphids have some parthenogenetic generations favours the suggestion that, irrespective of whether they live on herbaceous plants or trees, aphids are opportunists exploiting ephemeral habitats, which are continually being created (Dixon, 1987b).

As a consequence of parthenogenesis each genotype is represented by many individuals. As the individuals of a clone are likely to be widely dispersed within a habitat, a clone is much less likely to be wiped out by natural enemies or to fail to locate a suitable host plant. Competition between clones for resources is occasionally likely to be intense. This, together with the variation in the ability of clones to locate and feed on certain plants, avoid death from natural enemies and withstand periods of stress, determines which genotypes survive. Although it may be useful to equate reproduction with sexuality, and growth with asexuality, nevertheless, there is a crucial difference between 'reproducing' to make two aphids and growing to make one aphid twice as large. Reproduction involves passing through a single-cell stage, therefore a mutation present in the genetic material of that cell can affect development to cause reorganization of the body. Growth does not involve passing through a single-cell stage, and therefore a mutation can only affect a small group of cells and cannot cause reorganization of the body (Dawkins, 1982). However, in treating a clone as an 'evolutionary individual' it should not be forgotten that cyclical parthenogenesis creates other evolutionary possibilities. A clone can simultaneously experience different environments and hence different evolutionary forces, and it can recombine simultaneously with numerous genotypes in the population, including itself (Mackenzie and Guldemond, 1994).

5.1 Telescoping of generations

As a parthenogenetic egg does not require fertilization, it can begin to develop as soon as it is ovulated. Under congenial conditions aphids take approximately 1 week to develop from birth to maturity, whereas other similar sized insects take approximately 3 weeks. However, if one takes into consideration that an aphid starts developing inside its grandmother, then the actual development time is 2.5 times longer than it takes an aphid to develop from birth to maturity, i.e. approximately 3 weeks. Comparison of the generation times of organisms of a wide range of sizes clearly indicates that the larger and more complex organisms have much longer generation times than smaller organisms (Bonner, 1988). This trend suggests that organisms the size of aphids should have generation times in the order of 1 month and mites in the order of 1 week.

Therefore, there appears to be a minimum 'time' required for development, which is a function of the size/complexity of the organisms. Parthenogenesis has enabled aphids to start developing at ovulation and more importantly inside immature or even embryonic mothers. Then given that there is a constraint on the rate of development, there are great advantages in telescoping generations (Kindlmann and Dixon, 1989a) which has given aphids approximately a three-fold reproductive advantage (Dixon, 1990a).

In this way aphids have been able to achieve rates of population increase normally associated with much smaller organisms like mites.

The short generation time also enables aphids to track very closely the seasonal changes in their resources. Individuals in each generation must be able to survive the worst conditions they are likely to experience, which in long-lived individuals is likely to constrain their performance when conditions are favourable. In addition the seasonal sequence of short-lived generations show generation-specific strategies, by which they can anticipate in terms of morphology and physiology the seasonal changes in conditions (p. 161). This close tracking in time and matching in morphology and physiology the seasonal changes in resources has been a major factor determining the great abundance of many species of aphids relative to their resources.

5.2 Intra-clonal variation

Cognetti (1961) claimed that genetic recombination occurs during parthenogenesis and provides a significant and continued source of variation. Endomeiosis, as he called the process, could result in progeny differing genetically from their parents. However, most of the histological evidence indicates that parthenogenetic eggs are produced by mitosis during which no chiasmata are formed (Suomalainen, 1950; Blackman, 1979; Suomalainen *et al.*, 1980). The stability of aphid clones heterozygous for colour (Müller, 1962) and enzymes (Blackman, 1979; Suomalainen *et al.*, 1980; Tomiuk and Wöhrmann, 1981), and DNA fingerprinting (Carvalho *et al.*, 1991) indicate that no significant amount of recombination occurs during the development of parthenogenetic eggs.

The experimental evidence presented by Cognetti and others in support of endomeiosis should be treated circumspectly as it involved selecting for features like apterousness, which is under environmental control (p. 104). In selection experiments in which environmental conditions were rigidly controlled, there was no significant change over periods of 18–170 generations in the percentage of individuals showing the character selected for (Ewing, 1916; Lees, 1966; Blackman, 1979).

Blackman (1979) attributes the undoubted variation that occurs in laboratory 'clones' to contamination, mutation, mitotic recombination and variability of gene expression. Contamination is important and the longer a clone is maintained the greater the risk. Mutations that result in a marked change in fitness, as, for example, resistance to an insecticide or ability to colonize another host plant, are likely to occur as frequently in parthenogenetic as in sexually producing populations. In addition, genetic changes involving dissociation or rearrangement of chromosomes may also occur and may even be quite common (Blackman and Takada, 1975). Mitotic recombination whereby crossing over can occur and heterozygous genes can segregate in somatic cells is a rare event and has only been conclusively demonstrated in some Diptera.

In a given lineage an aphid can be winged or unwinged, parthenogenetic or sexual, hence different genes are active at different times. The sensitivity to environmental cues can be determined by the genotype, for example, the different response of green and pink forms to crowding (Sutherland, 1969a) and geographical variation in the photoperiodic responses and developmental threshold in the pea aphid (Campbell *et al.*, 1974; Lamb and MacKay, 1983). Chance may also be important in determining whether an embryo develops into a particular morph, especially when conditions do not significantly favour one or other course of development, and may account for the mixed families observed under these conditions.

Thus, there are several ways in which variability could occur within aphid clones. Although genetic changes occur relatively rarely, the extremely large populations of many aphid species substantially increase the likelihood of such changes and then parthenogenetic **reproduction** increases the likelihood of incorporating the changes at the organ level.

5.3 Adaptability of aphid clones

Blackman (1981) recounts that in 1953–54 a parthenogenetic female of the spotted alfalfa aphid (*Therioaphis trifolii* forma *maculata*), of Mediterranean origin, was introduced and established a clone in New Mexico. By 1956 it had spread through most of the southern US. It quickly developed resistance to organophosphorous insecticides and in 1958 attacked a previously resistant variety of alfalfa.

Before 1960 the aphid overwintered as parthenogenetic females in the southern US and migrated north each summer. Then, in the autumn of 1960, a population in the centre of Nebraska produced viable sexual morphs and passed the winter there in the egg stage. By 1964 the new variant had colonized Nebraska and extended the range of the species northward into South Dakota, an area previously not reached by the annual migration from the south, and another population had also started to reproduce sexually in Wisconsin. Today the aphid is found throughout the US and reproduces sexually, overwintering as eggs over most of the northern part of its range. In the south-western states where sexual reproduction has not been recorded, clones capable of colonizing previously resistant cultivars of alfalfa continue to appear. Therefore, the lack of genetic recombination in parthenogenetic reproduction has not prevented this aphid adapting to new conditions (Blackman, 1981).

Aphid population size can be very large, e.g. 2.5×10^6 aphids on a 20 m sycamore tree, 4×10^8 cereal aphids/hectare and 1.7×10^{11} aphids in one valley in California. With mutation rates of from 10^{-5} to 10^{-9}/gene/generation, even a rare mutation is likely to have occurred at least 1.7×10^{-2} times per generation in the Californian valley (Dickson, 1962). Therefore provided mutations occur, change becomes a matter of certainty rather than chance in such large populations. As an individual with an advantageous mutation may pass it to all its descendants, which can be as many as a hundred, the mutation is likely to spread very rapidly.

The expectation is that exclusively parthenogenetic reproduction should lead, despite compensating mutation (Wagner and Gabriel, 1990), to rapid extinction on an evolutionary time scale, with the clonal species that do evolve not giving rise to higher taxonomic categories (Suomalainen *et al.*, 1976, 1987). As pointed out by Maynard-Smith (1988) and Williams (1992) there are notable exceptions in the exclusively parthenogenetic rotifer order Bdelloidea and gastrotrich order Chaetonotoidea that each consist of several families and many genera. Within the Aphidoidea there is also an apparent exception to this notion, which is the wholly asexual Tramini. This tribe consists of four genera and 32 species (Eastop, 1977), and it would be of considerable interest to know more about this group's taxonomic status and evolutionary history. In addition molecular markers indicate that asexual clones have coexisted with sexual clones of *Rhopalosiphum padi* for nearly a million years (Simon *et al.*, 1996).

5.4 Genetic structure

Host adapted races in aphids have been known for more than 140 years (Walker, 1850) and almost half of the 36 insects pest species known to have biotypes are aphids (Tomiuk, 1990). Yet allozyme studies have revealed surprisingly little genetic polymorphism in aphids (Tomiuk and Wöhrmann, 1980). However, genetic fingerprinting is revealing genetic diversity (Carvalho *et al.*, 1991). The latter confirms what the breeding of plants resistant to aphids and the study of colour polymorphism in aphids has indicated. At least four biotypes of the European raspberry aphid (*Amphorophora idaei*) differentially infest resistant varieties of raspberry. The four aphid biotypes result from segregation at two gene loci and the plant resistance to aphids from different combinations of 14 genes. Surprisingly, the aphid strain that can colonize most raspberry varieties is consistently the least common. This has been attributed to its low sexual fertility (Briggs, 1965; Knight and Alston, 1974). Currently 90% of the UK raspberry hectarage is planted with cultivars containing one or more resistant genes (Jones, 1988). Associated with this has been a shift in the biotype constitution of the aphid population. In the 1960s, 97% of the aphid population was made up of the avirulent biotype 1 (Briggs, 1965), whereas in 1990–91, 77% of the population was made up of two resistance breaking biotypes: 2 and *x* (Birch *et al.*, 1994). Similar changes in the biotype constitution of greenbug, *Schizaphis graminum*, populations have produced devastating increases in host range and breakdown of genetic resistance over a period of 10 years or less in the

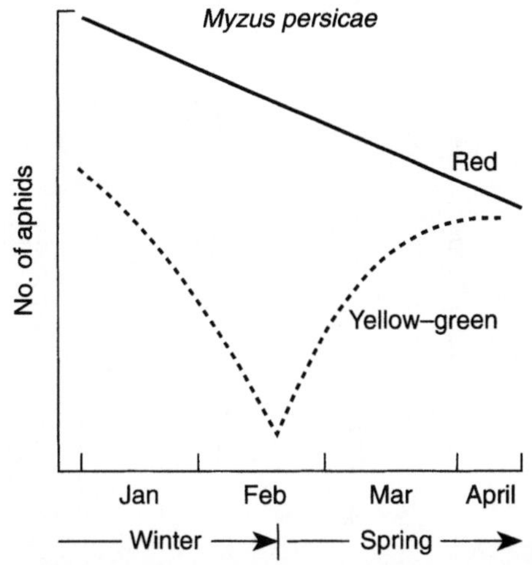

Figure 5.1 Population trends of red and yellow–green strains of *Myzus persicae* on Chinese mustard in the field from January through to April in 1978. (After Ueda and Takada, 1977.)

US (Puterka and Peters, 1990). That is, gene-for-gene relationships exist between aphids and their host plants, and there is a great genetic potential in aphids for plant resistance breaking even in mainly asexual species like the greenbug.

The pea aphid (*Acyrthosiphon pisum*) in north-west and central Europe is predominantly holocyclic and is made up of a number of strains that are each differently adapted to living on a range of species of leguminous plants (Müller, 1962). The green strains of this aphid introduced into North America, and more recently Australasia, lack the red allele common in European populations. The green strains are particularly well adapted to living on peas and have been assigned to the subspecies *Acyrthosiphon pisum destructor*. Similarly, only a portion of the Old World *Therioaphis trifolii* genome was introduced into North America (p. 85).

Several species of aphid have stable colour morphs that are genetically determined, e.g. the red allele is dominant to the green in the pea aphid (Müller, 1962) and in *Myzus persicae* (Takada, 1981). How do these strains manage to coexist? In certain polyphagous aphids, the individuals of one colour are better adapted to living on certain plant species within the aphid's host range, as in the pea aphid (Müller, 1962; Markkula, 1963; Ueda and Takada, 1977). For the same species it has also been claimed that colour polymorphism is maintained by the different susceptibility to parasitism and predation of the colour morphs (Losey *et al.*, 1997). In *Myzus persicae* red anholocyclic strains (which do not reproduce sexually) are better able to survive winter conditions than the green strains; then in the spring the green strains become predominant, increasing in abundance relative to the red

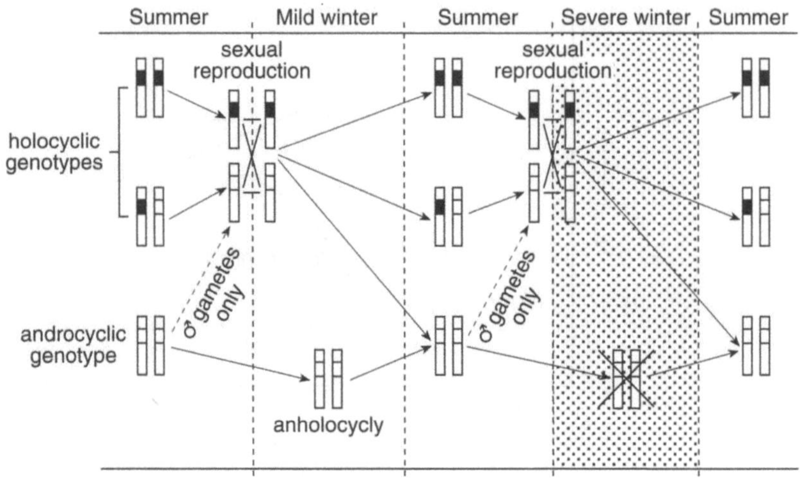

Figure 5.2 Maintenance of life cycle variation in *Myzus persicae* through three seasons. Androcyclic clones are generated afresh by the breeding system each year, even after a severe winter when parthenogenetic morphs outdoors are eliminated. (After Blackman, 1976.)

(Figure 5.1). Thus in addition to having different host ranges the strains may differ in their selective advantage during the course of a year. However, the significance of the colour (brown, yellow, red, etc.) to each of the strains of these aphids is unknown. There is also evidence for density-dependent differences in genotypic fitness. Changes in particular gene frequencies with increasing population density in the rose aphid (*Macrosiphum rosae*) indicate that the reproductive rate of one genotype decreases and another increases when population density itself increases (Tomiuk and Wöhrmann, 1981).

The range of genotypes present in aphid populations makes the breeding of aphid-resistant plants particularly difficult. In screening for resistant varieties of plants as great a range as possible of the naturally occurring aphid genotypes should be used. A rare resistance-breaking genotype could increase only too rapidly after the introduction and widespread planting of a new 'resistant' variety of crop plant.

Some species tend to become anholocyclic. A genetic change results in failure to produce sexual forms, and may result in the coexistence of holocyclic and anholocyclic forms of the same species, as in *Myzus persicae*. In response to short photoperiods, some clones produce oviparae and males (holocyclic); most produce some males and viviparous females (androcyclic) and a few do not produce sexual morphs at all (anholocyclic). These life cycle traits are inherited and determined monofactorially, with androcycly recessive to holocycly. Androcyclic and anholocyclic clones are likely to be killed in severe winters. However, they can be generated afresh from the holocyclic clones (Figure 5.2) (Blackman, 1976). The proportion of holocyclic, androcyclic and anholocyclic strains in the population is likely to vary from year to year in response to winter conditions, which are a major factor in determining the relative fitness of the strains. When winters are consistently mild and the environment is uniform and predictable, it is likely that an aphid species will become anholocyclic.

5.5 Sex

Both sexual females and males are produced parthenogenetically in response to external and/or internal cues (p. 114). The sexual females have two X chromosomes as do the parthenogenetic females. In order to develop into a male, however, oocytes have to shed half the sex chromatin of the parent female in the course of their single maturation division.

The X chromosomes undergo what Orlando (1974) refers to as a 'mini meiosis' in which they pair and then separate; the chromatids that remain on the equator of the spindle separate with the autosomes (Figure 5.3) and give rise to an XO male. Males are potentially capable of producing two types of sperm but during meiosis those without an X chromosome (Figure 5.3) degenerate. As all viable sperm carry an X chromosome, fertilized eggs always give rise to females. An XX/XY system of the type seen in some other insects would be an impossibility because most species of aphids exist for a large proportion of a year as all-female populations, and during this period the Y chromosome would have 'nowhere to go' (Blackman, 1995).

5.6 Why reproduce sexually?

As anholocyclic clones of aphids have been able to develop resistance to insecticides and colonize previously resistant cultivars of crop plants relatively quickly, and above all have a higher rate of increase than holocyclic strains, it is surprising that so few aphid species, only about 3%, are totally parthenogenetic (Blackman, 1980). What is the adaptive significance of sex for aphids?

Aphids predominate in the temperate regions of the world and to overwinter in a cold-resistant resting stage, the fertilized egg, is an adaptation to temperate conditions. This only accounts for egg laying in temperate regions, not sex. Pseudosexual reproduction in which females produce eggs without fertilization occurs in the closely related Adelgidae and Phylloxeridae and several other groups of animals. Therefore, in addition to asking why most aphids reproduce sexually one can also ask why aphids combined parthenogenesis with viviparity and sexual reproduction with oviparity.

How can sex be maintained in the short term by selection? After all, the greater rate of increase of an anholocyclic mutant could lead rapidly to the competitive elimination of holocyclic strains, but it does not.

The habitat occupied by an aphid species is not uniform but consists of a spatial–temporal mosaic of many different patches, each with its own

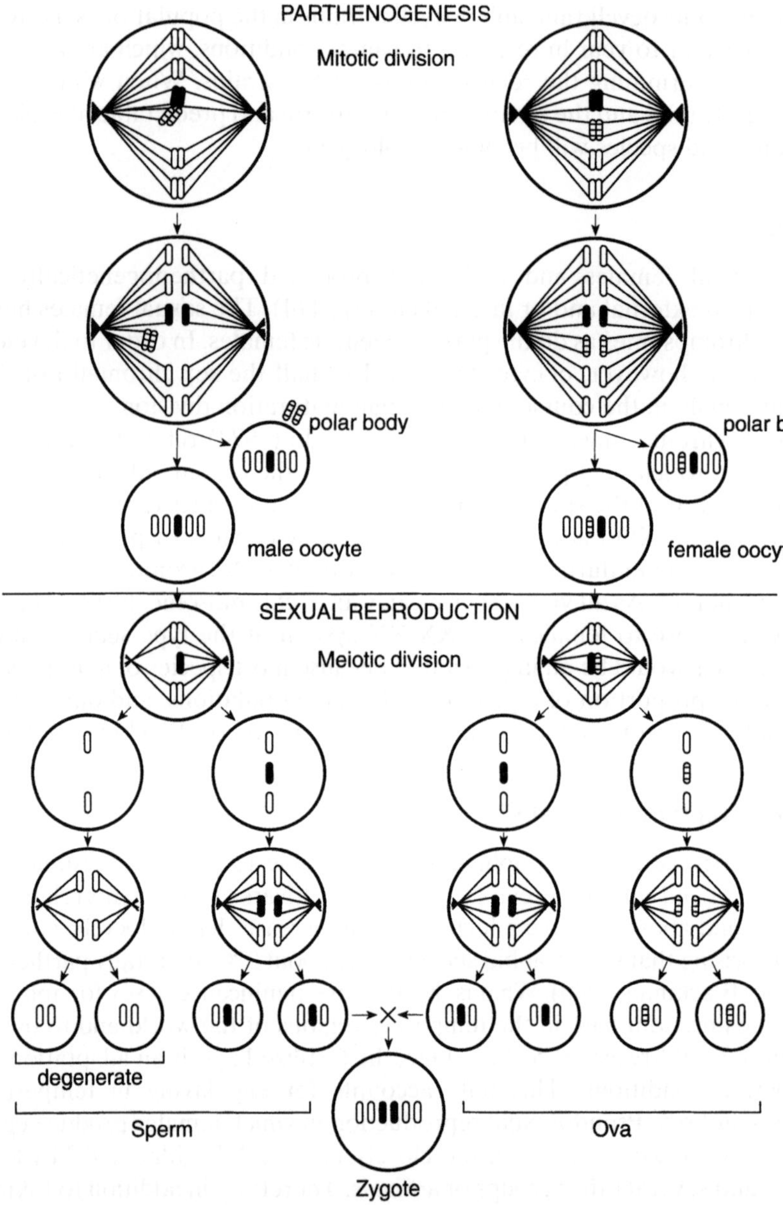

Figure 5.3 Diagram showing the behaviour of the chromosomes during the mitotic division leading to the development of male (XO) and female (XX) oocytes and the meiotic divisions in male and female aphids that result in the development of sperm and ova only from germ cells that carry an X chromosome. (After Blackman, 1980.)

complement of organisms and resources. An individual's exact location often determines its fitness. Thus one possible short-term advantage of sex is that it generates sibs with a range of genotypes and a range is more likely to include the fittest genotype for a particular patch than the single genotype of an asexual sibship would do. This is the basis of the 'lottery model' proposed by Williams (1975). The genetically similar individuals of an asexual sibship have the same ecological requirements and will compete severely with one another. A sexually reproducing parent reduces this competition by producing genetically dissimilar offspring. Thus genetically diverse sibs could have more 'elbow-room' as they are potentially capable of exploiting more than one kind of patch (Young, 1981).

It is possible that both 'lottery' and 'elbow-room' mechanisms operate simultaneously, but it remains to be demonstrated that either is relevant to aphids. However, both models assume environmental heterogeneity and imply that sex would be at a disadvantage in uniform habitats. Support for this comes from the tendency of many pest aphids not to reproduce sexually, even in temperate climates. Crop environments have far fewer interacting biological components, and are more uniform than natural habitats in which host plants, natural enemies and competitors are all generating variability. Possibly the single most important factor is the reduced variability of the host plant with a few genetically uniform cultivars being planted over great areas. Thus the habitat of many pest species has become more uniform and predictable and there is less advantage to continual adjustment of the genotype. This is especially true for aphid pests of glasshouse crops.

Blakley (1982) challenged the idea that genetic variation among host plants is important for the retention of sex in aphids. The criterion he used to establish the existence of 'host-specific superior aphid genotypes' surprisingly excludes the raspberry aphid (p. 86) and his survey does not include Ueda and Takada's (1977) work on *Myzus persicae*. However, even if it were unequivocably established that genetic heterogeneity among plants is not a major factor in the retention of sex, there are other environmental factors that generate habitat heterogeneity for aphids.

In parthenogenetic strains, deleterious or non-functional alleles will accumulate by what H. J. Muller (1964) has likened to a ratchet mechanism. Individuals with many mutant alleles will be selected against, but there is no way in which the overall load of deleterious genes can be reduced. Eventually, the size of the load will severely limit the possibility for further genetic change. However, the genetic recombination that occurs in sexual reproduction can expose the recessive deleterious mutants to selection in the homozygous form. Therefore, although a parthenogenetic clone may respond adequately to environmental change in the short term, its long-term evolutionary prospects are bleak because it has no means of effectively exposing recessive deleterious mutations to selection. That asexual and sexual clones of *Rhopalosiphum padi* have coexisted for nearly a million years (Simon *et al.*, 1996) indicates that the time scale is likely to be very long. Nevertheless, because of at least the two-fold reproductive advantage of parthenogens (*cf.* Newton and Dixon, 1988) over sexually reproducing species, it is difficult to account for the maintenance of sex by individual selection (Maynard-Smith, 1984).

Another factor, which might account for the maintenance of sex in aphids, is that many aphids show seasonally related generation-specific reproductive strategies (p. 161). If this is to be effective aphids need to reset their internal clock(s) each year, and going through an egg stage and/or sex could trigger this event. Those species that are exclusively parthenogenetic tend to live in situations where seasonality, so characteristic of wild plants, is greatly reduced or absent.

5.7 Timing of sex

Sexual reproduction precludes paedogenesis – an aphid that must mate cannot begin to mature its embryos before it is born. This means that a sexually reproducing female must either lay eggs or delay parturition until its offspring can mature. In autumn in temperate regions further increase in numbers is restricted by shortage of food and low temperatures, and most aphids enter a resting stage at this time. The production of overwintering eggs provides an excellent opportunity to reproduce sexually, since an egg need not be in an advanced stage of maturation to survive. Aphids have availed themselves of this opportunity: in all holocyclic species the sexual generation is the last in the season – that which produces the resistant eggs (Ward *et al.*, 1984).

5.8 Sex ratios

Fisher (1930) demonstrated that in the absence of inbreeding or competition between offspring of one sex for a mate the optimum investment ratio between male and female offspring is unity. This also applies if aphids are regarded as hermaphrodite and egg and sperm production is limited in the main by the same resources (Maynard-Smith, 1978). However, the relatively few sex ratios that have been recorded for aphids indicate a preponderance of sexual females over males.

Host alternation in aphids (p. 132) can be categorized as monomorphic or dimorphic on the basis of whether the remigrants that return to the primary host in autumn consist of one or two distinct morphs (Figure 7.3). The Anoeciidae, Hormaphididae and Pemphigidae are monomorphic, with the autumnal flight to the primary host being undertaken by a single morph, the sexupara. On the primary host the sexuparae give birth to both males and mating females. Because the sexuparae give birth to both sexes very quickly and they tend to stay together, a high degree of inbreeding is assured. Where there is local mate competition, parents are selected to economize on sons and female-biased sex ratios result (Hamilton, 1967).

This idea has been extended to aphids and in particular to situations where the sexuparae founding a local breeding group differ in their investment in offspring. The model developed by Yamaguchi (1985) predicts that each sexupara with more than a specified minimum investment in offspring should make the same investment in sons. The optimum number of sons depends on the size of the group of sexuparae and their average investment in offspring. It is difficult to see how an aphid could assess the size and total (or average) fecundity of the group of sexuparae it joins (May and Seger, 1985), especially as its offspring are nearly fully developed and their sex determined before it flies to the primary host.

Alternatively, the average size of the groups of sexuparae could be dependent on aphid abundance. In years when the aphid is abundant the size of the groups can be very large (Pergande, 1912; Marchal, 1928). In aphids, as in many insects, average adult size decreases as population density increases (Way and Banks, 1967). The gonads of aphids are bilaterally symmetrical and consist of a number of germaria and associated ovarioles arranged in pairs connected to a common genital pore. In the Pemphigidae the number of pairs of germaria that develop or ovulate depends on the size of the aphid and each germarium ovulates only one egg. The sex of each egg is possibly determined by the titre of juvenile hormone circulating in the mother at the time of ovulation (Hales and Mittler, 1983; Blackman and Hales, 1986) and the titre changes after the first pair, or two pairs, of germaria have ovulated. The exact instant of this change determines fitness and is thus subject to natural selection. As a result the first one or two pairs of germaria produce males and all subsequent germaria mating females (Lampel, 1969). Thus the small aphids produced when aphid population density is high produce a high proportion of males, and the large aphids produced when the aphid population density is low produce a relatively larger proportion of females. This model (Kindlmann and Dixon, 1989b) similarly predicts a constant absolute investment in sons. However, sexuparae do not assess the size or total investment of sexupara groups. The sex ratio is simply controlled through the affect of population density on size. At high aphid densities the size of the sexupara groups are likely to be very large and tend to fuse. In these circumstances outbreeding is more likely to occur and, as expected, the primary sex ratio approaches equality. However, in most years the aphids are less abundant, the sexupara groups small and the sex ratio female biased. Why the sexuparae of some species each give birth to two males and those of other species four males, etc., awaits a better understanding of the bionomics of aphids and their host plants.

In host-alternating species that are dimorphic and in autoecious species there is a greater certainty of outbreeding. In the host-alternating species it is virtually certain as the males and females return separately to the primary host. In the autoecious species the degree of inbreeding is likely to depend to a large extent on the size and length of life of the host plant, with large plants populated by more clones than small plants. However, even the host-alternating species, where outbreeding is certain, show very markedly female-biased sex ratios.

Most aphid clones are hermaphrodite (p. 114). Sex allocation by dimorphic host-alternating species, however, can best be described as sequential hermaphroditism: gynoparae are produced before males (Figure 5.5; Dixon and Glen, 1971). Resources are committed first to future ova, then to future sperm, i.e. the individuals in a clone undergo sex reversal from female to male. This analogy with sequential hermaphroditism in one-bodied individuals raises three questions (*cf.* Charnov, 1982) that must be answered if we are to understand the evolution of sex allocation in this group of aphids (Ward and Wellings, 1994):

1. Which sex should be first?
2. When should the clones change sex?
3. Should all clones change at the same time?

In many species the switch from parthenogenesis to investment in sex is cued by declining day length and temperature. Males or gynoparae are born on the summer host, develop to maturity and migrate to the primary host, where the gynoparae give birth to the mating females. These must reach maturity before they can mate and oviposit. There is thus a lag of one generation from cue to copulation if the clone produces males, and a two-generation lag between cue and oviposition if it produces gynoparae. In the bird cherry-oat aphid, *Rhopalosiphum padi*, the details of the cue mean that there is just enough time for gynoparae and oviparae to mature before leaf fall; the deadline for oviposition (Ward *et al.*, 1984). It is likely that this deadline applies to all other aphids in this group.

A game-theoretic model developed by Ward and Wellings (1994) predicts that the combination of the ecological deadline and the difference between the delays in reproduction through male and female functions means that gynoparae should be produced before males. As the deadline nears, there comes a time after which gynoparae are valueless but males still have time to mature, migrate and mate (switch (A) in Figure 5.4). Conversely, males born at the start of the sexual phase must wait until both the gynoparae and mating females have matured before they can mate, so are likely to suffer heavy mortality. That is, the fitness of the two sexes is changing in time and it is this that determines when clones should change sex.

When realistic values are used for the length of the delay, i.e. with mating female development time greater than 10% of the time between cueing of sex and the deadline for oviposition, the evolutionarily stable strategy is that all clones switch sex at the same time. If there is no lag the model predicts that half the clones should invest in males and the rest in 'gynoparae', i.e. it predicts Fisherian equality of investment in males and females. Whether the accumulated investment in sexes is male biased, equal or female biased depends on the rate of decline in abundance/mortality of the virginoparae

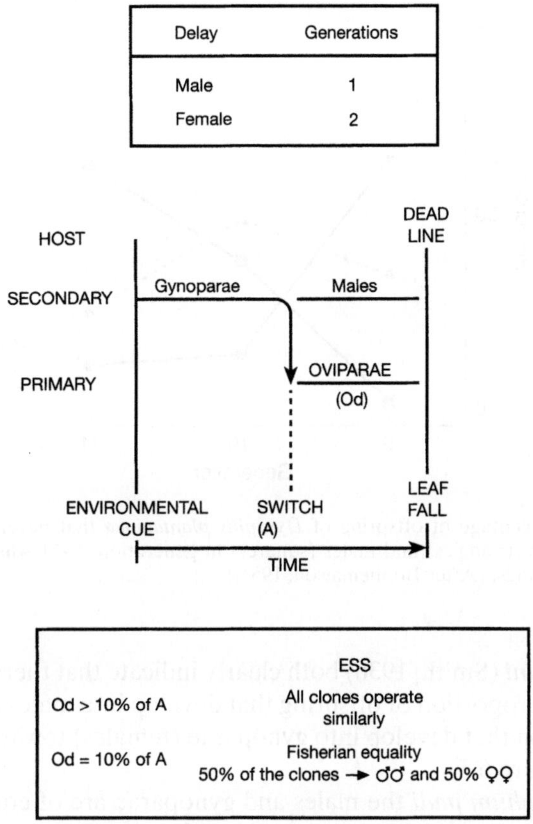

Figure 5.4 The delay in terms of generations required for the production of males and females, the environmental cue initiating the production of the sexuals on the primary and secondary hosts relative to the dead line (leaf fall), and the evolutionary stable strategy when the developmental time of the oviparae (Od) is greater than 10% or instantaneous relative to the time from the environmental cue initiating sexual production and the time of the switch to the production of males. (After Ward and Wellings, 1994.)

on the summer host and the length of the delay. The longer mating females take to mature, the longer early-born males must wait for mates, so the later the stable switch time becomes, the greater the investment in gynoparae. The more rapidly the source population on the secondary host declines, the fewer virginoparae survive to produce males. This should result in clones changing sex earlier, but the effect of this change is outweighed by the decrease in the number of virginoparae. The qualitative prediction is that they should invest less in males than in gynoparae.

What empirical evidence is there to support the idea of a switch from female to male production towards the end of the season in host-alternating aphids? Studies on *Dysaphis plantaginea* (Bonnemaison, 1951) and

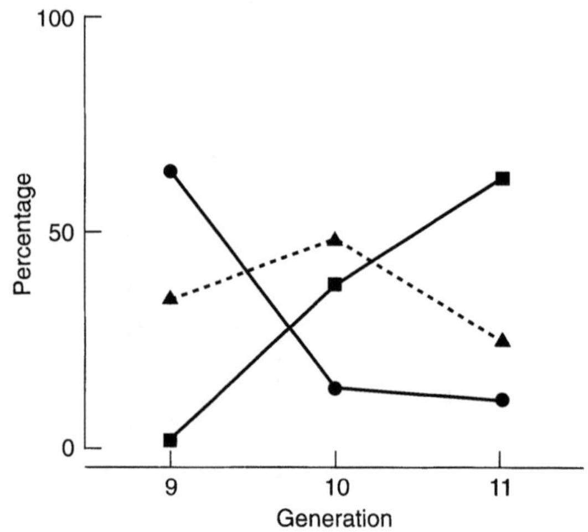

Figure 5.5 The percentage of offspring of *Dysaphis plantaginea* that develop into apterae (circles), gynoparae (triangles) and males (squares) in generations 9–11 when reared under autumnal photoperiods. (After Bonnemaison, 1951.)

Hyalopterus pruni (Smith, 1936) both clearly indicate that there is a marked increase in the proportion of offspring that develop into males and decrease in the proportion that develop into gynoparae (females) towards the end of the season (Figure 5.5).

In *Rhopalosiphum padi* the males and gynoparae are of equal weight at birth so it is not unreasonable to assume that the sex allocation ratio reflects the ratio of investment in the sexes. Suction-trap catches of the aphids migrating between the secondary and primary hosts indicate that in the field the investment is strongly female biased in virtually all years. In Sweden (Wiktelius, 1987) males made up 7.9% of the remigrants (range 5.8–37.7) and in Scotland 14.9 (range 4.8–82.4).

Autoecious species like the sycamore aphid, *Drepanosiphum platanoidis*, also end a season by producing males and mating females. These must mate and oviposit before leaf fall, but not be produced too early in a year since the switch to sex terminates the rapid reproduction of the parthenogenetic forms. There is thus a short mating season and an environmental deadline for successful sexual production. Unlike the above case the delay for the two sexes appears to be the same. In common with the above case, however, the fitnesses of the males and mating females are time dependent. In a population at ESS (evolutionary stable energy), sex allocation is initially male biased, then female biased and there is a final stage during which no males are produced. That is, the sycamore aphid should be protandrous and produce most of its males earlier rather than later or at the same time relative to the mating females. The overall

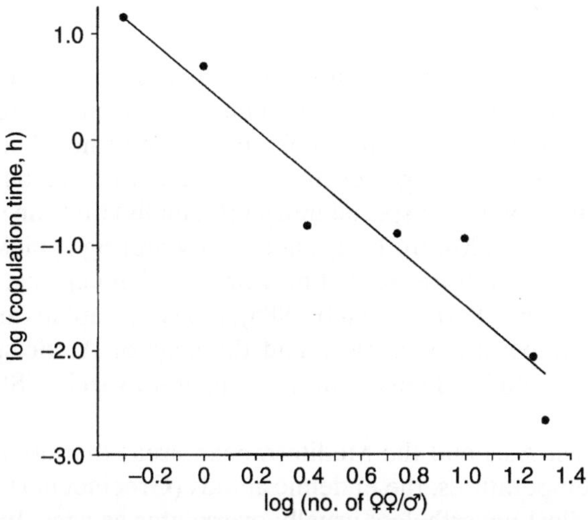

Figure 5.6 The relation between the time spent in copulation and the number of females to males in seven species of aphids.

allocation should be female biased, particularly if male mortality is low (Ward and Dixon, in prep.). The field data conforms very closely to this prediction.

5.9 Mate guarding

In *Pemphigus spyrothecae* males have been seen to push other males off females and in the presence of a second male to copulate for 3 times longer (Foster and Benton, 1992), i.e. the males of this aphid appear to be mate guarding.

As the numerical sex ratio varies greatly between species the expectation is that in those species where it is markedly female biased, males should spend most time searching for females and little time in copulation and *vice versa*. So far little attention has been given to studying the behaviour of males in aphids, possibly for a very good pragmatic reason, as they are only available for a relatively short period each year. The few results there are on copulation times and numerical sex ratios indicate a strong inverse relationship between copulation time and the number of females per male. Species like *Eriosoma ulmi*, where males out-number females 2:1, the males spend 14 h in copulation/mate guarding, whereas in *Chromaphis juglandicola* where there are approximately 18 females to every male the males only spend 30 s copulating with each female (Figure 5.6), i.e. the empirical data supports the prediction that 'mate guarding' should be most striking in those species where males out-number females and absent in those species where females greatly out-number males.

5.10 Why lay eggs?

Genetic recombination in animals and plants is often associated with a resting stage, with the new genotype(s) emerging from the resting stage at a time and place where they are likely to encounter novel conditions. Not all aphids overwinter as eggs even in the temperate regions: some overwinter viviparously or as a special morph (hiemalis) that can survive winter without feeding (p. 120). Both eggs and overwintering aphids in temperate regions contain an anti-freeze that may protect them against low temperatures (Sömme, 1969; Parry, 1979a,b, 1985). However, aphids are not as resistant to low temperatures as eggs, and the eggs of Arctic species have a higher level of cold hardiness than the temperate species (Strathdee *et al.*, 1995a).

In south-east Asia and the Mediterranean areas that do not experience low winter temperatures, the endemic aphids (Greenideinae, Hormaphidinae and Fordini) nevertheless usually overwinter as eggs. In the Mediterranean region the Fordini spend alternate winters in the egg stage and as young parthenogenetic females. Thus if the whole population is synchronized in this way, an egg stage would not appear to be essential for survival.

If, however, plant foliage is largely absent or of very poor quality for a time and/or the physical environment is harsh, aphids are in danger of desiccation especially if they live on the above-ground parts of plants. In these circumstances an egg could be an advantage. In this respect it is interesting to note that the special overwintering morphs produced by a few species overwinter in the soil, where humidity is likely to be high.

Thus the timing of the sexual stage in the life cycle, prior to the onset of harsh conditions, generates a new range of genotypes from which the fittest will be selected when conditions become favourable again. It then becomes a question of how best to survive the adverse conditions. On above-ground parts of plants desiccation may be the most important factor, and an egg the best means of combating it. In temperate regions there is the additional advantage that eggs are more resistant than other stages to very low temperatures.

Summarizing, aphids show cyclical parthenogenesis. Although there is no significant amount of genetic recombination during the development of parthenogenetic eggs, the large size of many aphid populations combined with parthenogenetic reproduction increases the likelihood of a mutation occurring and becoming established. This accounts for some of the adaptability of aphid clones. In addition the sexual reproduction that usually occurs once a year, just prior to the production of eggs, generates genetic diversity that further improves the chances of survival of the species in an heterogenous and changing environment. In addition, sex possibly, serves to reset the internal clocks that control the expression of the seasonally keyed generation specific strategies.

Those species in which the autumnal migrants found local breeding groups as theory predicts, economize on sons and produce, in terms of biomass, markedly female-biased sex ratios. However, even in those species where out-breeding is certain the sex ratio is also female biased. This appears to be due to the relative fitness of the sexes changing dramatically within a season. Those species in which there are numerically relatively few females per male spend more time in copulation, 'mate guarding', than those species where there is a marked preponderance of females.

6 Resource tracking: mechanism – polyphenism

Aphid populations are made up of clones, the 'evolutionary individuals' in Janzen's (1977) terminology. How a clone allocates resources to particular functions is likely to determine its fitness (Dixon, 1977). Individuals in each clone are involved to a varying degree in defence, dispersal, reproduction and aestivation/hibernation (Figure 6.1). However, specialization in one or other of these functions imposes constraints in terms of resource allocation, for carrying out the other functions. At certain times particular functions are more important than others for the overall fitness of a clone and this has resulted in the evolution of a division of labour within a clone that is reflected in its polyphenism. This implies that altruism is common in aphids. However, the general view is that only the soldier caste is altruistic. That most soldiers are sterile and are likely to die defending and safeguarding the survival of their clone mates supports the claim that they act altruistically. Winged forms similarly have a greatly reduced fecundity and a very low probability of surviving to reproduce, but in incurring this cost they benefit the overall fitness of the clone. It is also difficult to visualize how this benefit to the clone could be undermined by cheating, i.e. unrelated individuals profiting from the behaviour but making no contribution in return. Thus it would appear that winged forms also act altruistically. If this line of argument is accepted, then contrary to Hamilton's (1987) claim altruism is common in aphids.

Each clone goes through a seasonal life cycle made up of a sequence of morphs that differ in their behaviour, physiology and structure. In a few species there can be as many as eight morphs that differ at least in their external morphology. This polyphenism is characteristic of aphids and reaches its greatest development in host-alternating species (p. 128). The ancestors of aphids were winged, as is still largely the case for many species in the family Drepanosiphidae, in which only the sexual females lack wings. However, in many aphids the parthenogenetic females show alary dimorphism, i.e. they can be either winged (alate) or unwinged (apterous). Most individuals reproduce either asexually or sexually. The striking exception is the sterile soldier caste that has been recorded in species of the families Hormaphididae and Pemphigidae (Aoki, 1975, 1977).

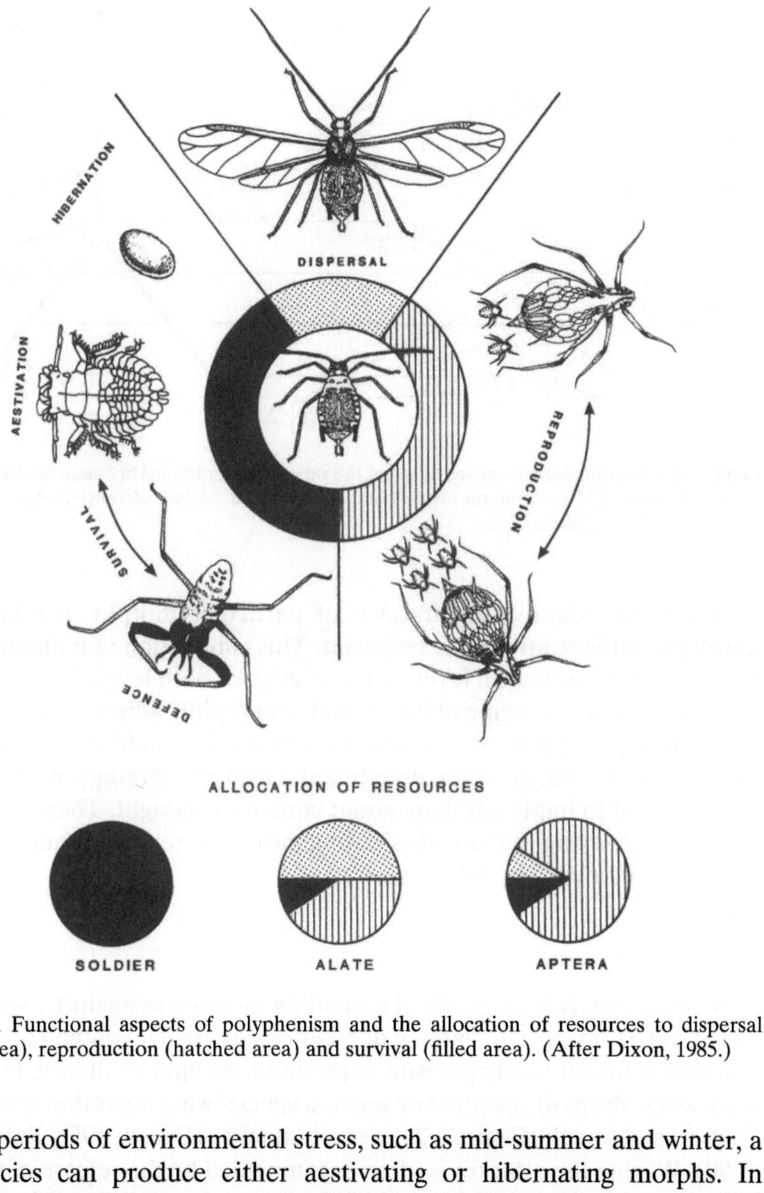

Figure 6.1 Functional aspects of polyphenism and the allocation of resources to dispersal (dotted area), reproduction (hatched area) and survival (filled area). (After Dixon, 1985.)

During periods of environmental stress, such as mid-summer and winter, a few species can produce either aestivating or hibernating morphs. In autumn most aphids produce sexual forms: usually wingless females and winged or apterous males, with a few species producing both apterous and alate males. After mating, sexual females lay eggs and this enables many species to survive the winter. The eggs resume development and hatch the following spring. Mating and the egg stage appear to be important in initiating a sequence of generation-specific strategies, which make up the annual life cycle.

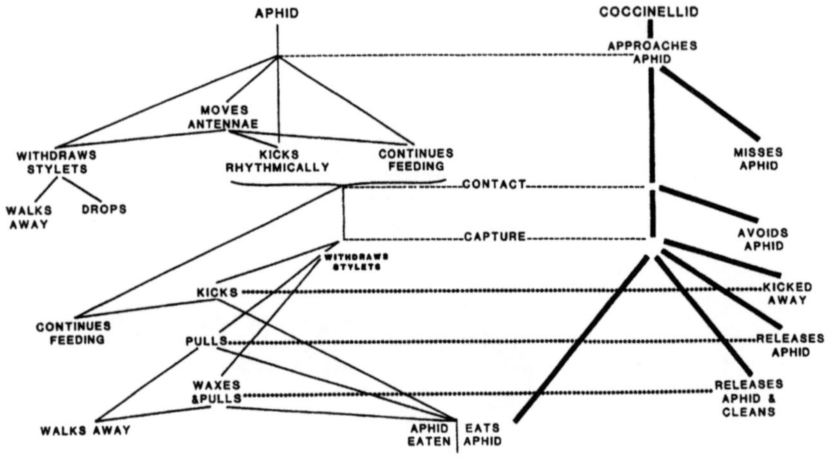

Figure 6.2 Diagrammatic representation of the possible interactions between the behaviour of the nettle aphid (*Microlophium evansi*) and the ladybird (*Adalia decempunctata*) during an encounter. (After Dixon 1958.)

Below the adaptive significance of particular morphs and how their development is controlled is reviewed. This can be done on either an ecological or a physiological level. A knowledge of both levels is necessary for a complete understanding of how aphids can rapidly adjust to, and to a very large extent anticipate, environmental change. It should be noted that the words anticipation, decision, investment, optimum, strategy and tactics are not intended to imply anything about conscious thought. They are a shorthand way of saying that aphids showing particular responses and structures have been favoured by selection.

6.1 Defence

Individual aphids are capable of defending themselves against a wide range of natural enemies, with varying degrees of success. Depending on the relative size of the attacking parasite or predator, an aphid will kick, attempt to walk away, drop off the plant or smear a special waxy secretion it can exude from its siphunculi (Figure 6.2) over an attacker (Dixon, 1958; Foldi-Hope, 1990). Waxing the attacker is often accompanied by the release of an alarm pheromone that alerts nearby aphids. There are several alarm pheromones. The most widespread interspecifically is *trans*-β-farnesene (Pickett and Griffiths, 1980). Aphids respond to these volatile substances by either swivelling their bodies, or leaving the area by walking, flying or jumping. However, in some aphids the alarm pheromone only serves to heighten the aphid's response to mechanical and visual stimuli associated with the presence of a parasite or predator (Dixon and Stewart, 1975; Nault and

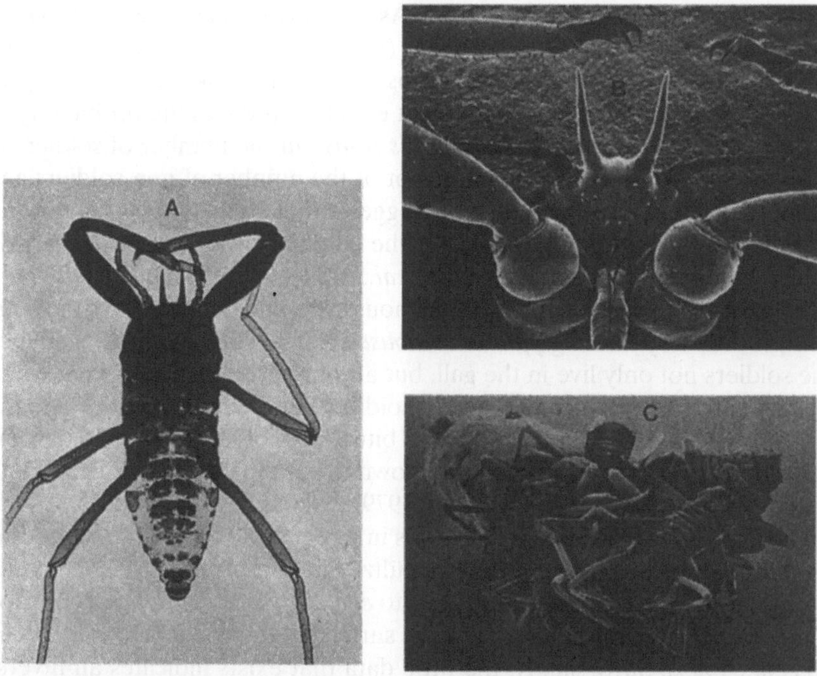

Figure 6.3 Soldiers of *Pseudoregma alexanderi*: (A) shows the enlarged prehensile forelimbs and frontal horns, (B) a ventral view of a soldier's head and (C) a group of soldiers attacking a syrphid lava. (Photographed by Dr S. Aoki.)

Montgomery, 1977). As adjacent aphids are likely to be of the same clone this altruistic act increases the probability of the clone surviving.

These defensive responses and the associated morphology may not be equally well developed in all morphs, especially in species that show a high degree of polyphenism. For example, only the first instar fundatrices of certain gall-forming aphids have the heavy sclerotization that affords them some protection against attack by other individuals of the species (Aoki and Makino, 1982). However, if siphunculi are to be used effectively for defence then an aphid has to be able to see and manoeuvre. This poses problems within a gall, particularly a closed gall (p. ??), and an alternative form of defence is needed. The existence of 'soldiers' in some mainly gall-forming species (Aoki, 1975, 1977) is a good example of an alternative form of defence and of the division of labour within a clone, with some individuals specialized for defence. Soldiers can make up 13% of a colony, are very short-lived, do not feed or reproduce, and defend the colony against insect enemies (Figure 6.3). In the sugarcane woolly aphid, *Ceratovacuna lanigera*, the soldiers respond to alarm pheromone emitted by reproductives, not by dispersing, but by aggregating and attacking any object contaminated by siphuncular exudate (Arakaki, 1989). As a colony is likely to consist mainly of the descendants of a founder aphid (i.e. it is a clone), a soldier, in using its well-developed prehensile forelimbs and frontal spike, or even its proboscis, to seize and kill potential insect enemies, reduces the probability of a clone becoming extinct. In *Cerataphis fransseni* the number of soldiers in a gall is proportional to the square-root of the number of non-soldiers and to the surface area of the gall. This suggests that soldiers are produced to defend an area, e.g. the perimeter of the colony or the surface of the gall, rather than individual aphids (Stern *et al.*, 1994).

Approximately 50% of the 100–200 thousand aphid inhabitants of the large galls produced by *Ceratoglyphina bambusae* on snowball trees are 'soldiers'. The soldiers not only live in the gall, but also on the gall and adjacent twigs. When a gall is handled or shaken the soldiers fall or stream off the gall and can inflict painful bites on man. These biters possibly afford a colony some protection against squirrels that are known to eat the contents of aphid galls (Geiler, 1956; Aoki *et al.*, 1977; Aoki, 1979; Aoki and Kurosu, 1991).

The investment in soldiers fluctuates in a regular fashion both within the life of a colony (Sakata *et al.*, 1991; Schültze and Maschwitz, 1991; Akimoto, 1992), seasonally (Sunose *et al.*, 1991; Ito *et al.*, 1995) or in direct proportion to the surface area of the colony or the surface of the gall (Stern *et al.*, 1984; Ito *et al.*, 1995). Surprisingly, the little data that exists indicates an inverse relationship between the numbers of predators (syrphid larvae) and the proportion of soldiers in aphid colonies (Ito *et al.*, 1995). However, this may arise because the soldiers are effective at decreasing predation by syrphids (Tanaka and Ho, 1995). The proportion of biters or soldiers in a colony, and whether they are sterile, possibly ultimately depends on the probability of attack by natural enemies and the constraints on defence imposed by structure and feeding biology.

6.2 Dispersal

As even trees eventually die or are killed, it is advantageous for aphids to disperse. Individuals of most morphs are capable of moving to other parts of their host plants or to adjacent plants. In some species, like the sycamore aphid, adults of most generations possess wings and are highly mobile as they retain the ability to fly throughout adult life. In species that show alary dimorphism, movement over short distances within and between adjacent plants is mainly undertaken by apterae, but alatae disperse or migrate over greater distances. Migration is most noticeable when aphids are abundant, especially in summer.

6.2.1 What constitutes a crowd?

In the vetch aphid (*Megoura viciae*), it is the tactile stimulation associated with crowding that induces the development of winged forms. When several apterous mothers that have been reared in isolation are confined together in a small glass tube for 24 h and are then placed individually on bean plants they give birth to alatae. Apterous mothers confined individually in glass vials for the same period give birth only to apterous offspring (Figure 6.4). This happens even when the offspring are reared in isolation and is not influenced by the quality of the host plants their mothers fed on. Likewise, crowding the offspring born to either crowded or uncrowded mothers, does not increase the proportion that develops into alatae, i.e. the response to crowding in the vetch aphid is maternally mediated (Lees, 1967) and two aphids can constitute a crowd.

In other species, like *Aphis fabae* (Shaw, 1970a–c), *Rhopalosiphum padi* (Dixon and Glen, 1971) and *Sitobion avenae* (Watt and Dixon, 1981), both mothers and offspring respond to crowding. The more intense the crowding, the higher the proportion of offspring that develop into alatae, especially if both mothers and offspring experience crowding (Figure 6.5).

6.2.2 Host quality

The quality of the food available to an aphid is important in determining its size, survival and reproductive rate. High-quality food is usually available to aphids feeding on actively growing or senescing plants or parts of plants and often declines rapidly when plants cease growing. Therefore an ability to respond to changes in the quality of phloem sap and to anticipate the onset of adverse nutritional conditions is of great adaptive significance.

During summer the English grain aphid (*Sitobion avenae*) feeds on the flower heads of several species of grasses that are rich feeding for only a relatively short period. Winged forms develop in response to both pre- and postnatal crowding and to changes in the nutritional quality of a host plant associated with the ripening of the seed (Figure 6.6). When mothers reared on the ears of grasses at flowering are crowded in glass vials for short periods, they give birth to proportionally fewer winged offspring than mothers similarly treated but reared on mature ears (Watt and Dixon, 1981). Similarly the pea aphid (*Acyrthosiphon pisum*) shows a heightened response to crowding when developing on mature leaves and, like the rosy apple aphid (*Dysaphis devecta*), can even produce alatae solely in response to changes in host quality (Sutherland, 1969b; Forrest, 1970).

Spring migrants of *Rhopalosiphum padi* develop on the primary host, bird cherry, in response to both crowding and changes in host quality. When *Rhopalosiphum padi* is reared in isolation on maturing bird cherry leaves, an increasing proportion develops into alatae in subsequent generations, but no alatae develop if *Rhopalosiphum padi* is reared in isolation on young leaves (Figure 6.7). However, when the same species is crowded on young leaves it readily produces alatae (Dixon and Glen, 1971).

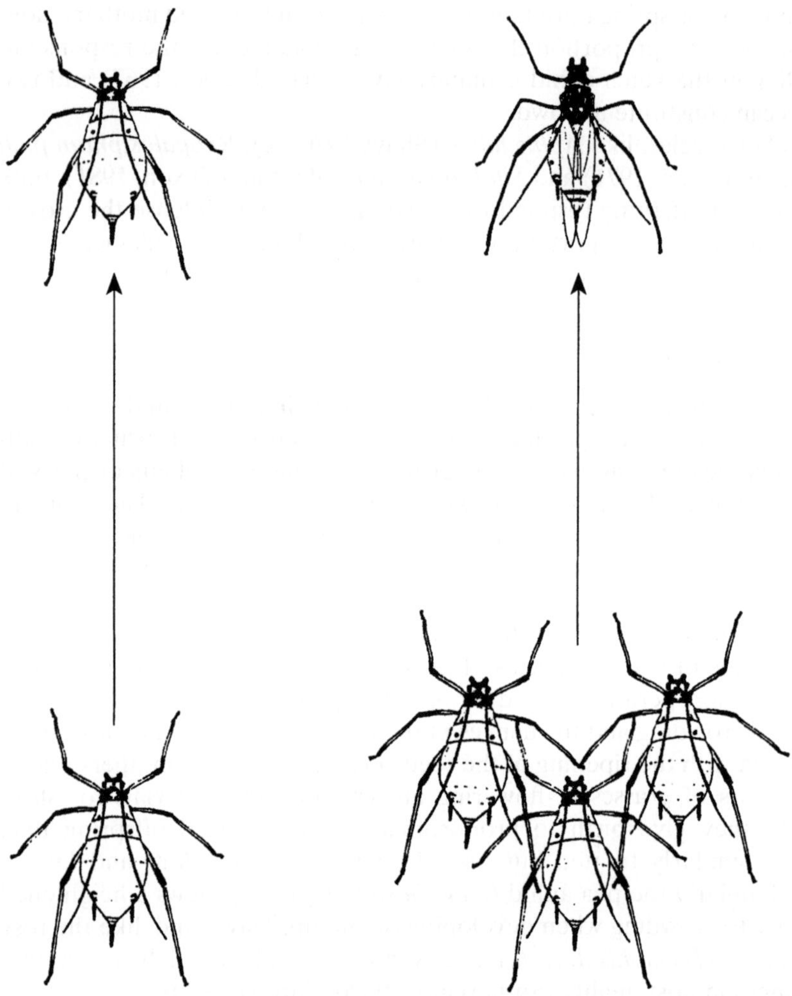

Figure 6.4 Diagram of the effect of rearing the vetch aphid (*Megoura viciae*) in isolation (a) or in crowds (b) on the production of alatae.

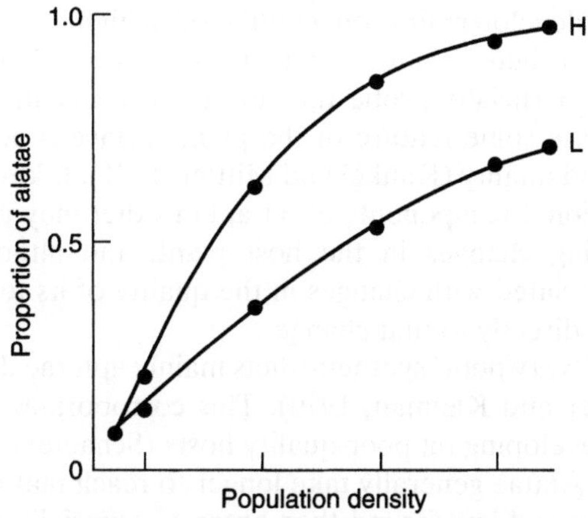

Figure 6.5 Proportion of the progeny of *Aphis fabae* that develop into alatae in relation to the population densities at which they were reared, for mothers themselves reared under low (L) and high (H) levels of crowding. (After Shaw, 1970.)

Rearing nymphs and adults of *Myzus persicae* in isolation on a synthetic diet has revealed that the composition of the food only partially controls alary dimorphism in this species (Sutherland and Mittler, 1971). Although

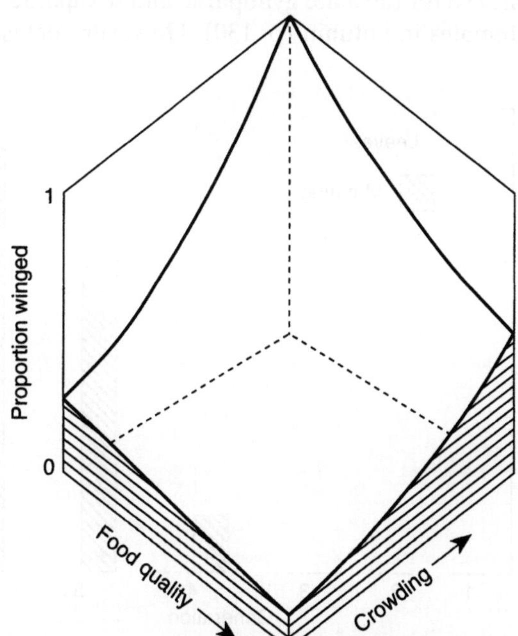

Figure 6.6 The proportion of the English grain aphid (*Sitobion avenae*) that develop into alatae in relation to food quality and crowding. (After Watt and Dixon, 1981.)

apterae tend to develop mainly on 'good' food, a high proportion of apterae develops on poor diets if, before giving birth, their mothers are allowed to walk on and superficially probe the surface of a nutritionally good host plant. In this case some feature of the plant surface is acting as a 'token stimulus' of good quality (Kunkel and Mittler, 1971). Likewise it is possible that non-nutritional components of an aphid's diet may also act as token stimuli indicating changes in the host plant, and although an aphid's response is associated with changes in the quality of its food, perhaps it is not responding directly to that change.

However, on 'very poor' synthetic diets mainly apterae develop in *Myzus persicae* (Mittler and Kleinjan, 1970). This corroborates observations on other aphids developing on poor-quality hosts (Schaefers and Judge, 1971; Dixon, 1972a). Alatae generally take longer to reach maturity (p 74; Table 6.1), are smaller and less fecund than apterae, especially in poor habitats. Therefore, in very harsh conditions it is possibly advantageous for a clone to develop apterae better able to survive until conditions improve, rather than allocate limited resources to risky dispersal (p. 148).

6.2.3 *Day length*

In several species, alatae are produced at a definite time of year (Figure 6.8). This is particularly so for the alate gynoparae and sexuparae that give rise to the males and females in autumn (p. 130). However, alates also appear at

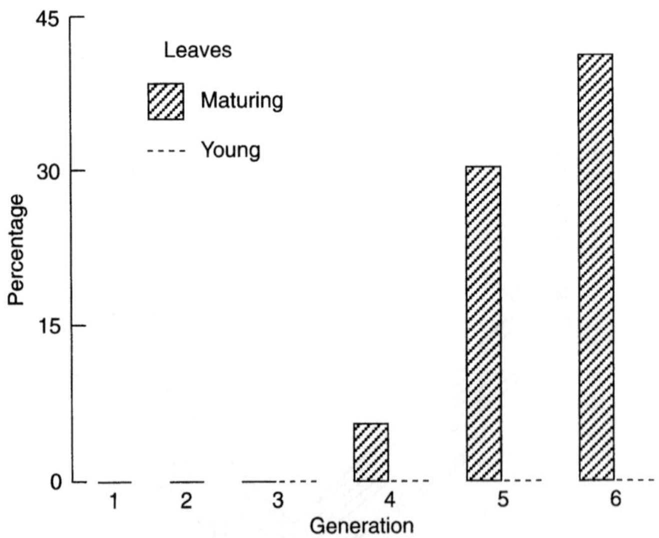

Figure 6.7 The percentage of emigrants developing when the bird cherry-oat aphid (*Rhopalosiphum padi*) is reared continuously in isolation for six generations, starting with the fundatrix, on maturing and young leaves of bird cherry. (After Dixon and Glen, 1971.)

Table 6.1 Duration of development of apterae and alatae

		Duration of development (days)			
Aphid	*Temperature (°C)*	*Apterae*	*Alatae*	*Percentage increase*	*Source*
Elatobium abietinum	15	15.6	18.0	15	Fisher (1982)
Metopolophium dirhodum	15	13.2	14.3	8	Thornback (1983)
Rhopalosiphum maidis	10	21.4	26.9	26	Noda (1960)
	25	5.0	6.0	20	
Rhopalosiphum prunifoliae	10	21.3	25.7	21	Noda (1960)
(*insertum?*)	25	4.7	5.5	17	
Sitobion avenae	10	21.4	26.9	26	Noda (1960)
	25	5.0	6.0	20	

specific times. In the green spruce aphid (*Elatobium abietinum*) the maturation and flight of the alatae coincide with bud burst of spruce. Although the proportion of green spruce aphids that become alate is influenced by crowding, host quality and temperature, the dominant factor is the increase in day length in spring. The maximum critical day length is 14 h, beyond which the proportion developing into alatae declines (Figure 6.8). This response is not influenced by the generation of aphid, being just as marked and occurring at the same time, whether the aphid overwintered viviparously or as an egg (Fisher, 1982). Increased alate production in response to changes in day length has also been reported for *Myzus persicae* (Figure 6.9) (Matsuka and Mittler, 1978) and *Cinara todocola* (Yamaguchi, 1976).

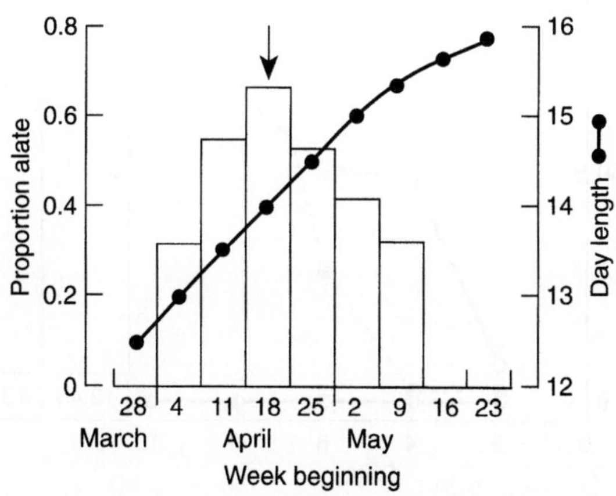

Figure 6.8 Proportion of the green spruce aphid (*Elatobium abietinum*) developing into alatae in relation to their time of birth and day length (↓ = maximum critical photoperiod). (After Fisher, 1982.)

Alate production and dispersal have been regarded as a means of regulating numbers, and in several species this is most marked when there are large populations, i.e. the incidence of alate production is density dependent. However, the production of alatae at a particular time of year, even when the aphids are reared in isolation, reveals that the primary role is dispersal. In many species there is possibly only a short period when dispersal is likely to be successful and the response to changes in day length enables the aphid to synchronize its development with this window. The reason why aphid populations commonly produce alatae in a density-dependent fashion is possibly because it is the best strategy for maximizing the number of migrants produced during the life of a colony (Kidd, 1990) and the seasonal life cycle of a clone.

6.2.4 *Maternal control*

It takes at least one generation to produce alate individuals. Therefore by responding to cues that indicate an impending change in habitat quality, aphids anticipate the change. In species in which there is a maternal switch mechanism, it is possibly under neural control and occurs just as an embryo is released from an ovariole or is actually passing down an oviduct (Lees, 1979), i.e. the 'decision' is made just before birth. In other species,

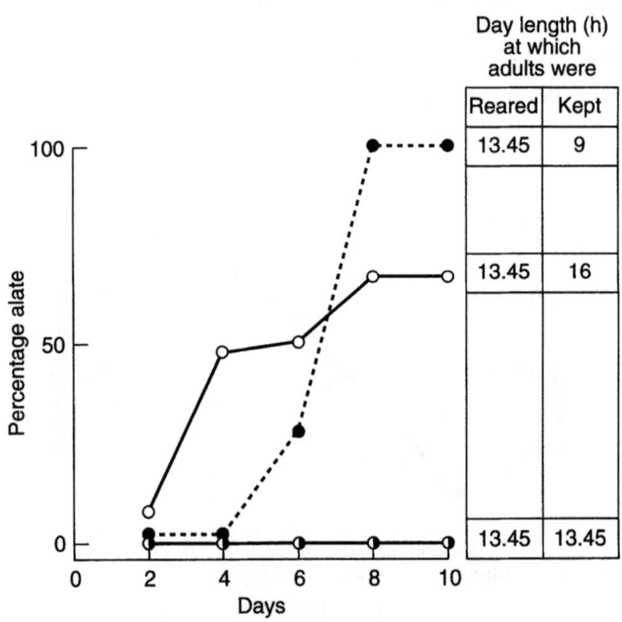

Figure 6.9 Percentage of alatae amongst the offspring produced every 2 days by *Myzus persicae* when reared and kept at the same and different day lengths. (After Matsuka and Mittler, 1978.)

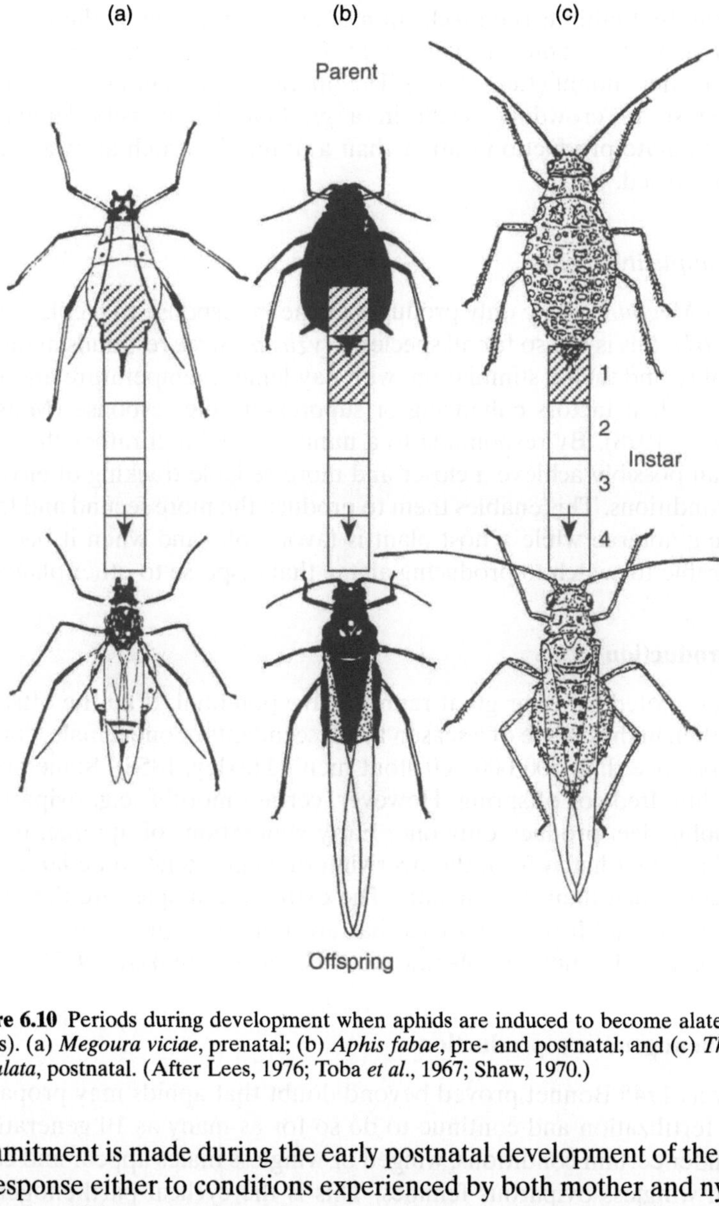

Figure 6.10 Periods during development when aphids are induced to become alate (hatched boxes). (a) *Megoura viciae*, prenatal; (b) *Aphis fabae*, pre- and postnatal; and (c) *Therioaphis maculata*, postnatal. (After Lees, 1976; Toba *et al.*, 1967; Shaw, 1970.)

commitment is made during the early postnatal development of the nymph, in response either to conditions experienced by both mother and nymph or to the conditions experienced by the nymphs only (Figure 6.10). As the rate of change in habitat quality is influenced by several factors and is therefore variable, then the later in development an aphid commits itself to the switch to alate development, the more likely it will be that alates are produced at the optimum time.

Colonies founded by alatae initially contain only apterous offspring, and later an increasing proportion of alatae develop. There is a biological clock mechanism that inhibits the production of alatae, not only by the founding alate but also by her apterous daughters. This refractory phase is time, not generation, dependent (Lees, 1966). The operation of an interval timer and the response to crowding result in a gradual, i.e. density-dependent, increase in alate production, rather than a dramatic switch after a certain time has elapsed.

6.2.5 *Multiplicity of cues*

Although *Megoura viciae* only produces alatae in response to tactile stimuli (Lees, 1967), this is not so for all species. *Myzus persicae* responds mainly to food quality and tactile stimulation, with day length, temperature and non-nutritional plant factors enhancing or suppressing the response (Matsuka and Mittler, 1978). By responding to a number of stimuli rather than one, aphids can possibly achieve a closer and more reliable tracking of environmental conditions. This enables them to produce the more fecund and faster developing apterae while a host plant is favourable, and when it becomes unfavourable to switch to producing alatae that disperse to other plants.

6.3 Reproduction

Aphids are noted for their great reproductive potential. If all the offspring survive, then in the course of a season the descendants of one female 'contain more substance than 500 000 000 stout men' (Huxley, 1858). Some aphids produce hundreds of offspring. However, certain morphs, e.g. oviparae of the Pemphigidae, produce only one. Early generations of apterae, in particular those that hatch from the overwintering eggs, tend to be larger and more fecund than their descendants. The extreme examples are the fundatrices of some gall-forming species that are nothing more than 'awkward bags' packed with embryos (Mordwilko, 1935; Shaposhnikov, 1977).

6.4 Parthenogenetic reproduction

As early as 1745 Bonnet proved beyond doubt that aphids may propagate without fertilization and continue to do so for as many as 10 generations. Then, under certain conditions, winged or wingless males appear and copulate with wingless oviparous females. This is the cyclical parthenogenesis characteristic of most aphids.

At egg hatch in spring, food is abundant and of high quality, and natural enemies are scarce. Under these conditions parthenogenesis has a strong selective advantage over sexual reproduction (Dixon, 1987b) and in particular life-history theory predicts that small progeny will be favoured. When food quality is poor, or small aphids are likely to be killed by natural enemies, then a parent tends to produce a few large progeny that are better able than small progeny to survive the harsh conditions (Williams, 1966; Stearns, 1976). This occurs in aphids not only interspecifically, but intraspecifically as well. Aphids of each of the generations that make up a life cycle have specific reproductive strategies possibly adapted to the conditions they are most likely to encounter. The first generation individuals of *Aphis fabae* and *Cavariella aegopodii* produce smaller offspring relative to their size and have relatively smaller lipid reserves than individuals of subsequent generations. Changing the size of aphids by rearing them on better or poorer quality hosts than they normally experience does not affect the trend (Dixon and Dharma, 1980b). In some species these trends are reflected in programmed changes in the numbers of ovarioles in their gonads from generation to generation (Wellings *et al.*, 1980), i.e. by means of generation-specific reproductive strategies aphids possibly anticipate the predictable seasonal trends in habitat quality (Dixon and Wellings, 1982; p. 161).

Whether an aphid has or lacks wings affects its reproductive rate. When the viviparous generations are all winged, as in the sycamore aphid (*Drepanosiphum platanoidis*), the individuals usually have relatively low reproductive rates, compared with the wingless viviparous individuals of more polyphenic species. Alatae of aphids that do show alary dimorphism take from 8 to 26% longer than apterae to reach maturity (Table 6.1), and although they may initially have a reproductive rate comparable to apterae (e.g. *Aphis fabae*; Dixon and Wratten, 1971), they are more likely to have a lower overall reproductive rate (Wratten, 1977). Alatae also give birth to smaller offspring than apterae, which may be an adaptation to the generally better quality of the plants alatae colonize than leave. However, if alatae produced offspring as large as apterae do, the reproductive rate of the alatae would be even further reduced if the same total allocation of resources was given to reproduction. Thus in order to colonize a new habitat the rapid production of offspring by alatae may be more important than their size at birth, and this could have contributed to the evolution of the small size of their offspring.

Winged individuals of species that show alary dimorphism autolyse their wing muscles shortly after settling on a plant (p. 172). This prevents further dispersal and possibly has the advantage of making a small amount of protein available for embryo development (Dixon, 1971b). However, the brachypterous alate of *Drepanosiphum dixoni*, which differs from the macropterous morph in lacking indirect muscles and not being able to fly, is 32% more fecund (Dixon, 1972b). Therefore the maintenance of wing muscles is costly, and by autolysing or not developing them, aphids free resources for the development of offspring.

The shorter developmental time and greater reproductive rate of apterae, compared with alates, also gives them a higher rate of increase. A first step in the evolution of apterousness was possibly a simple increase in fecundity

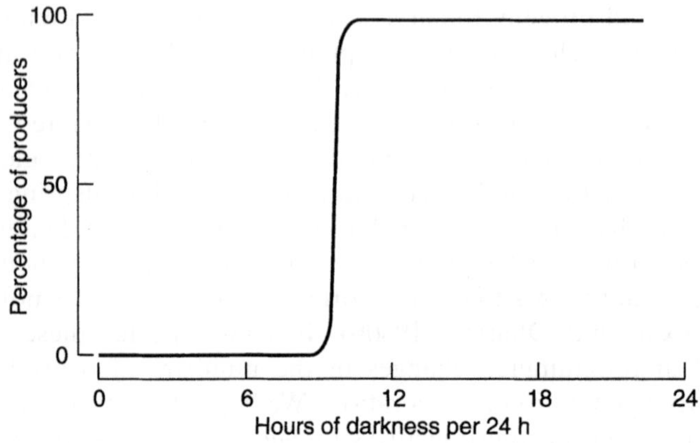

Figure 6.11 The percentage of vetch aphids (*Megoura viciae*) producing sexual females (*oviparae*) in relation to the number of hours of darkness per day they were reared under. (After Lees, 1959.)

with the appearance of brachypterous forms or as a result of wing muscle autolysis. Later apterous development could then give the additional advantage of a shorter developmental time (Table 6.1) and possibly lower temperature thresholds for development (Noda, 1960).

6.5 Sexual reproduction

Continued parthenogenetic reproduction was thought to result in aphids becoming effete. Sexual reproduction supposedly invigorated a clone, and the 'spermatic force', which weakened in its passage between successive generations of cells, determined the number of generations and proliferation of a clone. Huxley (1858) was the first to show that if aphids were kept warm and supplied with food they could reproduce parthenogenetically without deterioration, apparently indefinitely. This is also supported by the existence of many anholocyclic species of aphids. It was not until 1924 that Marcovitch demonstrated the role of photoperiod in the induction of sexual forms, shortly after the discovery of the effect of photoperiod on flowering in plants, i.e. aphids show environmentally determined sex.

6.5.1 Extrinsic factors

At the onset of autumn in temperate regions many species of aphids switch to the production of sexual forms; with each clone (evolutionary individual) producing both egg-laying females (oviparae) and males, i.e. a clone is hermaphrodite. Although associated with short days, the switch is actually triggered by long nights (Figure 6.11) (Lees, 1973).

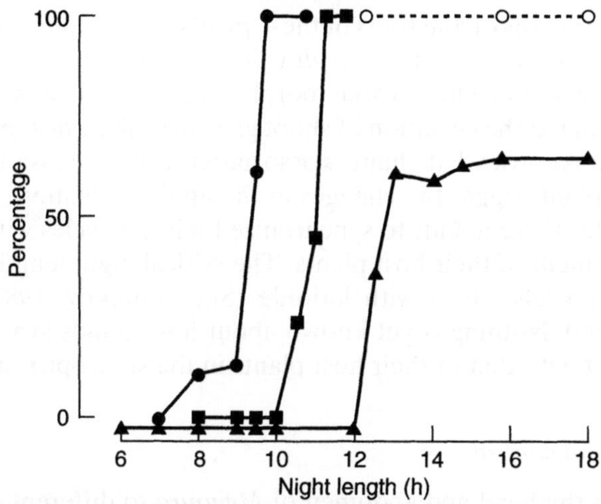

Figure 6.12 Percentage of the bird cherry-oat aphid (*Rhopalosiphum padi*) developing into gynoparae and males in relation to the night length they are reared under, at 10°C (filled circles), 14°C (filled squares) and 18°C (filled triangles; open circles, third generation). (After Dixon and Glen, 1971.)

However, aphids live in a multivariate environment in which the time of leaf fall and the coolness of the autumn vary from year to year. Parthenogenetic reproduction continues as late as is compatible with successful egg laying. Optimally, sexual forms are produced only when the interval before leaf fall is equal to the time required for development, migration, mating and oviposition. As aphids are poikilothermic they need longer to complete the transition in cooler autumns. It is also likely that the average temperature in autumn affects the time of leaf fall.

The mean temperature in July can be used as a positive predictor of the mean temperature in August (Ward *et al.*, 1984). Therefore by producing sexuals at shorter and shorter night lengths at lower temperatures (Figure 6.12), aphids can initiate sexual development earlier in the years when there is a high likelihood of a cool autumn. Field results for several species indicate that the initiation of sexual forms is advanced more than is necessary to compensate for the extra time needed for development at the lower temperatures. It is possible that plants also respond to predictors of an early onset to winter, as sometimes they shed their leaves early. Thus the early initiation of sexual forms in cool autumns possibly compensates for both the extra time necessary for development as well as an earlier leaf fall (Ward *et al.*, 1984).

In several aphid species sexual production occurs in response to changes in the host plants. In temperate regions the short-day conditions in autumn induce plants to cease growth and become dormant. Aphids living in continuous darkness, at relatively constant temperatures and often at considerable depths on the roots of these plants, still produce sexual morphs as the plants become dormant. *Aphis farinosa* and *Dysaphis devecta* live above ground, but produce sexual morphs when night length is at its shortest, responding to the cessation of shoot growth of their host plant (Forrest, 1970). It is unknown what changes associated with the cessation of growth of the host plant trigger the changes in the aphids. Whatever the mechanism, it enables these aphids to synchronize their life cycles with the growth and development of their host plants. The critical night length for the production of sexuals varies with latitude (Shaposhnikov, 1981; Smith and Mackay, 1990). Nothing is yet known about how aphids synchronize their development with that of their host plants in the subtropics and topics.

6.5.2 Maternal control

By exposing the head and abdomen of *Megoura* to different photoperiods and destroying groups of cells in the brain by microcautery, Lees (1964) and Steel and Lees (1977) have revealed that neurosecretory cells in the pars intercerebralis, and tissue lateral to them, control sex determination. The light receptor and probably the photoperiodic clock are located in the lateral area, and the neurosecretory cells are the endocrine effectors producing a virginopara determinant. This determinant may be delivered direct to the embryos via the thoracic ganglion or released into the blood as a neurohormone (Figure 6.13). However, in the gynoparae of *Aphis fabae* the photoperiodic clock controls nymphal development through the corpus allatum, which produces the juvenile hormone, which influences both the external form and the type of gonads. The amount of juvenile hormone is also thought to affect the proportion of males produced by apterous virginoparae of *Myzus persicae* kept under long-night conditions (Lees and Hardie, 1981; Hales and Mittler, 1983).

6.5.3 Aphid clocks

Although many aphids hatch from eggs early in the year and are present in the spring when night length is as long as in autumn, the sexuals rarely appear in the field before autumn (Figure 6.14). This is because of the operation of an intrinsic timing mechanism referred to as a 'facteur fundatrice' (Bonnemaison, 1951) or as an 'interval time' (Lees, 1960, 1966).

In *Megoura* no oviparae or males are produced by young clones founded by fundatrices for 80–90 days, even if they are kept in a constant long-night regime at 15°C. This response is independent of generation, as it has been shown that a lineage of the early-born mothers of each generation will pass through four generations before ovipara producers develop, whereas only two generations are required in a parallel lineage derived from the last-born (Figure 6.15). Although the timing mechanism(s) is temperature

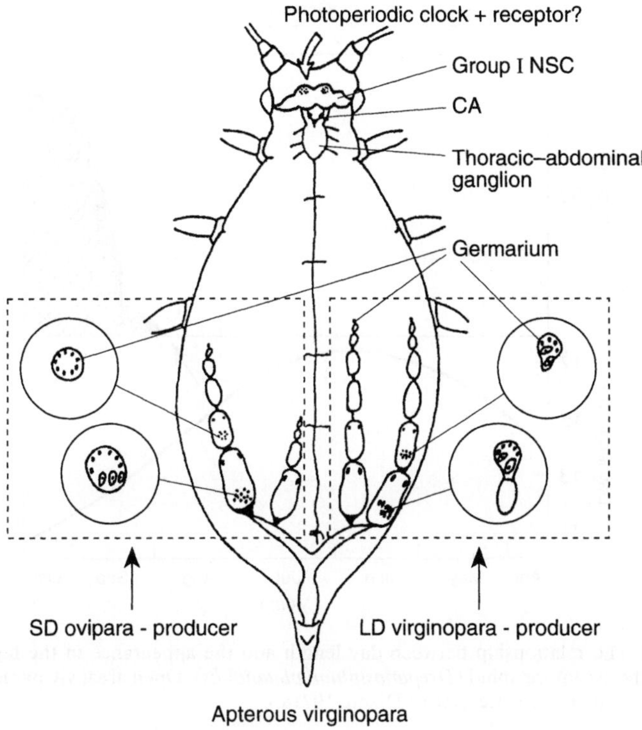

Figure 6.13 Representation of the adult apterous virginopara of *Megoura viciae*, showing some of the internal structures. The ovarioles contain chains of embryos the largest of which under long-day conditions (right) contain functional germaria, which have already produced one or two embryos (encircled), and under short-day conditions (left) the embryos are destined to become oviparae and the germaria produce yolky haploid eggs. (NSC, neurosecretory cells; CA, corpus allatum). (After Lees and Hardie, 1981.)

dependent, the slight variations in the time interval are unimportant provided the period of delay extends into early summer when the short night length will take over the role of the inhibitor.

However, in the lime and sycamore aphids the interval timer mechanism does not function independently of the environment while it is 'running out'. With the passage of time the ratio of the number of oviparae to virginoparae produced by aphids in successive generations gradually increases, but does so more rapidly in short, than in long, photoperiods (Figure 6.16). Incomplete inhibition of the interval timer in autumn permits clones to continue reproducing parthenogenetically in the short days when the senescent foliage of sycamore provides a rich source of food. This is advantageous in those autumns when leaf fall is late (Dixon, 1971a, 1972c).

Similarly, in the short season and continuous daylight of the high arctic, aphids show a gradual switch to the production of sexuals controlled by an

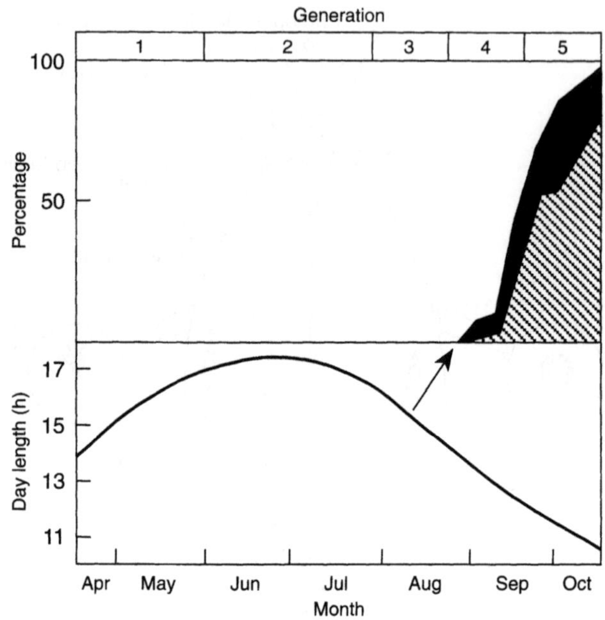

Figure 6.14 The relationship between day length and the appearance in the field of sexual morphs of the sycamore aphid (*Drepanosiphum platanoidis*). Open area, viviparae; solid area, males; shaded area, oviparae. (After Dixon, 1971a.)

interval timer. For example, the offspring of the first generation of *Acyrthosiphon svalbardicum* consist of 83% sexuals and those of the second generation 100% sexuals (*cf.* Strathdee *et al.*, 1993). In favourable years the offspring of the second generation are likely to survive to maturity and egg production is increased by five-fold over those years when only the offspring of the first generation reach maturity (Strathdee *et al.*, 1993).

6.6 Aestivation and hibernation

In summer, many plants, especially shrubs and trees, cease growing and, until their leaves become senescent at the onset of autumn, are poor hosts for some aphids. Host-alternating aphids seek out and colonize plants that are still growing during this period, whereas several autoecious species of aphids aestivate. The sycamore aphid anticipates the onset of adverse conditions associated with the cessation of growth of its host; its summer morph matures with poorly developed gonads and a well-developed fat body and may not reproduce for up to 8 weeks (p. 153). Another aphid also living on sycamore, *Periphyllus testudinaceus*, aestivates as a first instar nymph. Poor nutrition induces this aphid to produce small flattened offspring whose

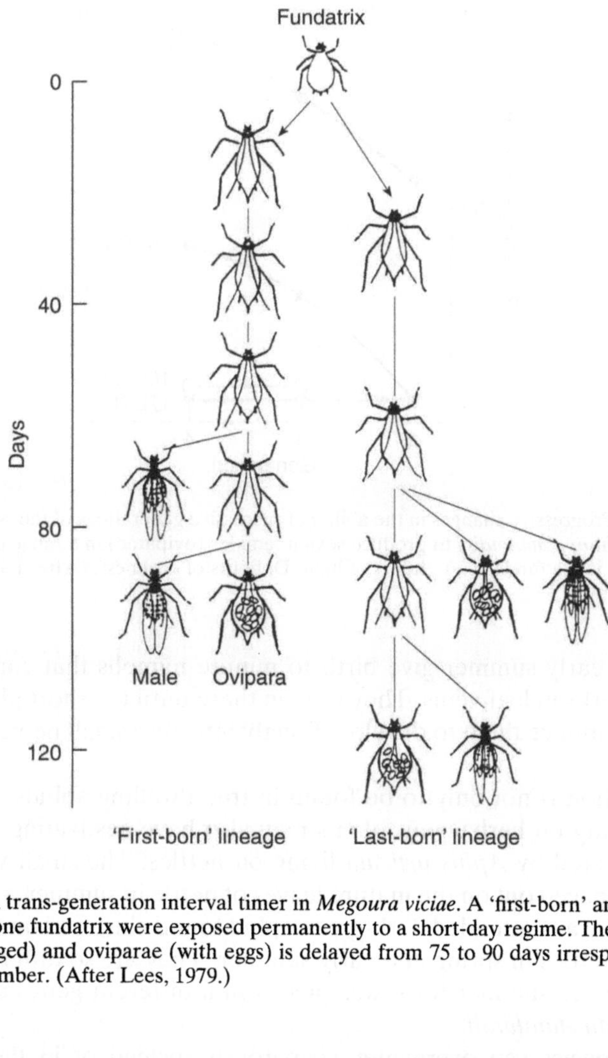

Figure 6.15 A trans-generation interval timer in *Megoura viciae*. A 'first-born' and 'last-born' lineage from one fundatrix were exposed permanently to a short-day regime. The appearance of males (winged) and oviparae (with eggs) is delayed from 75 to 90 days irrespective of the generation number. (After Lees, 1979.)

bodies are covered and fringed with minute plates giving them the appearance of a tortoise, and so their specific name. If transferred from mature to young, or from young to mature leaves, it take approximately 4 days at 15°C (Figure 6.17) for a mother to respond and give birth to the appropriate nymphs: aestivating nymphs on mature leaves and normal nymphs on young leaves. Aestivating nymphs attach themselves very closely to the leaves and are very difficult for predators to remove or for wind brushing the leaves together to dislodge. The short photoperiods of autumn induce the aphids to resume growing and develop into sexuparae that give birth to males and oviparae (Bonnemaison, 1956; Shearer, 1976). *Thelaxes dryophila*, living on oak, has a similar life cycle. The alatae that are produced when the leaves

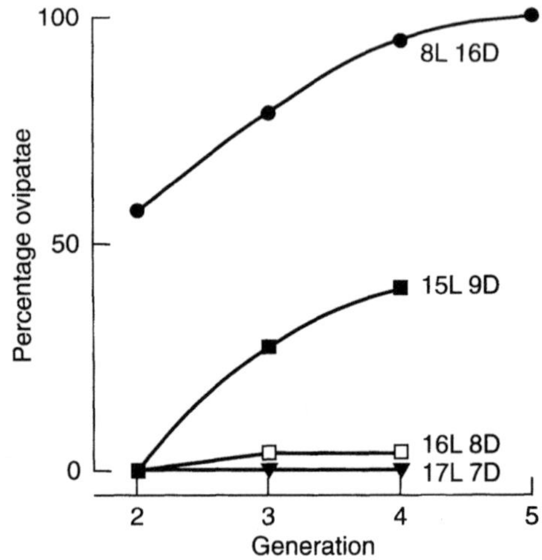

Figure 6.16 Progressive changes in the ability of successive generations of the sycamore aphid (*Drepanosiphum platanoidis*) to produce sexual females (oviparae) in a range of day lengths. Generation I is the fundatrix, L, hours of light, D, hours of darkness. (After Dixon, 1971a.)

mature in early summer give birth to minute nymphs that conceal themselves at forks in leaf veins. They remain there until the short photoperiods in autumn induce them to develop directly into very small neotenous males and females.

Aestivation is not only to be found in tree-dwelling aphids. Autoecious species living on herbaceous plants may also have aestivating forms, as is well illustrated by *Aphis urticata* living on nettles. The small yellow form that lives spaced out on the mature leaves of nettle in summer is very different from the ant-attended dark-green aphid found clustered on the growing points of nettles in spring and early summer. They are so different that for some time the summer form was placed in a different genus and species, *Pergandeida stanilandi*.

Many species can overwinter viviparously instead of in the egg stage (Müller, 1971b). In some a special overwintering form, called a hiemalis, is induced to develop in response to environmental cues. About a fifth of the autumn apterae of the lettuce root aphid (*Pemphigus bursarius*) can survive without a host plant for 48 weeks at 3°C (Dunn, 1959). They have poorly developed gonads and a well-developed fat body (Judge, 1967; Sutherland, 1968). The environmental cue that induces the development of the hiemalis in this root-feeding aphid is low temperature; at 10°C, 95% develop into hiemalis that overwinter in the soil and the remainder develop into sexuparae that return to the primary host, poplar. At this temperature the hibernating aphids do not reproduce for 103 days, which is long enough for low winter temperatures to further delay their development until the following spring.

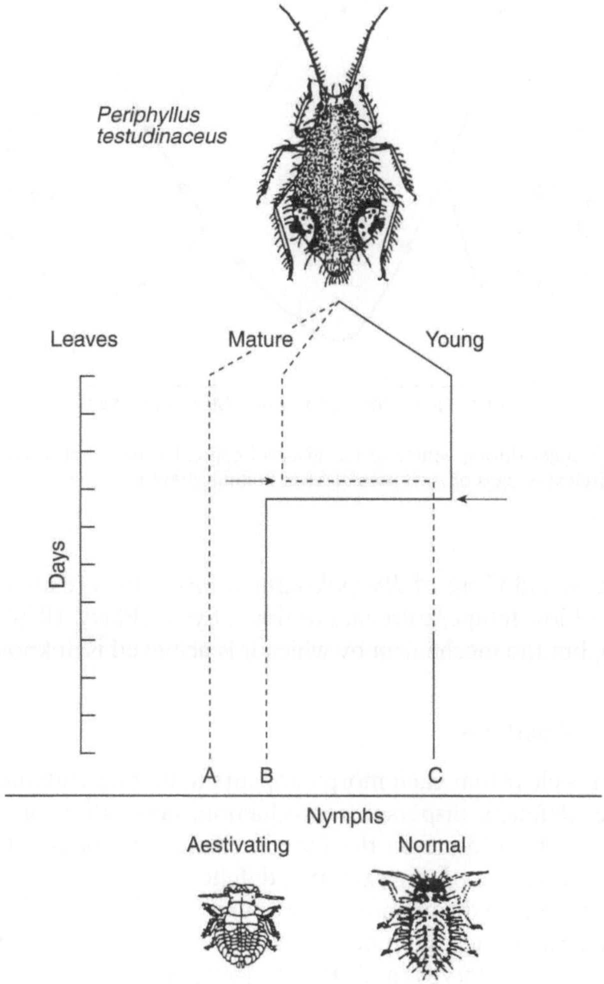

Figure 6.17 Production of aestivating (broken line) and normal (solid line) offspring by apterous fundatrigeniae of *Periphyllus testudinaceus* when (A) kept as adults on mature leaves for 10 days, (B) kept on young leaves for 3 days and then transferred (←) to mature leaves or (C) kept on mature leaves for 3 days and then transferred (→) to young leaves. (After Shearer, 1976.)

Eggs too are well adapted to survive periods of stress. Some can survive temperatures lower than –40°C (James and Luff, 1982). This cold-hardiness of eggs (Sömme, 1969), like that of the active overwintering stages of the green spruce aphid (*Elatobium abietinum*) (Parry, 1979a), is associated with changes in the levels of certain polyhydric alcohols (glycerol and mannitol) (Figure 6.18). However, the cold-resistance of the eggs of *Cinara pilicornis* is not related to their glycerol or mannitol content. Cold-hardiness of the eggs

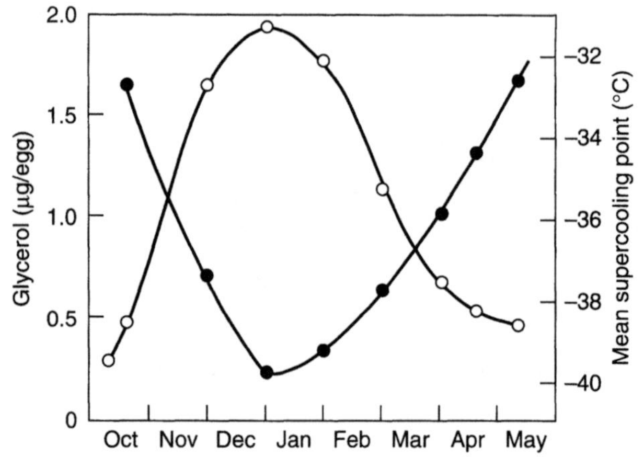

Figure 6.18 Changes during winter in the glycerol content (open circles) and supercooling point (solid circles) of eggs of *Apis pomi*. (After Sömme, 1969.)

of this species and those of *Rhopalosiphum insertum* is related to previous experience of low temperature and to time of year (Parry, 1979b; James and Luff, 1982), but the mechanism by which it is achieved is unknown.

6.7 Ecological patterns

Although it is clear that each morph appears to be more adapted for a particular role: defence, dispersal, reproduction, aestivation or hibernation, little work has been done on the functional reasons for the different patterns in the incidence of, for example, defence or dispersal in aphids. One approach is to regard the phenotypes we currently observe as limited by that of the ancestral species, i.e. phenotypes are not determined only by optimization criteria (Harvey and Pagel, 1991; Stern and Foster, 1996a). Although, this is undoubtedly true in that all aphids share the same basic design, the extent to which phylogeny has determined the incidence of particular phenotypes at the level of family, tribe or genus is debatable. This approach is dependent on a good phylogeny of the taxa under consideration, which is not always available. Alternatively, the historical transitions can be ignored and species regarded as independent evolutionary events (Charnov, 1993). Below, the incidence of defence and dispersal in aphids will be used to illustrate this debate.

6.7.1 Defence by 'soldiers'

This topic has been reviewed by Stern and Foster (1996). Of the 4400 species of aphids, approximately 1% have soldiers and these are confined to two of the 10 families of aphids – the Hormaphididae and Pemphigidae. These families are very closely related, have similar life cycles and are an ancient group having probably originated in the Cretaceous over 65 million years ago (*cf.* Figure 1.2) (Heie, 1987). The phylogenies of these two families indicate that soldiers have evolved a minimum of 6–9 times (Stern and Foster, 1996).

Does the very restricted distribution of soldiers reflect a phylogenetic constraint? In particular is the high incidence of wandering, fighting and offensive and defensive structures in the early instars unique to these two families? The young and adults of aphids of other families frequently move between leaves and even plants (Dixon and McKay, 1970; Dixon, 1970a; Harrington and Taylor, 1990) are aggressive to one another and space out by kicking (Dixon and Logan, 1973), have thick defensive cuticles (e.g. *Atheroides*, *Brachycaudus*) or defensive/offensive spines (Greenideinae), and even defend themselves against insect predators by kicking and waxing (Dixon, 1958). They have even been recorded as using their stylets to attack and kill eggs of predators and other species of aphids (Banks *et al.*, 1968). Therefore, the supposed phylogenetically restricted behavioural and/or morphological features are not restricted to the Hormaphididae and Pemphigidae.

Another feature associated with soldier-producing species is their tendency to induce closed galls on their primary host. The form of a gall is a consequence of an interaction between an aphid and its host plant in which the aphid is the architect, whose pattern of probes to a very large extent determines the structure of a gall (Dixon, 1983). Unfortunately, closed galls are not produced by aphids other than those belonging to the Hormaphididae and Pemphigidae.

Providing the closed gall forms quickly, it is likely to enclose a monoclonal population of aphids, which is defendable. Migration between galls has been reported in the Pemphigidae, with as many as 25% of the aphids in a gall not of the same genotype as the fundatrix (Setzer, 1980). In other Pemphigidae, *Melaphis rhois*, however, there appears to be little or no intergall migration (Herbert *et al.*, 1991) as also appears to be the case for the only soldier-producing species and members of the Hormaphididae that has been studied, *Ceratovacuna nekoashi* (Fukatsu and Ishikawa, 1994). In addition, there is no evidence that soldiers can discriminate between kin and non kin (Foster, 1990; Aoki *et al.*, 1991). Therefore, it is unlikely that soldiers evolved as an effective means of reducing interclonal competition, which is more effectively achieved by the rapid production of a closed gall. It is likely, however, that soldiers evolved to defend monoclonal gall populations against predators.

Optimality models developed by Akimoto (1995) and Stern and Foster (1996) indicate ecological features that may have promoted the evolution of soldiers in species that produce monoclonal gall populations. Gall duration and predator pressure are interrelated as the longer a gall lasts and the more predators there are, the more likely the galls are to be attacked; this is particularly so where the gall is not completely closed. As the population in a gall increases, intraclonal competition for resources is likely to increase and birth rate to decline. The models indicate that investment in soldiers should increase with decline in birth rate, i.e. there is then more need to defend the clones' investment in long-term survival, the reproductive individuals. Why it is the early instars, rather than adults, which function as soldiers in dimorphic species is possibly not only because they can be produced quickly but because they do not require resources for growth and therefore do not compete with the reproductive individuals for the limited resources.

The above cannot account for the occurrence of soldiers in colonies of some of these species when on their secondary host plants, as the individuals in these colonies are not enclosed by a gall and are therefore less likely to be genetically identical. Although, optimization criteria can account for the evolution of soldiers where these colonies are enclosed in galls, the question of why closed galls are not induced by other groups of aphids remains unresolved. Is this the constraint that resulted in the evolution of soldiers in this lineage but not in other lineages of aphids?

6.7.2 Dispersal

Species in which all the virginoparae are winged (macropterous) only occur in the Drepanosiphidae and are restricted to the Drepanosiphinae and Phyllaphidinae, two ancient and closely related lineages (Heie, 1987). However, even within the Phyllaphidinae the incidence of macroptery varies greatly between genera. All 50 species of *Tuberculatus* live on *Quercus* and are macropterous, whereas only in one species of the five species of *Diphyllaphis* and four species of *Mexicallis*, all of which also live on *Quercus*, has a macropterous virginopara been reported. Eighty-four percent of the 37 species of *Myzocallis*, which live on species of Fagaceae, and a third of the 15 species of *Calaphis*, which all live on species of Betulaceae, are macropterous. Interestingly one of the polyphenic species in each of these two genera, *Myzocallis myricae* and *Calaphis myricae*, lives on *Myrica*, which is shrub. One common feature of the macropterous species is that they all live on climax forest trees.

Nearly all species of Drepanosiphidae, both macropterous and polyphenic, have an apterous morph, the sexual female. As the genetic machinery for apterousness exists in these species it is difficult to accept that macroptery can be non-adaptive and maintained by constraint.

6.7.3 Macroptery

In winged aphids the optimal partitioning of resources between soma, fuel and gonads is dependent on the timing and frequency of migration (Dixon *et al.*, 1993a). This is an optimization problem concerned with the risk associated with migration, the resources necessary to fuel the flight, the optimum time for migration and whether an aphid should autolyse its wing muscles. The following is an illustration of the possible mechanisms.

The number of individuals in a population at a particular site at time t is $N = N(t)$ and the risk connected with migration is expressed as the probability p that an average individual will survive the migration. The risk is likely to be inversely proportional to the degree of habitat fragmentation. The reproductive success of an individual at a particular site is defined by $rxf(N)$, where x is the number of its offspring and r the potential rate of increase of the offspring under optimal conditions, i.e. in high-quality sites and in the absence of density dependence. The function f describes all the negative influences that either the environment or the number of individuals present (density dependence) can have on the potential rate of growth of the population. The function f may also be dependent on site and time, as there may be differences between sites both in space and in time, t, as the quality of the site undergoes, e.g., seasonal changes. However, this is neglected here.

The migratory strategy is expressed in terms of the proportion q of a clone that stays and reproduces at the original site, and the proportion $(1 - q)$ that attempts to locate and reproduce at a new site. The fitness of the clone under such assumptions can be expressed as:

$$F = qrf(N) + (1 - q)prf(M) \tag{6.1}$$

where M denotes the number of individuals of this species already present at the new site. The necessary condition for fitness to be maximized is $dF/dq = 0$, which gives:

$$f(N) = pf(M) \tag{6.2}$$

Before plant quality deteriorates we assume N increases with time. If only a small proportion of sites are occupied then a migratory individual is likely to arrive at an empty site ($M = 0$) and if f is a decreasing function of N, which is the usual case, then equation (6.3) defines the critical density, N_{crit}, for migration:

$$f(N_{crit}) = pf(0) \tag{6.3}$$

For illustration one can assume a logistic growth in equation (6.3) and look for the dependence of the interval between flights on habitat fragmentation, represented by $1/p$ in Figure 6.19. From this figure it is evident that when $1/p$ is low, insects should fly often, and when $1/p$ is large and increasing the time to migration should increase dramatically. Thus species living in non-fragmented habitats, like forests dominated by one species of tree, where p is high, i.e. the cost of flight is low, should always be winged but invest relatively little in fuel and retain the ability to fly throughout adult life (i.e.

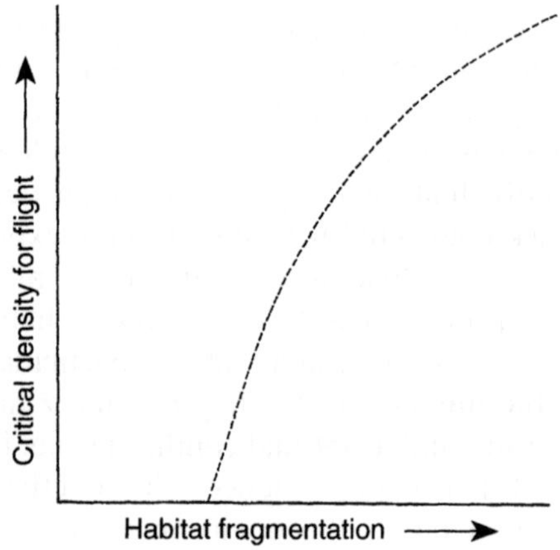

Figure 6.19 Relationship between the critical density for flight and habitat fragmentation.

retain their wing muscles), even though it results in a reduced fecundity and lower population growth rate. That is, as the cost of host (patch) transfer is low it is frequently advantageous to locate another host (patch) than to stay. However, in those species that occupy very fragmented habitats (most herbaceous plants), i.e. the cost of flight is high, flight is only advantageous when the patch becomes very overcrowded. The optimum interval between flights in this situation may exceed the length of an individual's life. In such species there are advantages in a high investment in fuel but, on colonizing an empty habitat, to go in for wing muscle autolysis, use all the energy for reproduction, and produce unwinged offspring, i.e. switch from a migratory to a reproductive mode.

As predicted, nearly all the aphids that are fully winged live on abundant forest trees. In addition, most of the aphids that live on herbaceous plants, which are often spatially unpredictable resources, tend to go in for wing muscle autolysis after locating a suitable host plant, whereas those that live on trees that are spatially predictable and abundant resources, retain their wing muscles. Associated with wing muscle autolysis is a dramatic and often very rapid increase in the size of the gonads. As wing muscles compete with the gonads for resources (p. 73; Newton and Dixon, 1990), there are advantages in wing muscle autolysis and investing all the resources in reproduction. Wing muscle autolysis has the additional advantage of preventing an aphid accidentally being displaced from its host plant, the diffusive losses of Roff (1986).

The abundance per unit of resource of aphids that live on predictable and abundant plants (forest trees) is also likely to differ from those that live on herbaceous plants (p. 207). This could also influence the evolution of macroptery. When living on abundant trees, aphids are abundant and the ability of these species to avoid the more frequently experienced intraspecific competition by flying has possibly been another factor counterbalancing the cost of being able to fly and of flying. The aphids living in such habitats are frequently fully winged, although living in a permanent habitat. A good example is that of the tree-dwelling Drepanosiphidae, which mainly produce only winged virginoparae. However, in the case of the uncommon *Drepanosiphum dixoni*, which only lives on the leaves of *Acer campestre* that are in deep shade, i.e. in a very fragmented habitat, the virginoparae are mostly brachypterous, with the aphid switching to the production of the less fecund fully winged individuals only when it becomes abundant (Dixon, 1972a,b).

When the host plant is uncommon and its distribution very fragmented then the losses incurred in migrating are likely to be very high and the aphid only infrequently abundant. Although the host plants here are most likely to be herbaceous and temporary habitats, nevertheless it is in such habitats that apterousness is likely to be favoured. On colonizing such vacant habitats, there are advantages for the colonist in wing muscle autolysis and producing apterae in order to exploit the resource and later generations switching to alate production when the plant eventually becomes overcrowded.

Summarizing, their clonal structure has enabled aphids to develop a division of labour in which each morph is adapted more for either defence, dispersal, reproduction or survival, i.e. altruism is common in aphids. They track the seasons, producing the various morphs in response to a multiplicity of environmental cues and to changes in their internal clocks. The extent to which phylogeny has determined the incidence of particular phenotypes is debatable. Although phenotypes are undoubtedly constrained by that of the ancestral species, nevertheless the patterns in the distribution between taxa of particular phenotypes can largely be accounted for in terms of optimization criteria.

7 Resource tracking in time

Aphid life cycles are of two types: autoecious (host-specific) and heteroecious (host-alternating) (*cf.* Figure 3.1). Autoecious species of aphids live on one or a few species of a particular genus of plants (Figure 7.1), and heteroecious species spend autumn, winter and spring on a woody host and summer usually on herbaceous plants (Figure 7.2). The primary and secondary hosts belong to different families of plants (*cf.* Table 6.1), and the aphids are classified as polyphagous. However, many of these aphids are sequentially monophagous as they live on one plant species at a time, e.g. the damson hop aphid alternates between damson and hop (Figure 3.1).

Members of four of the 10 subfamilies of the Aphidoidae (the Aphididae, Anoeciidae, Hormaphididae and Pemphigidae) show host alternation (Eastop, 1977). However, even within a genus some species may be heteroecious and others autoecious, and occasionally the same diversity occurs within a species at the subspecific level (Shaposhnikov, 1981).

7.1 Host alternation

Complex life cycles, in which a regular seasonal shift in habitat occurs, have evolved in numerous taxa. Ten percent of aphids seasonally transfer between unrelated host plants. This phenomenon, know as host alternation or heteroecy, has long attracted interest. Bonnet (1745) suspected it existed and it was first described by Walker (1848). The typical life cycle involves migration between a woody host, on which sex occurs and eggs are laid (the primary host), and an herbaceous plant, where the reproduction is solely parthenogenetic (the secondary host).

Clearly, the twice yearly host transfer with associated mortality is a cost, especially in a taxon with limited host location abilities. Yet many aphid species are rigidly fixed to a system which would appear risky and dangerous. Why then, do they do it? Throughout the literature of the last century, two themes are presented as answers: (1) the adaptive hypotheses and (2) the maladaptive. The development of these ideas is reviewed and a synthesis presented. It is concluded that such obligate resource switching occurs when changing hosts is predictably more profitable than remaining on the

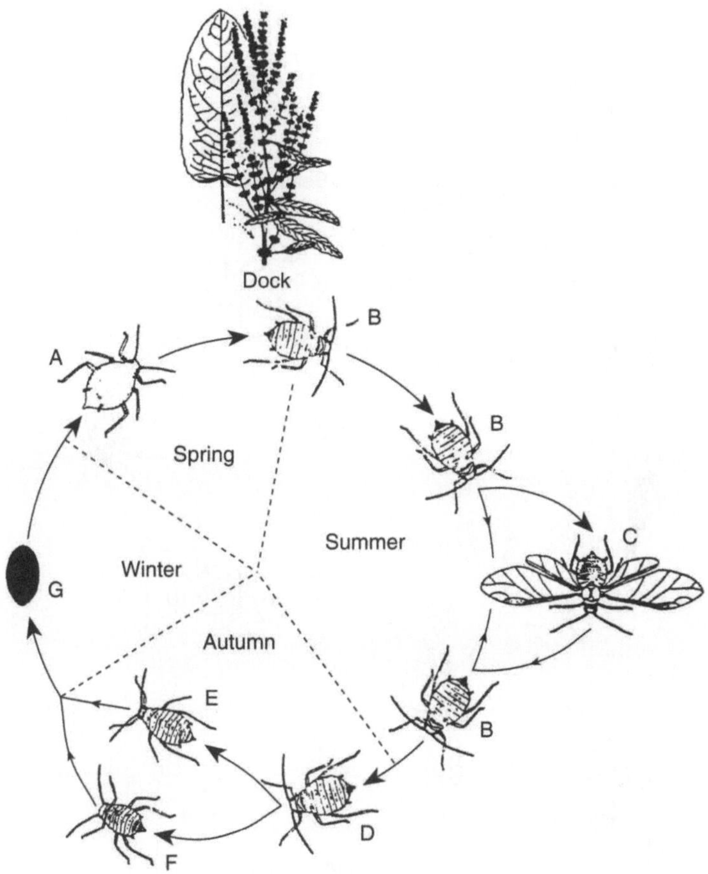

Figure 7.1 Life cycle of the autoecious dock aphid, *Aphis rumicis*. (A) Fundatrix or stem mother, (B) aptera, (C) alate, (D) sexupara, (E) ovipara or mating female, (F) male, (G) egg. (After Jones, 1942a.)

same plant species. The major cause of this benefit is where alternative sympatric hosts display asynchronous phenologies.

7.2 The historical debate

Why aphids host alternate was first addressed by Weed (1896) in a general article 'The hibernation of aphids' where he argued that these aphids benefited in at least three ways: (1) escape from natural enemies, (2) more succulent food and (3) lessening of the injury to, or even saving from destruction, their host plants. Mordvilko (1900, 1901, 1928, 1934), however, was the first to discuss the problem in detail. He noted that:

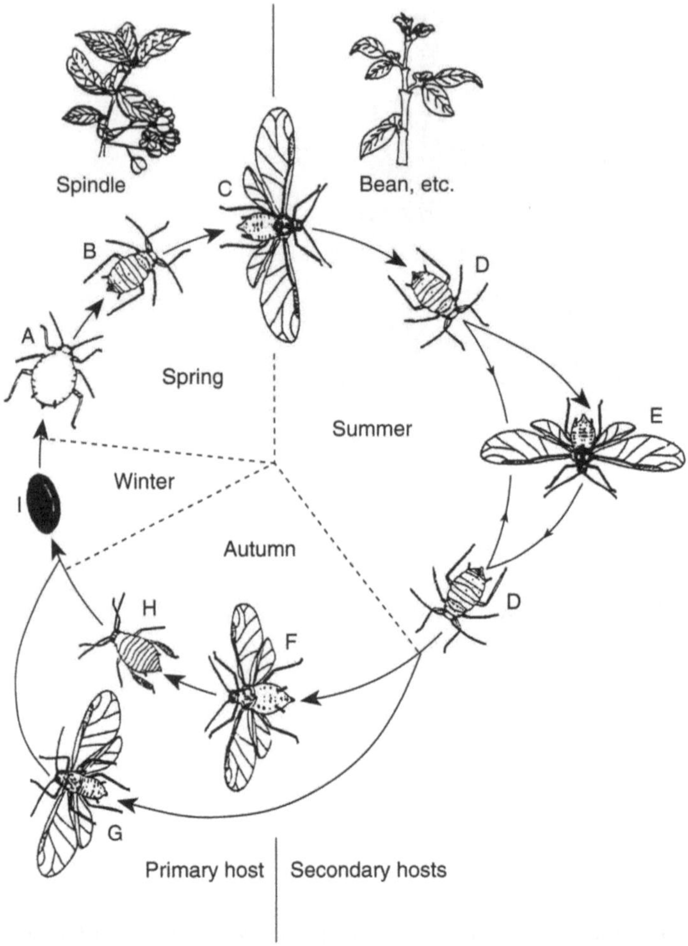

Figure 7.2 Life cycle of the heteroecious black bean aphid, *Aphis fabae*. (A) Fundatrix or stem mother, (B) fundatrigenia, (C) emigrant or spring migrant, (D) apterous exule, (E) alate exule or summer migrant, (F) gynopara, autumn migrant or remigrant, (G) male, (H) ovipara or mating female, (I) egg.

1. The trees and shrubs on which most host-alternating aphids overwinter as eggs are often phylogenetically older than their summer hosts and therefore designated the woody plant as the 'primary' host.
2. In temperate regions, trees and shrubs are generally nutritionally unfavourable for aphids during the summer.
3. The evolution of herbaceous plants, which generally continue to grow and remain nutritionally favourable throughout summer, presented the opportunity for host alternation.

These observations led Mordvilko (1928) to reach two conflicting hypotheses accounting for the evolution of host alternation:

1. It enables aphids to exploit the complementary growth patterns of herbaceous and woody plants. Aphids can do this because they can produce many generations in the course of a season.
2. It allows the exploitation of herbaceous plants, which he regarded as a higher quality food.

Mordvilko greatly favoured the notion that aphids should move from more ancient taxa to phylogenetically younger and nutritionally more favourable plants. However this hypothesis suggests that the most successful life cycle is autoecy on an herbaceous plant. Why, then, should heteroecy persist if a shift of all life cycle stages to the secondary host is advantageous? The answer, he thought, was that some of the morphs had undergone such extensive morphological and physiological changes that it became impossible for them to live on other plants. In some heteroecious species, the fundatrix (the parthenogentic female hatching from the bisexually produced egg) is obviously morphologically different from other morphs. Thus, he decided that the adaptations to the primary host constrained the evolution of secondary autoecy. This fundatrix-constraint hypothesis, accounting for the maintenance of host alternation in the face of the proposed disadvantages, has been both exhumed and attacked since Mordvilko proposed it.

Börner (1939) proposed that aphids were initially polyphagous and have become increasingly specialized, such that the summer and winter hosts of a heteroecious species may be regarded as polyphagy spread over time. Börner does not consider that this system is shaped by constraints. It should be noted that the pattern of host utilization in aphid taxa provides clues as to the evolution of heteroecy. In many aphid taxonomic groups, the hosts (especially the primary hosts) are closely related. This suggests an ancestral association of a particular lineage with particular hosts, followed by either diversification or specialization. This is illustrated by species of *Cavariella* which all hibernate on species of *Salix* and spend the summer on Umbelliferae (Hille Ris Lambers, 1950). Clearly the precursor of these species possessed this pattern of host alternation. Whilst Börner's model of increasing specialization may hold, it does not account for this pattern.

Shaposhnikov (1951, 1956, 1987) suggested evolutionary host shifts occurred due to environmental changes which, by altering the physiology of the host plant, weakened the aphids' specificity and made possible the colonization of a new host. The new host is chosen from the available flora and hence secondary hosts are unrelated to the primary hosts. Thus the poor nutritional quality of woody plants in summer may initiate a move to other plants during this period. Similarly, Kennedy and Stroyan (1959) suggest that the summer forms of aphids, which are produced in large numbers, are less specialized and settle out randomly over the entire plant cover

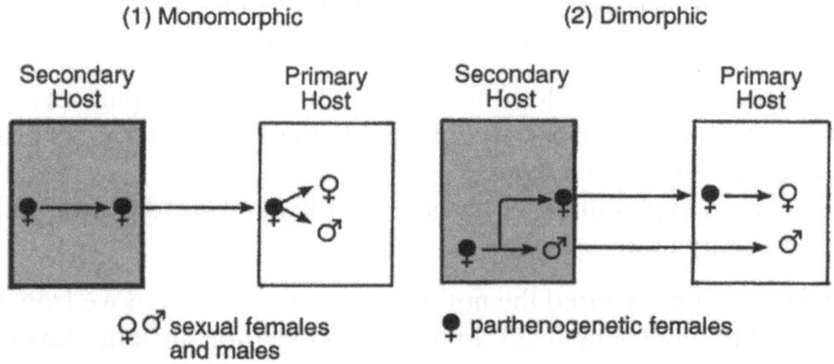

Figure 7.3 The two types of heteroecy (host alternation) in aphids.

(Kennedy and Booth, 1951; Kennedy, 1953) and are the potential initiators of a host shift. Kennedy and Stroyan consider that the fundatrix may develop close associations (in the exteme, galling) with a primary host, which could constrain the evolution of secondary autoecy.

A detailed analysis of aphid life cycle patterns was performed by Remaudière (1953), who noted that holocyclic taxa have open to them a number of different evolutionary pathways, but that assessing the evolutionary history of a given species, which is critical in this debate, is often impossible on the available evidence. This is an important point; if we are to make substantial progress in understanding host race and life cycle evolution, we must have dependable genetic phylogenies.

Hille Ris Lambers (1953) argued convincingly on the basis of his studies on *Cryptomyzus* that heteroecy in aphids should not be considered as an evolutionary dead end, because of the occurrence of several species autoecious on herbaceous plants. Szelegiewicz (1978) attributes the conflict between Hille Ris Lambers and Mordvilko on this issue to the existence of at least two types of heteroecy (Figure 7.3). Monomorphic heteroecy is characterized by the autumnal flight to the primary host being undertaken by a single morph, the sexupara. On the primary host the sexuparae give birth to both males and mating females. Because the sexuparae give birth to both sexes inbreeding is assured, except when population density is very high (Kindlmann and Dixon, 1989b). Szelegiewicz (1978) suggests that the high incidence of inbreeding might account for the low species diversity in this group of aphids. Loss of the association with the primary host in these taxa almost always appears to result in a species becoming anholocyclic, i.e. producing exclusively by parthenogenesis. Dimorphic heteroecy occurs in the Aphididae and is characterized by an autumnal flight to the primary host of two types of migrants, males and gynoparae, the latter giving birth to the mating females on the primary host. As the two morphs return to the primary host independently of each other and at different times, the chance

	Life cycles exhibited	Galls on primary host	Number of species
Pemphigidae Anoeciidae Hormaphididae	heteroecious (monomorphic type), autoecious and holocyclic on woody plants, autoecious and mostly only parthenogenetic on herbaceous taxa	often complicated	469
Drepanosiphidae	autoecious, mostly on woody plants	none	446
Aphididae	hereroecious (dimorphic type), autoecious and holocyclic on both woody and herbaceous plants	simple or none	2229

Figure 7.4 A simplified phylogeny of aphids. (Data from Heie, 1987; Eastop, 1977.)

of outbreeding is increased. This is associated with the production of simple or no galls on the primary host. The outbreeding is thought to account for the high diversity and morphological differentiation of the species in this group. In the Aphididae, breaking connection with one of the hosts does not necessarily result in a species becoming anholocyclic, but the evolution of secondarily autoecious and holocyclic species. These patterns are summarized in Figure 7.4, a simplified phylogeny of the major aphid families.

Considering the general phenomenon of the evolution of complex life cycles, Istock (1967) suggested that they are unstable and tend to break down to the phase conferring greatest advantage. He contended (Istock, 1970) that complex life cycles may only persist in the face of a genetic constraint. Slade and Wassersug (1975) and Wilbur (1980) criticize Istock's thesis and suggest that the evolution of complex life cycles will be favoured where sympatric high-quality ephemeral resource patches occur predictably. Evolutionary models of resource diversification cover two areas of intra- and interspecific interactions, which are relevant to a consideration of heteroecy, competition and 'enemy-free space' (Strong *et al.*, 1984; Futuyma and Moreno, 1988). As intraspecific competition becomes more intense, the relative risk of migration declines. It has been shown that aphid populations on trees in summer are particularly sensitive to density variation (Dixon, 1966). Thus, the move onto an alternative summer host may be promoted by such competition. The concept that habitat shifts may be in response to the predator and parasite burden (Kennedy and Booth, 1954; Gilbert and Singer, 1975; Bernays and Graham, 1988) has been suggested as a mechanism for the evolution of aphid heteroecy by Weed (1896) and Way and Banks (1968).

A prerequisite of a site where eggs are to be laid is the predictable availability of a food source in the spring. Long-lived perennials provide this, as the probability of the plant dying over the winter is negligible, and buds, where a high-quality food source is guaranteed, are laid down in the previous autumn. Thus Moran (1983) suggests that heteroecy is maintained by a balance between the advantages of feeding on herbaceous plants in the summer and the greater winter survival and spring fecundity attained on the woody host.

Recently, Moran (1988, 1990) has reopened the issue of host alternation in aphids and forcefully argued that it exists because generations on the ancestral host, particularly the fundatrix, 'are more constrained in their host affinities and less able to shift to newly available nutritionally superior hosts' (Moran, 1988, p. 694), i.e. like Mordvilko and Istock, she sees host alternation as a consequence of a fundatrix constraint and, like Mordvilko and Shaposhnikov, sees aphid evolution as a series of steps progressing to autoecy on herbaceous plants.

Mackenzie and Dixon (1990), Dixon and Kundu (1994) and Kundu and Dixon (1995), on the other hand, favour the view that for insects that can complete many generations in the course of a season, host alternation is one way of optimizing the use of plants that show complementary changes in quality and of exploiting a wide range of plants, i.e. they favour the other of Mordvilko's original hypotheses – the optimal use of the complementary nature of the growth of herbaceous and woody plants. Moran's belief that all aphid evolution is pressing inexorably though unsuccessfully towards secondary autoecy leads her to consider these mechanisms irrelevant (Moran, 1990).

Moran (1988) believes any theory based on the optimal use of host plants must address the following features of aphid life cycles: (1) the acquisition of new primary hosts, (2) secondary autoecy, (3) broader host plant range of summer morphs, (4) the evolutionary infidelity of summer morphs in heteroecious taxa, (5) the presence of 2-year cycles in some species, (6) the alleged alternation between hosts that are not complementary in growth. The optimization arguments for each of these is outlined below. In addition, an ecological explanation of the rarity of host alternation (7), galling and the specialization of fundatrices (8), evidence for complementary plant growth (9) and evidence against aphid evolution being in pursuit of recent plant taxa (10) are presented.

7.2.1 Acquisition of primary hosts

If host alternation is a transitional phase in the shift from a primary host to a nutritionally more favourable 'secondary host' then one would expect the primary hosts of a particular group of aphids to be closely related. Or if heteroecy is superior to year-round feeding on herbaceous plants, then species on herbaceous hosts should sometimes acquire woody hosts for spring and autumn generations (Moran, 1988).

In late Eocene (40–45 million years bp) Baltic amber, the commonest fossil aphid is *Germaraphis dryoides* (Heie, 1987), which very closely resembles *Prociphilus bumeliae*, an extant taxa host alternating between ash, *Fraxinus excelsior* (the primary host), and firs, *Abies*, The presence of large numbers of only first instar aphids in pieces of amber led Hille Ris Lambers (1979) to state it was as if 'a procession of the progeny of emigrants was caught in a blob of resin rolling down the trunk of the amber tree', i.e. they had been caught in the act of host alternating (Hille Ris Lambers, 1979). That this group has had a long association with conifers is not doubted (Heie, 1987). European species of *Prociphilus* overwinter on *Crataegus*, *Fraxinus* and *Lonicera*, and spend the summer on the roots of *Pinus*, *Abies* and *Picea*, respectively. A similar migration to the roots of Coniferae is found in the closely related *Pachypappa*, which all overwinter on *Populus*. Thus it is likely that the species of *Prociphilus* on *Crataegus*, *Fraxinus* and *Lonicera* evolved from aphids that host alternated between *Populus* and the roots of Coniferae (Hille Ris Lambers, 1950). That is, *Prociphilus* has captured new primary hosts, not from within the Dileniidae to which *Populus* belongs but from the more highly evolved Rosidae (*Crataegus*) and Asteridae (*Fraxinus* and *Lonicera*).

The complexity of the host plant relationships of *Aphis fabae* has largely been revealed by Müller (1982, 1985b). What was previously regarded as a highly polyphagous species is now seen as a species complex with each 'subspecies' exploiting only part of the host range of the original *Aphis fabae*. The host relationships of the *Aphis fabae* complex are illustrated in Figure 7.5. As in *Cryptomyzus* cited below, the sympatric coexistence of secondarily autoecious *Aphis f. armata*, host-alternating *Aphis f. fabae*, *Aphis f. solanella*, *Aphis f. cirsiiacanthoidis* and *Aphis f. mordwilkowi*, and autoecious species on the primary hosts, *Aphis f. euonymi*, *Aphis f. philadelphi* and *Aphis viburni*, argues strongly for 'balancing' selection and against herbaceous plants being generally more favourable. It also cannot be argued that the autumn migrants are rigidly constrained to *Euonymus europaeus*, which belongs to the Rosidae (Celastraceae), since *Aphis f. fabae*, *Aphis f. cirsiiacanthoidis* and *Aphis f. mordwilkowi* have captured between them two other primary hosts, *Viburnum opulus* (Asteridae, Caprifoliaceae) and *Philadelphus coronarius* (Rosidae, Hydrangeaceae) (Müller, 1982; Thieme, 1988). Not only are the fundatrices of this aphid moving between families of plant within a subclass, but also like *Prociphilus* between subclasses.

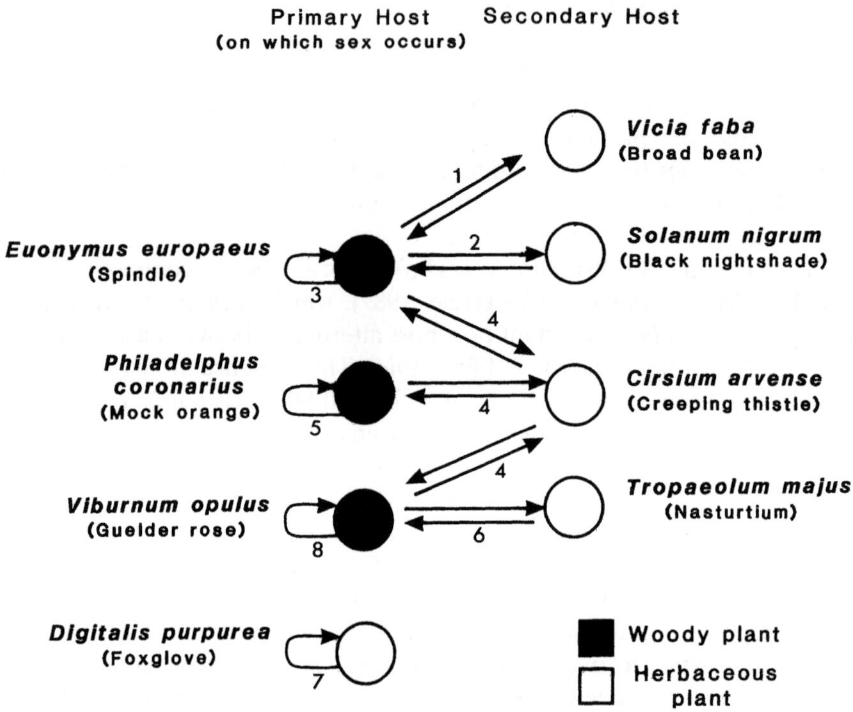

Figure 7.5 Host plant relationships in the aphid species group *Aphis fabae*. The host races/subspecies are numbered as follows: (1) *A. fabae fabae*, (2) *A. fabae solanella*, (3) *A. fabae euonymi*, (4) *A. fabae cirsiiacanthoidis*, (5) *A. fabae philadelphi*, (6) *A. fabae mordwilkowi*, (7) *A. fabae armata*, (8) *A. viburni*. (After Müller, 1982; Stroyan, 1984; Heie, 1986; Thieme, 1988.)

In addition, the primary hosts of the host-alternating species of *Aphis*, at least in the UK (Stroyan, 1984), belong to several plant families. Caprifoliaceae (three species), Celastraceae (one species), Cornaceae (one species), Grossulariaceae (two species), Hydrangeaceae (one species) and Rhamnaceae (two species). Similarly, the primary hosts of the sibling species *Aphis gossypi* and *Aphis glycines* belong to the Rutaceae and Rhamnaceae, respectively (Zhang and Zhong, 1990). That is, *Aphis* also appears to have been very good at capturing primary hosts from a wide range of families from two subclasses of plants. Interestingly, it is also debatable in species like *Aphis f. solanella*, which host alternates between *Euonymus* (Rosidae) and *Solanum* (Asteridae), whether it is the primary or the secondary host that is phylogenetically older, and in cases like *Aphis sambuci* the primary host (*Sambucus*, Asteridae) is in a phylogenetically younger taxon than the secondary hosts (*Rumex*, *Dianthus*, *Silene*, Caryophyllidae).

A powerful argument for the optimal nature of seasonal shifts in host usage is that it does not occur exclusively between woody and herbaceous plants. *Acyrthosiphon pisum* host alternates between perennial vetches and the annual *Pisum sativum* (Mordvilko, 1928; Müller and Steiner, 1985), and *Uroleucon gravicorne* and *Uroleucon olivei* between perennial *Solidago* and *Aster* and annual *Erigeron* (Moran, 1983, 1987). In all cases, sex occurs on the perennial plant, which offers a high probability of continuity in time. The annual plant is more favourable for reproduction during summer. This indicates that aphids can exploit the complementary nature of the growth of plants wherever it occurs, even between herbaceous plants, and emphasizes yet again that it is not just complementary nutritional quality that is important, i.e. the favoured host is not merely something fed on, it is something lived on (Kennedy, 1953).

From the examples presented above, it is apparent that in many taxa the fundatrix morph has made the evolutionary shift from one host to another, unrelated plant. In most cases, though not all, the new primary host is a woody plant. If woody plants are indeed a poor food source, then breaking of the association with the first woody primary host should not be followed by the establishment of a new relationship with a woody plant. This is a clear rebuttal of the predictions of Moran's (1988) model.

Furthermore, the host relationships of *Prociphilus* and many Aphididae reveal that Mordvilko's view that woody or primary hosts were the first to be captured is not always true. However, as the terms primary and secondary are now widely used, primary should imply nothing more than being the host on which eggs are laid and the life cycle starts and ends.

7.2.2 Secondary autoecy and the maladaptive nature of heteroecy

The suggestion that the evolution of autoecy in taxa that formerly were heteroecious indicates that heteroecy is not advantageous (Moran, 1988, p. 697) can be challenged. It is mistaken to suggest that because two related species possess different characters, one must be maladaptive. This can be illustrated by reference to the host distribution of the various species and forms of *Cryptomyzus* (Hille Ris Lambers, 1953; Guldemond, 1991). The host-alternating habit of *Cryptomyzus galeopsidis*, *Cryptomyzus heinzei*, *Cryptomyzus korschelti*, *Cryptomyzus maudamanti*, *Cryptomyzus ribis* and *Cryptomyzus stachydis* is regarded as the original way of life of the genus, and the autoecious habit of *Cryptomyzus alboapicalis*, on what is regarded as a secondary host plant, to have evolved later (Hille Ris Lambers, 1953; Szelegiewicz, 1978) (Figure 7.6).

The presence of genetically determined forms of *Cryptomyzus galeopsidis* that remain all year on the primary hosts, *Ribes nigrum* and *Ribes rubrum* (Guldemond, 1987, 1990b), indicates that evolution proceeds in both directions with species becoming secondarily autoecious on either the primary or the secondary host. This raises the question: what are the factors that favour host alternation or autoecy?

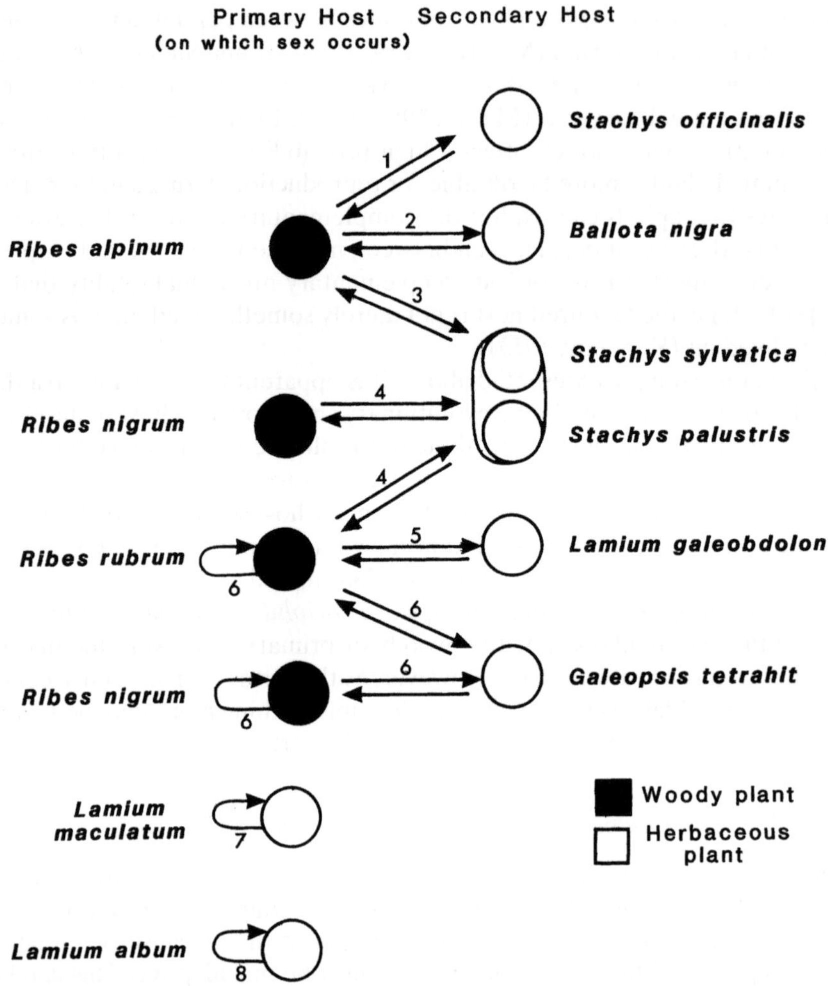

Figure 7.6 Host plant relationships in the aphid genus *Cryptomyzus*: (1) *C. heinzei*, (2) *C. ballotae*, (3) *C. korschelti*, (4) *C. ribis*, (5) *C. maudamanti*, (6) *C. galeopsidis*, (7) *C. ulmeri*, (8) *C. alboapicalis*. (After Guldemond, 1990a.)

7.2.3 Broader host plant range of summer morphs

In general, heteroecious species utilize more plant species during the summer months than during their sexual phase. If we view the secondary host plant range of heteroecious species in the context of the diet breadth of related autoecious species, it is apparent that host-alternating species are in general much more polyphagous than other species (Figure 7.7). Why are autoecious species specialists and heteroecious species generalists?

Evolutionary theories of specialization may provide us with some clues. Futuyma and Moreno (1988) outline a number of factors that may be

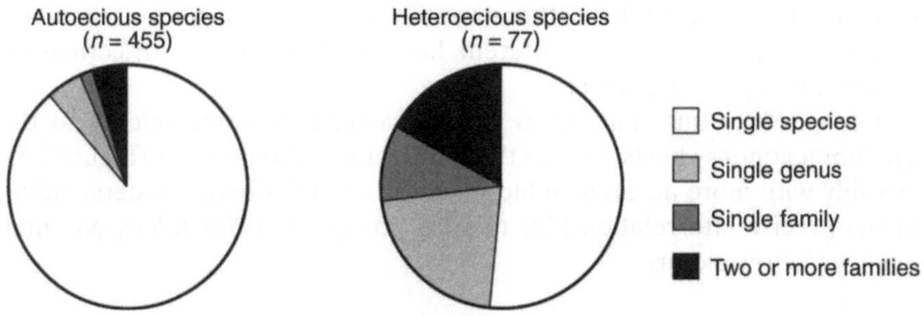

Figure 7.7 Summer host range of European members of the Aphididae.

important in the determination of a species' specialization, in particular: (1) environmental constancy – the more predictable the environment, the more readily it may be partitioned, and (2) mating rendezvous – finding a mate is of paramount importance to any sexual species and concentrating all individuals on a single resource is one way of achieving this. Ward (1987a, 1991) emphasizes the importance of mating rendezvous in aphid life cycle evolution. Heteroecious species returning to a single primary host species clearly have no problem in locating mates, so they may diversify across a broad host plant range during the summer, avoiding the risk of an obligate association with a single herbaceous species.

Habitat permanence has long been recognized as an important selective force in insect life-history evolution (Southwood, 1977). Herbaceous plants, especially annuals, are liable to dramatic changes in density, even over short time periods, whilst woody plants have far more stable population structures. Thus, in specializing on a constant resource (the woody primary host) and diversifying over a range of stochastically varying herbaceous species, the heteroecious species has the best of both worlds. As Ward (1987a) states, only in heteroecious species 'is the host range free to expand to include all nutritionally, phenologically and architecturally suitable hosts'.

7.2.4 Evolutionary infidelity of summer morphs

Hille Ris Lambers (1950, 1979) and Moran (1988) have drawn attention to the fact that, especially in the Pemphigidae, summer morphs are generally much more evolutionarily flexible in their choice of host taxa than the sexual stages. Moran (1988) construes this as evidence that secondary autoecy is optimal, on the basis that secondarily autoecious species would commonly capture new primary hosts if heteroecy were adaptive. She claims the move from autoecy is 'never reversed' (Moran, 1990, p. 136). However, Mackenzie and Dixon (1990) indicate that new primary hosts are regularly captured in the Aphididae and within *Prociphilus* in the

Pemphigidae. In general, whether or not the preceding stage was secondary autoecy or heteroecy on a different host, is impossible to discern from current phylogenetic evidence.

It is worth noting that *Aphis f. cirsiiacanthoidis* shows fidelity to the *Cirsium* secondary hosts, but has three unrelated primary hosts (Figure 7.5). Possibly with more accurate aphid phylogenies and a better understanding of the precise interrelationships of such groups as *Aphis fabae*, one may reach a clearer picture.

7.2.5 Two-year life cycles

Moran suggests that biennial life cycles, which occur in the Fordini and Hormaphidini, 'should provide evidence useful in evaluating hypotheses proposed for (heteroecy) evolution' (1988, p. 688). Unfortunately, she neglects to explain how this may be, except to say 'none of the optimization arguments is consistent with . . . the existence of 2-year cycles' (1988, p. 697). The following illustrates how the life cycle pattern of a typical species fits with the complementary growth hypothesis.

In *Baizongia pistaciae* (Fordinidae), eggs are laid on the shrub *Pistacia palaestina*. Fundatrices hatch from March to mid April and immediately commence feeding on the smallest embryonic compound leaves at the apex of a shoot. Their feeding and that of their descendants induce a large banana shaped gall up to 10 cm long. The gall opens from the end of October through November and releases up to 8000 alatae. These emigrants fly off and colonize the roots of various Poaceae usually at the end of a long dry season. The following spring, alate sexuparae develop on grass, return to *Pistacia* and give rise to the sexuales, the females of which lay the eggs that remain dormant until the following spring, when they hatch somewhat prior to the arrival of the sexuparae. *Baizongia* exploits *Pistacia* while it is actively growing in spring, and extends this favourable period during summer and early autumn by galling (p. 46) (Wertheim, 1954). The emigrants that leave *Pistacia* in autumn do so as the autumnal rains terminate the long summer dry season, the grasses begin to grow and *Pistacia* sheds its leaves and becomes vegetatively dormant (Davatchi, 1958). This species would appear to fit the complementary growth hypothesis very well (Figure 7.8). The presence of the extended egg diapause is irrelevant, but probably accounted for by the necessity of the fundatrices to co-ordinate hatching precisely with the onset of growth in the shrub, to enable successful gall initiation, which Wool (1984) indicates is critical.

Moran (1988, p. 700) also notes that in some species, of which *Baizongia* is an example (Wertheim, 1954; Wool and Manheim, 1986), host alternation occurs in parallel with continued parthenogenesis on the secondary host. Such facultative host alternation (Remaudière, 1953; Müller, 1961, 1966) is widespread in aphids (Lathrop, 1928; Savary, 1953; Stroyan, 1963; Judge,

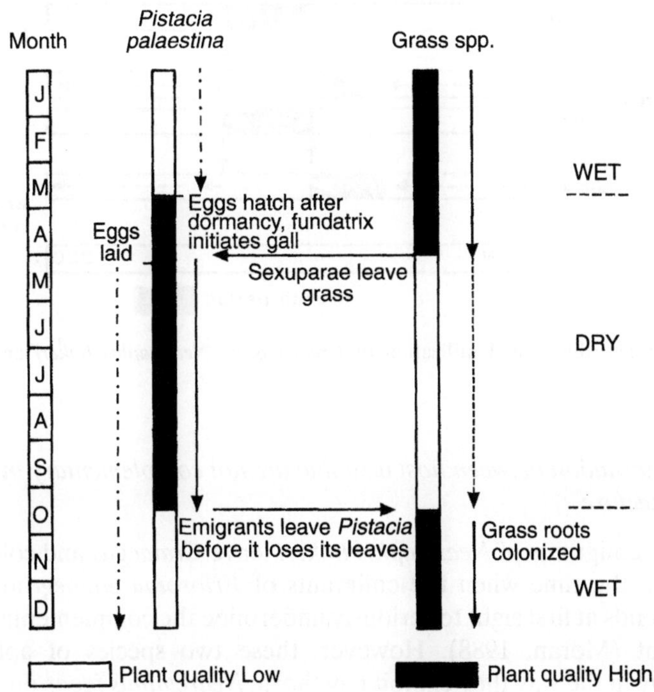

Figure 7.8 The life cycle of *Baizongia pistaciae* (Hormaphididae).

1967; Kolesova, 1974; Smith and Denmark, 1984). Moran implies that the system of mixed heteroecy and parthenogenesis continues due to the long-term advantages of sex, and despite the disadvantages of utilizing the primary host, one would expect all of the population to shift, rather than a proportion of it. However, if the environment is subject to stochastic variation, then either the secondary host might be 'unexpectedly' rewarding or the primary host unexpectedly poor. If the aphid is unable to predict the climate, then a mixed strategy is optimal. It should be emphasized that natural selection in aphids takes place at the level of the clone, not the individual aphid. It has been indicated (Müller and Steiner, 1985; Tatchell and Parker, 1990; Guldemond, 1991) that a single clone may produce both sexual and parthenogenetic morphs in the autumn, showing that the mixed strategy is at the level of the genotype, not the population.

Thus the aphid *Baizongia pistaciae* would appear to be using the complementary growth patterns of *Pistacia* and grasses to extend the period available for reproduction and the long life of one of the host plants to increase the probability of survival of the sexually produced offspring. A similar explanation could be offered for the 2-year life cycle of Hormaphidini. It is not necessary to view such life cycles as a consequence of the specialization of the fundatrix tying the aphid to one of the host plants.

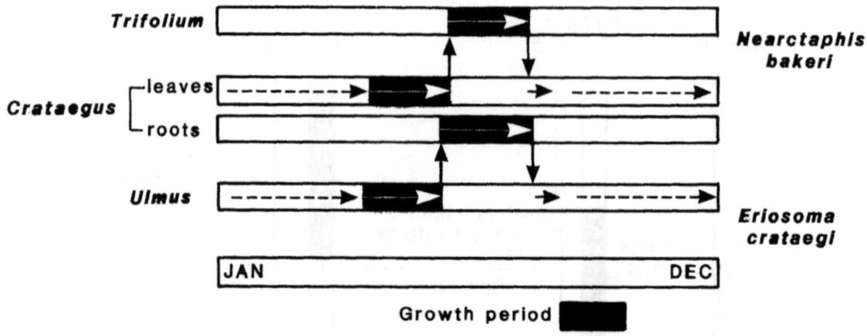

Figure 7.9 The differential utilization of *Crataegus* by *Nearctaphis bakeri* and *Eriosoma crataegi.*

7.2.6 Alternation between host taxa that are not complementary in growth or architecture

That the emigrants of *Nearctaphis bakeri* leave *Crataegus* and colonize *Trifolium* at the time when the emigrants of *Eriosoma crataegi* arrive from *Ulmus* tends at first sight to seriously undermine the complementary growth argument (Moran, 1988). However, these two species of aphid differ markedly in the way they exploit hawthorn; *Nearctaphis* feeds on the leaves and *Eriosoma* on the roots (Heriot, 1938). Leaf growth in shrubs and trees occurs earlier in the season than root growth. Therefore, these aphids exploit two markedly different resources when on hawthorn. For *Eriosoma crataegi* it is the growth of the roots of hawthorn, not the leaves, that show a complementary growth pattern with the leaves of its primary host, *Ulmus* (Figure 7.9). It should be emphasized that summer generations of all *Erisoma* species, like the majority of Pemphigidae, are root dwellers (Akimoto, 1983; Heie, 1980). The adaptations to root living are clearly highly specific. Hence a transition of *Nearctaphis bakeri* from hawthorn leaves to roots, which would clearly avoid some migration losses, is constrained by the inflexibility of the summer morphs.

7.2.7 Rarity of host alternation

If host alternation is advantageous then why is it so rare? Only about 10% of aphid species host alternate (Eastop, 1986), although some of them like *Aphis fabae* and *Myzus persicae* are very abundant. Two factors appear to be important, the availability of two suitable sympatric hosts and a sufficiently long season to complete the necessary development and reproduction.

In climax forest dominated in the summer by few species of tree, there is reduced opportunity for the evolution of host alternation. The closed canopy characteristic of such woodlands prevents the development of herbaceous plants during summer when the 'quality' of the trees for aphids is low. In contrast early successional trees, like *Populus*, *Prunus*, *Pyrus*, *Salix* and *Ulmus*, which have many heteroecious aphids associated with them, live in more open and temporary habitats where there is often a rich herbaceous flora in summer. Although both the boreal and Mediterranean forests are open, they are unlikely to afford a suitable habitat for host alternation. The growth of the plants in boreal forests tends to be synchronized because of shortness of the season, and in the Mediterranean forests the herbaceous vegetation tends to dry out in summer. Thus, opportunity in terms of common woody and herbaceous plants having complementary growth patterns and occurring close together, i.e. a suitable plant community structure, must have played an important part in the evolution of host alternation (Dixon, 1990b).

If the secondary host were to become rare in part or all of its common distribution with the primary host, then selection will favour genotypes that stay on the primary host during summer, and similarly, secondary autoecy would be selected in the absence of the primary host. Thus the relative abundance of the various plants will determine whether and how they are utilized by aphids (Figure 7.10). This factor has been largely ignored in accounting for the distribution of aphids between different species of plants (Dixon, 1987a).

The evolution and maintenance of host alternation is constrained by the time available each season for development and reproduction. The transfer between hosts needs a specialized winged morph, so there must be enough time to develop at least four generations. In addition the production of winged forms is costly in terms of rate of increase as they take longer to develop and have a lower reproductive rate (Mordvilko, 1928; Dixon and Howard, 1986). Thus, it is only advantageous to shift hosts if the cost of moving can be more than offset by multiplication on the secondary host, which is likely to require several generations. High altitudes and latitudes, where the season is very short, are unlikely to favour host alternation as a way of life. This may in part account for why an autoecious species of *Hormaphis* (with three generations per year) replaces the host-alternating species (with seven generations) at higher elevations in north central and north-western Virginia (von Dohlen and Gill, 1989), though secondary host availability also plays a role. The absence of host-alternating forms of *Cryptomyzus galeopsidis* in Iceland (Heie, 1964) is also evidence for a season length effect.

7.2.8 Fundatrix specialization

Central to the idea that host alternation exists mainly because of the constraints imposed by the close association of the fundatrix with the primary

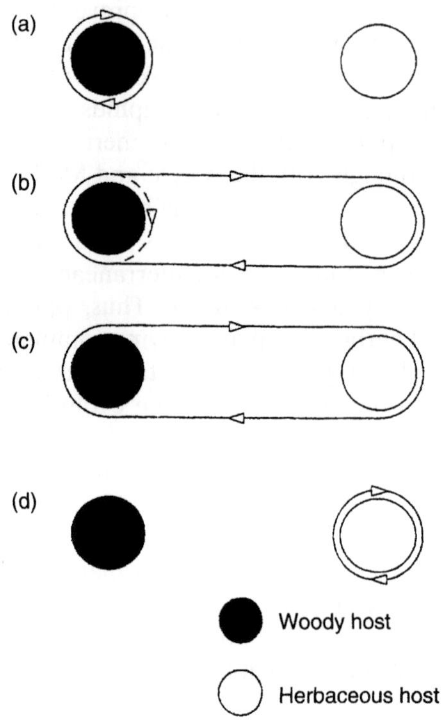

Figure 7.10 Schematic diagrams of an autoecious life cycle on a woody host (a), host alternation between a woody primary host and an herbaceous secondary host with some clones (dashed arrow) remaining on the primary host throughout the year (b), host alternation with all the clones transferring seasonally between the primary and secondary host plants (c), and secondarily autoecious life cycle on an herbaceous host (d)

host, is a high degree of specialization of the fundatrix morph in heteroecious species. The evidence presented in support of this is (1) the morphological differences that exist between fundatrices and subsequent generations, and (2) the tendency for the fundatrix to gall or deform its host in host-alternating species.

The morphology of the fundatrices of all species differ from that of subsequent generations. In most species the fundatrices differ in having shorter appendages and more egg tubes (ovarioles) in their gonads (Lees, 1966; Dixon, 1974b; Wellings *et al.*, 1980). However, in some species the fundatrices tend to be strikingly larger and globular in shape. Commonly, fundatrices of heteroecious species have a more extreme morphology (Figure 7.11), but there are exceptions to this (Moran, 1988: Figure 5, *Periphyllus*).

This difference in external morphology is associated with differences in the number of ovarioles. Species like *Thecabius affinis* (heteroecious) show big differences in morphology and ovariole number (50 in the fundatrix, 13 in the next generation), whereas *Lachnus roboris* (autoecious) shows little

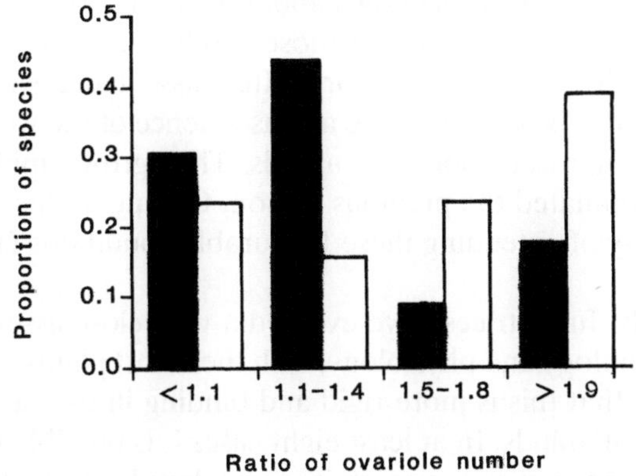

Figure 7.11 Relative fundatrix morphology in terms of the ratio of the number of ovarioles in the fundatrix to that in the summer apterae in heteroecious (open bars, $n = 13$) and autoecious species (filled bars, $n = 23$) of aphids.

or no difference in either (Tashev and Markova, 1982; Polaszek, 1987). As ovariole number is positively associated with reproductive rate (Wellings *et al.*, 1980; Leather and Wellings, 1981), the globular nature of some fundatrices only indicates they have a very high reproductive rate. Why should a high reproductive rate bind an aphid to a plant any more than a low reproductive rate? There is no reason to believe that a fundatrix should not survive on an herbaceous host because of its high fecundity. The increased fundatrix fecundity in heteroecious species is achieved at a cost of reduced mobility. This results in a limited ability to locate resources and a decreased predator avoidance capacity. This is particularly clear if the winged fundatrices of the wholly autoecious Drepanosiphidae are compared with the heteroecious, globular short-legged *Thecabius*. Heteroecious species are taking a risk, placing high spring fecundity at the highest priority. This is necessary because (1) to make use of the complementary growth of herbaceous plants, alatae must be produced within a relatively short time, and (2) due to migration losses, only a small proportion of emigrants will reach the secondary hosts (Taylor and Taylor, 1979; Ward *et al.*, 1998). As aphid alatae are relatively inept at finding host plants, a genotype can only benefit from heteroecy if large numbers of emigrants are produced. Thus, heteroecy may be viewed as a high-risk, high-reward strategy that utilizes extreme fundatrix morphology.

It has been argued that galling of woody hosts by fundatrices indicates a long association with a host plant (Mordvilko; 1928; Hille Ris Lambers, 1979; Szelegiewicz, 1978; Moran, 1988). However, aphids feeding on herbaceous plants frequently distort their growth and in the case of *Cryptosiphum artemisae*, *Eumyzus gallicola*, *Eumyzus impatiensae* and *Hayhurstia atriplicis* induce complicated and colourful galls (Buckton, 1879; Hayhurst, 1909;

Moritsu, 1983). What little is known about the induction of these galls indicates that they are very similar to those produced on trees by host-alternating aphids. That galling is more often associated with woody than herbaceous plants, is possibly more a consequence of the way woody plants grow than a long association with aphids. They grow rapidly, drawing on resources accumulated the previous season. Galling in this context may be viewed as a way of extending these favourable conditions (Kennedy, 1958; Forrest, 1971).

Undoubtedly, fundatrices have evolved a very close association, both in terms of morphology and physiology, with their host plants. However, there is no evidence that this is more rigid and binding in the case of host-alternating species of aphids. In at least eight cases it is possible to rear the fundatrix of a heteroecious species on its secondary host plant (*cf.* Table 7.2) (Mordvilko, 1928; Dixon and Dharma, 1980b; Dixon and Kundu, 1994). Their globular structure and galling habit seems to be related more to their ecology than to a long evolutionary association with a particular group of plants.

7.2.9 Nutritional quality of the secondary host plant

That secondary host plants should be generally more nutritionally favourable than the primary host, rather than complementary, is an important requirement of Moran's thesis (Moran, 1988). Many trees and shrubs start growing early in the season before most herbaceous plants.

Observations on host-alternating aphids tend to support the view that while on the primary host in spring these aphids achieve large size and high fecundity and those that remain there over summer are small and fare badly, whereas those that colonize herbaceous plants in summer thrive. This stresses the complementary nature of the relationship rather than one host being generally better than the other.

The little experimental work that has been done indicates that in the case of *Aphis*, *Cavariella*, *Dysaphis*, *Pemphigus* and *Uroleucon*, they do better in terms of fecundity per generation while on *Euonymus*, *Salix*, *Crataegus*, *Populus* and *Solidago*, respectively, than when on their secondary hosts. However, if they had remained on the primary host, they would have done worse (Kennedy and Booth, 1951; Goszczynski and Cichocka, 1978; Moran, 1983; see below), i.e. many herbaceous plants are a better nutritional resource during summer.

7.2.10 The progressional acquisition of more recently evolved taxa

Mordvilko (1928) and Moran (1988) contend that over evolutionary time aphids have progressively moved onto more recently evolved taxa, driven by the increasing food plant quality.

Table 7.1 Capture of more ancient host taxa by aphids

(a) Heteroecious species

Aphid	Primary host	Secondary host
Pemphigidae		
Melaphis rhois	Angiospermae	Bryophyta
Prociphilus spp.	Angiospermae	Gymnospermae
Pachypappa sp.	Angiospermae	Gymnospermae
Asiphum spp.	Angiospermae	Gymnospermae

(b) Autoecious species

All examples are from the Aphididae, which are regarded as ancestrally heteroecious with a Rosales (Rosaceae) primary host (Hille Ris Lambers, 1950)

Aphid	Host	Aphid	Host
Myzodium	Bryophyta	*Amphorophora ampullata*	Pteridophyta: Filicopsida
Aspidaphium	Bryophyta	*Idiopterus*	Pteridophyta: Filicopsida
Aphis equiseticola	Pteriodphyta: Equisetales	*Ephedraphis*	Gymnospermae: Gnetaceae
Sitobion equisetum	Pteridophyta: Equisetales	*Elatobium*	Gymnospermae: Pinaceae
Sitobion	Pteridophyta: Filicopsida	*Illinoia*	Gymnospermae: Pinaceae
		Sanbornia	Gymnospermae: Pinaceae

Yet there are numerous examples of aphid species that have moved from young to more ancient plant taxa. In some cases, the more ancient plant is a secondary host (Table 7.1a), whilst in others the evolutionary move to the more primitive host is complete and the aphid is autoecious on this plant (Table 7.1b). This would appear to be a direct contradiction of the above hypothesis, that aphid evolution is involved in the pursuit of the most recent plant taxa.

7.3 Experimental evidence

The aphids that hatch from the overwintering eggs (fundatrices) of eight species of host-alternating Aphidinae have been successfully reared on their secondary host plants (Table 7.2). This has mainly been achieved by transferring the aphid to the secondary host plant at egg hatch (Dixon and Kundu, 1994). In *Cavariella aegopodii*, however, the eggs were laid, hatched and the emerging fundatrices reared to maturity on the secondary host plant. Although the fundatrices do not survive as well on the secondary host as they do on the primary host at bud burst they do survive and grow as well or better than those reared on the mature foliage of the primary host (Kundu and Dixon, 1995).

Table 7.2 The species of host-alternating Aphididae whose fundatrices have been successfully reared on a secondary host

Species	*Host alternates between*
A woody primary and herbaceous secondary host	
Aphis fabae	spindle and bean
Cavariella aegopodii	willow and Umbelliferae
Cavariella pastinacae	willow and Umbelliferae
Cavariella theobaldi	willow and Umbelliferae
Metopolophium dirhodum	rose and grasses
Myzus persicae	*Prunus* and species of over 40 different plant families
A perennial herbaceous primary and annual herbaceous secondary host	
Acyrthosiphon pisum	perennial vetches and annual *Pisum* spp.
Uroleucon gravicorne	perennial *Solidago* and annual *Erigeron*

That is, contrary to the prediction of the fundatrix specialization hypothesis there does not appear to be a physiological barrier to some host-alternating Aphididae transferring their whole life cycle over to the secondary host plant, i.e. becoming autoecious. Therefore, if host-alternating aphids are not constrained to this way of life, what are its advantages and under what conditions is it likely to have evolved?

Host alternation involves at least two host transfers each year in most species. Thus host-alternating clones need to be able to compensate for the large losses incurred in host transfer otherwise there will be a strong selection pressure to simplify the life cycle and for the aphid to remain (become autoecious) on either the primary or secondary host.

The population growth rates (Figure 7.12) and fecundities (Figure 7.13) recorded in the field for each generation of three host-alternating aphids reveals that the potential amplification in the number of individuals in each clone on the primary host in spring could compensate for losses incurred in the spring host transfer. Only as few as 1:100 are thought to make each transfer between hosts (Taylor, 1977). Thus these clones are better represented at the beginning of the season on the secondary host than they would have been if they had overwintered there. In two of these species the further amplification in numbers achieved on the primary host in autumn is probably not greater than it would have been if they had stayed on the secondary host (Figure 7.13). Thus the amplification in numbers in each clone achieved in spring on the primary host has not only to offset the cost of one host transfer, but of both host transfers. By virtue of the very high fecundities achieved on the primary host in spring these aphids can potentially compensate for the huge losses incurred in making the two transfers.

The remigrants of most host-alternating Aphididae that have been studied either do not or, judging by the relatively few offspring they produce, are thought not to feed on the primary host (Dixon and Kundu, 1994). This has been attributed to the short period remaining for ovulation and

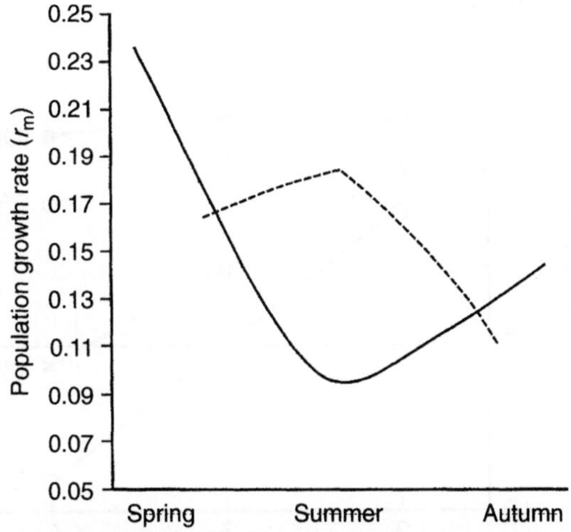

Figure 7.12 Seasonal changes in the population growth rate (r_m) of *Cavariella aegopodii* on willow (solid line) and carrot (dashed line). (After Kundu and Dixon, 1995.)

maturing additional embryos before leaf fall (Walters *et al.*, 1984). Those of the willow–carrot aphid, *Cavariella aegopodii*, however, are exceptional as they achieve a high fecundity, feed and mature half their embryos on the primary host. This difference in the reproductive behaviour of the remigrants is possibly associated with the pattern of leaf fall of their tree hosts.

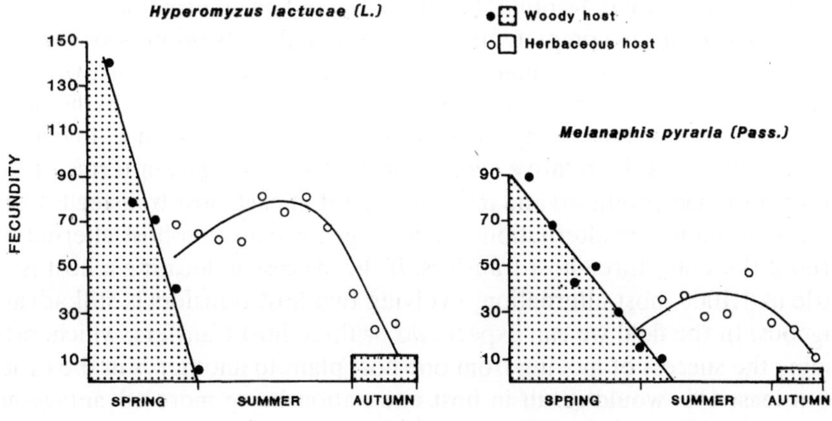

Figure 7.13 The fecundity and number of generations that *Hyperomyzus lactucae* and *Melanaphis pyraria* achieve on their primary (filled circles) and secondary (open circles) hosts in spring and summer, respectively, and the fecundity achieved by the autumnal migrants of these species on their primary hosts. (After Karczewska, 1976, 1979.)

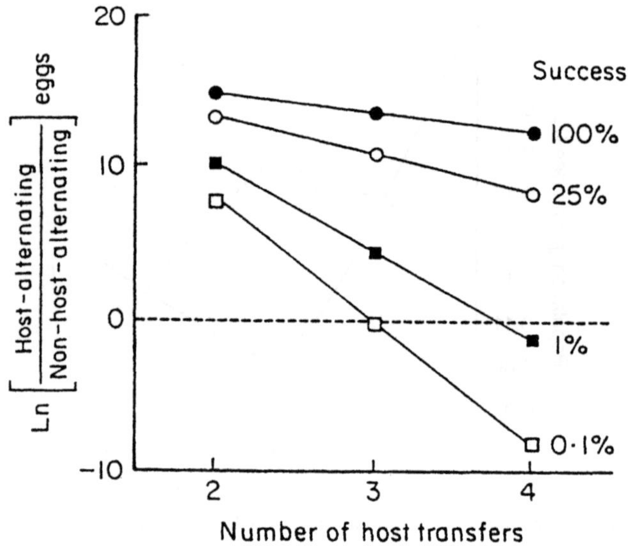

Figure 7.14 Relative fitness of host-alternating and non-host-alternating forms of *Cavariella aegopodii* in terms of eggs produced at the end of a season assuming different degrees of success in locating a new host and different numbers of host transfers for the host-alternating form.

Willow (*Salix* spp.), the primary host of *Cavariella aegopodii*, retains its leaves much later than do the primary hosts of other host-alternating aphids, e.g. *Malus*, *Prunus* and *Pyrus*, and as a consequence the aphid can further amplify its numbers on its primary host, i.e. the relative phenology of the two host plants will determine the advantage of host alternation.

As *Cavariella* can complete its life cycle on either its primary or secondary host it is possible to estimate the advantage of host alternation over non-host alternation assuming that all individuals of clones with the later strategy make one host change in the third generation between willow trees. Figure 7.14 gives the relative success of the two strategies in terms of the number of eggs produced and indicates that if at each host transfer 1% are successful then host alternation is advantageous over non-host alternation even if there are three host transfers. If the success in locating a host is as little as 1:1000, host alternation involving two host transfers is still advantageous. In the field we can expect two or three host transfers, which, providing the success of moving from one host plant to another is in the order of at least 1%, would result in host alternation being more advantageous than autoecy (Kundu and Dixon, 1995).

That is, the fitness of a clone is dependent on its success in maximizing the collective performance over the whole life cycle, the evolutionary unit of Janzen (1977). However, any factor that reduces performance on the primary or secondary host would change the balance and selection might then favour clones that transfer their whole life cycle to either the woody or the herbaceous host plant. In the case of the host-alternating Aphididae there is good experimental evidence in support of the optimization hypothesis – these aphids are exploiting the complementary growth patterns of woody and herbaceous plants. As observed by Szelegiewicz (1978), whether optimality or constraint is the favoured interpretation tends to depend on the group of aphids studied. The very close association between the Pemphigididae and their woody host plants, which is reflected in the complex galls induced by these aphids, makes this group of aphids very unsuitable experimental material. However, once it is accepted that the form of the fundatrix is an adaptation to the host quality it experiences, then the current host plant associations of these aphids tends to lend support rather than refute the idea that optimization criteria have been important in the evolution of host alternation generally.

The fundatrix constraint was proposed as the reason why only aphids amongst phytophagous insects show host alternation (Mosan, 1988). The above tends to cast doubt on it being the major factor. However, the above analysis does indicate that host alternation is only possible in insects that have very high rates of increase. Aphids are unique amongst insects in this respect and it is this rather than the fundatrix constraint that probably accounts for why only aphids amongst the insects host alternate.

7.4 Autoecious way of life

Most of the detailed studies on aphid life cycles have been on heteroecious species because they are often pests when on their secondary hosts and because of the interest in what in insects is a rare way of life. The autoecious way of life is equally interesting. Many of the species that are primarily autoecious, i.e. not thought to have evolved from heteroecious species, e.g. Drepanosiphidae, live entirely on trees, which are perceived to be nutritionally unfavourable hosts during the summer months.

7.4.1 Food quality

When aphid performance is measured in terms of the percentage of energy consumed that is assimilated and used for growth, aphids feeding on herbaceous plants appear to do better than those feeding on trees (Figure 7.15; Dixon, 1975a; Llewellyn, 1982). This has been cited in support of the notion that herbaceous hosts are nutritionally superior to trees as hosts for aphids. The magnitude of this assimilation/consumption ratio is dependent on the sugar content of phloem sap. In trees it is very high (Dixon, 1975a) and most of it is excreted by aphids. What is important is whether aphids can obtain sufficient quantities of amino-nitrogen per unit time to sustain high growth and reproductive rates. Aphids compensate for the low amino-nitrogen content of

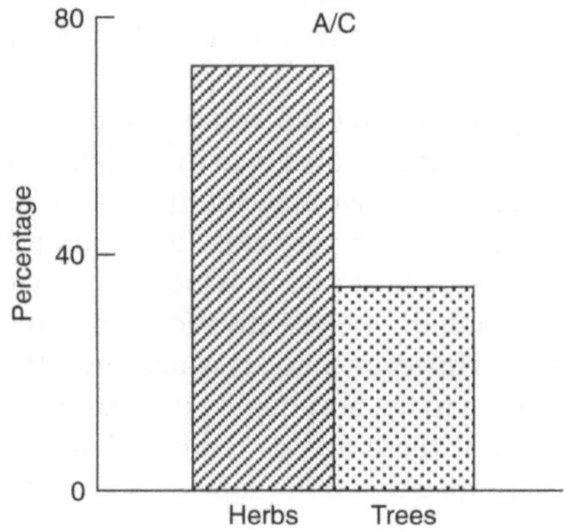

Figure 7.15 The percentage of the energy consumed (C) that is assimilated (A) for aphids that live on herbaceous plants and trees. (After Llewellyn, 1982.)

their food by processing large quantities per unit time and by using their symbionts to upgrade some of the amino acids (p. 19).

Some tree-living aphids like the lime aphid (*Eucallipterus tiliae*) and the Turkey-oak aphid (*Myzocallis boerneri*) are reproductively active all through the season (Figure 7.16). Others, like the sycamore aphid (*Drepanosiphum platanoidis*), enter reproductive diapause as an adult during the summer (Figure 7.17). Generally, their reproductive rates are

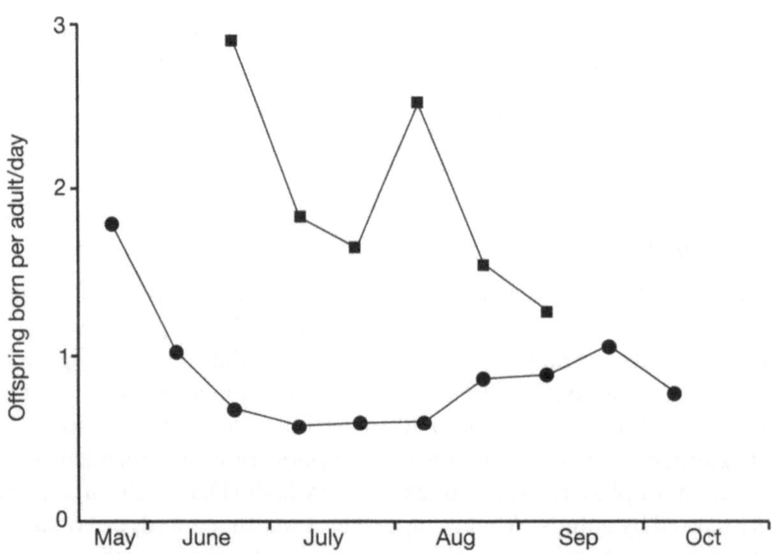

Figure 7.16 Seasonal changes in the daily reproductive rates of the lime (filled squares) and Turkey-oak (filled circles) aphids.

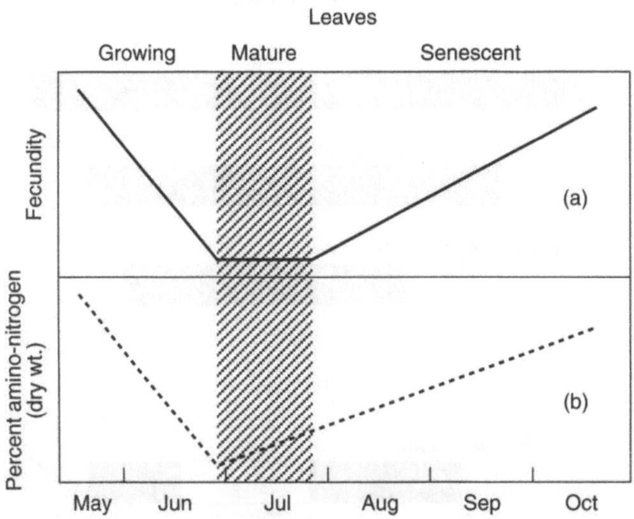

Figure 7.17 Seasonal changes in the fecundity of the sycamore aphid *Drepanosiphum platanoidis* (a) and the amino-nitrogen content of the leaves of sycamore *Acer pseudoplatanus* (b). (After Dixon, 1970a).

highest in spring, when leaves are actively growing and decline in summer when the leaves are mature, to zero (sycamore aphid) or to a third (Turkey-oak aphid) or to a half (lime aphid) of the spring rates (Figures 7.16 and 7.17). Some species, like the sycamore aphid, show an increase in reproductive rate with the onset of autumn (Figure 7.17).

On the same trees can live other species of autoecious aphids, even belonging to the same genus, which have different reproductive strategies. *Drepanosiphum acerinum* continues reproducing all through summer on sycamore, whereas *Drepanosiphum platanoidis* usually aestivates as an adult (Figure 7.18). *Myzocallis boerneri* similarly reproduces all through summer on oak, but *Thelaxes dryophila* aestivates as a first instar. *Eucallipterus tiliae* can maintain very high reproductive rates throughout summer on lime, whereas *Patchiella reaumuri* leaves lime and oversummers on the roots of *Arum maculatum* (Figure 7.18).

Below the temporal patterns in the life cycles of tree-living autoecious species and heteroecious species living on the same trees are reviewed. This temporal patterning could be secondary, but if it is not it could reveal a physiological constraint to optimization.

7.4.2 Temporal changes in environmental quality

Temperatures can reach high levels in summer, especially in the canopy of trees. In addition to the marked seasonal changes in the quality of food, tree-dwelling aphids, more than many other species, experience very high

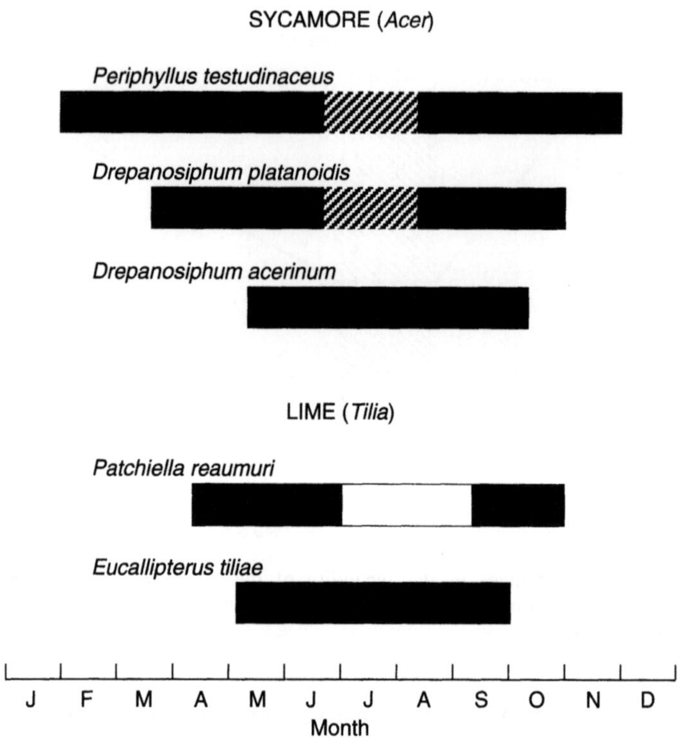

Figure 7.18 The months of the year when aphid species are either growing and reproducing (solid blocks) or aestivating (hatched blocks) on the tree hosts sycamore and lime, or leave to colonize herbaceous plants (open blocks).

temperatures during summer and, over the course of a season, are also exposed to a wide range of temperatures (Figure 7.19). The consequence of this for aphids is possibly revealed by the timing and pattern of the life cycles of various tree-dwelling aphids.

Of the three species of aphid that live on sycamore, *Periphyllus testudinaceus*, hatches early, in late January or early February; followed by *Drepanosiphum platanoidis* in late March and then *Drepanosiphum acerinum* in late April or early May (Figure 7.20). The earliest to hatch, *Periphyllus testudinaceus* and *Drepanosiphum platanoidis*, aestivate during summer, the former as a highly specialized first instar nymph (p. 118) and the latter as an adult. The last species to hatch, *Drepanosiphum acerinum*, does not aestivate, but reproduces all through the summer (Figure 7.18). Similar life cycle patterns occur on other trees. On lime the early-hatching *Patchiella reaumuri* leaves to spend the summer on *Arum maculatum* (Roberti, 1939), whereas *Eucallipterus tiliae* that hatches later, in early May, actively reproduces throughout summer on lime (Figures 7.16 and 7.18). On apple, the aphids *Rhopalosiphum insertum* and *Dysaphis plantaginea* hatch

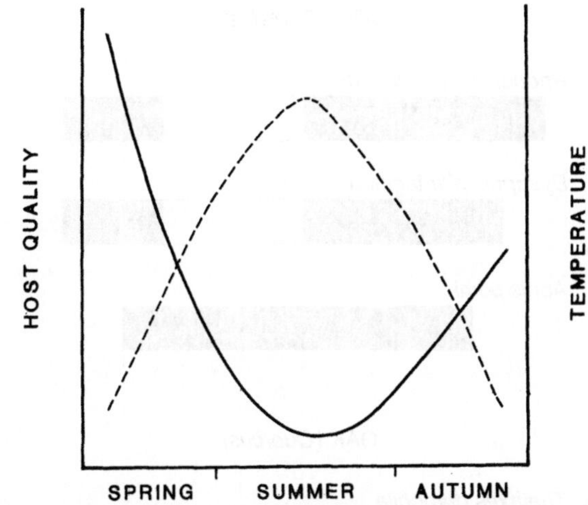

Figure 7.19 Seasonal changes in host quality (solid line) and temperature (dashed line) experienced by aphids living on trees.

early and then spend the summer on grasses and plantain, respectively, in contrast to the later-hatching *Aphis pomi* that reproduces all through summer on apple and starts producing its sexuals in early autumn, usually before the return of *Rhopalosiphum insertum* and *Dysaphis plantaginea*

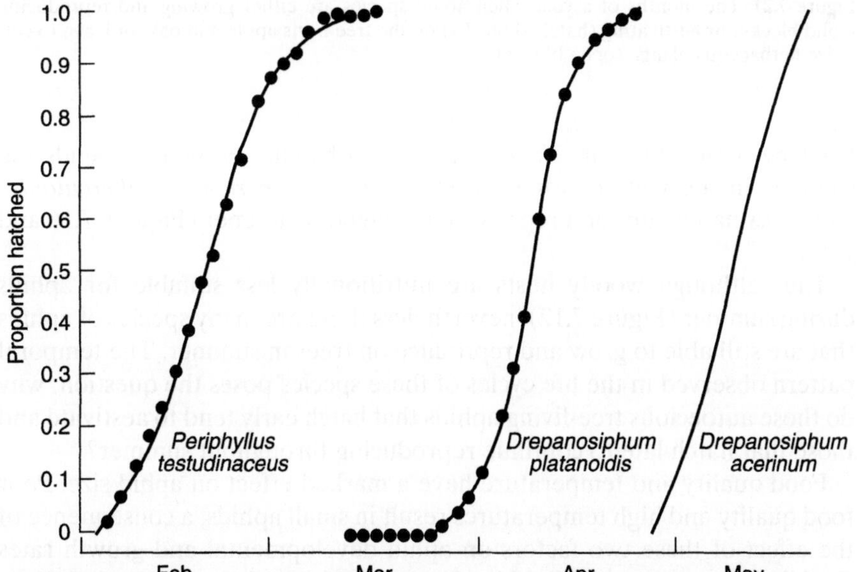

Figure 7.20 The proportion of eggs that have hatched in relation to time of year for *Periphyllus testudinacens, Drepanosiphum platanoidis* and *Drepanosiphum acerinum.*

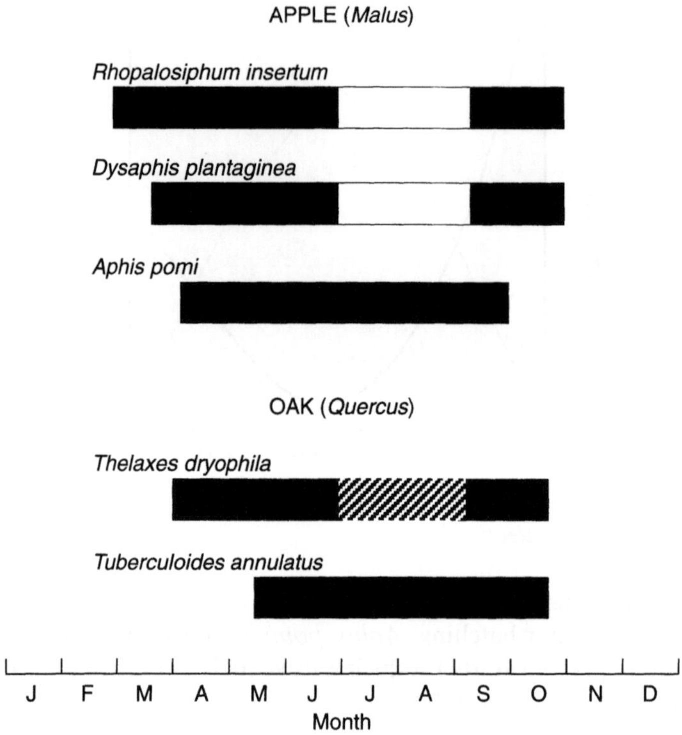

Figure 7.21 The months of a year when aphid species are either growing and reproducing (solid blocks) or aestivating (hatched blocks) on the tree hosts apple and oak, or leave to colonize herbaceous plants (open blocks).

(Massee, 1946). On oak *Thelaxes dryophila* hatches early and spends the summer in aestivation, whereas *Myzocallis boerneri* and *Tuberculoides annulatus* hatch later and reproduce throughout summer (Figures 7.16 and 7.21).

Thus although woody hosts are nutritionally less suitable for aphids during summer (Figure 7.12), nevertheless there are many species of aphids that are still able to grow and reproduce on trees in summer. The temporal pattern observed in the life cycles of these species poses the question: why do those autoecious tree-living aphids that hatch early tend to aestivate and those that hatch late to continue reproducing throughout summer?

Food quality and temperature have a marked effect on aphid size. Poor food quality and high temperatures result in small aphids; a consequence of the effect of these two factors on aphid developmental and growth rates (p. 60). Young sycamore aphids (*Drepanosiphum platanoidis*) taken from laboratory cultures and caged on sycamore trees in the field in summer suffer a high mortality and only achieve a third of the weight of the smallest

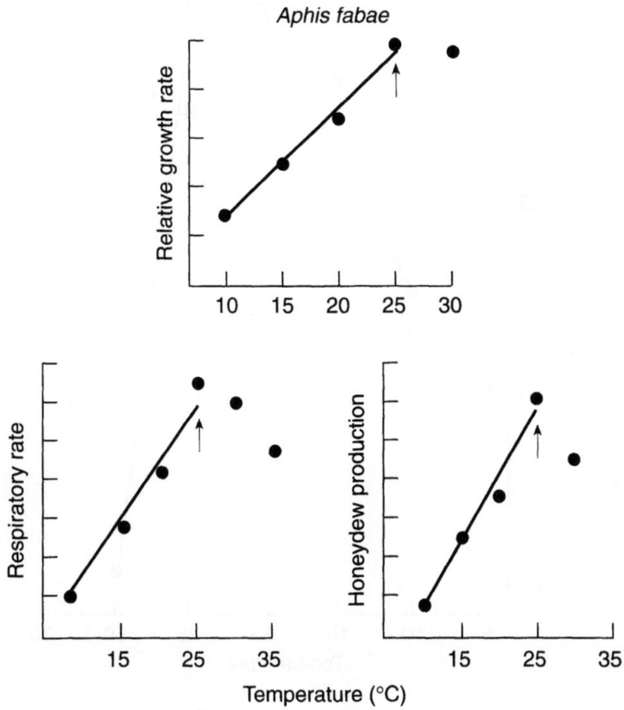

Figure 7.22 The increase in relative growth rate, respiratory rate and honeydew production of *Aphis fabae* in relation to temperature (arrow indicates temperature above which these rates decrease and mortality increases).

individuals recorded in the field. Such small aphids are unable to fly and give birth to very few offspring. However, should a leaf senesce in summer then aphids that exploit this high-quality food source grow large and achieve high reproductive rates. Thus, in the sycamore, aphid aestivation is in response to the poor nutrition and high temperatures of summer. In contrast the later-hatching *Drepanosiphum acerinum* nevertheless continues to develop and reproduce throughout summer.

For each species of aphid there is possibly a limited range of food quality and temperature over which it can increase in numbers. At high temperatures an aphid's relative growth rate declines as does its respiratory rate and honeydew production (Figure 7.22), and associated with this is an increase in mortality. However, in addition there also appear to be physiological constraints that prevent an aphid from thriving at both low and high temperatures, with species like *Lipaphis erysimi* thriving at higher temperatures than species like *Macrosiphum euphorbiae* and *vice versa* (Figure 7.23). The combination of poor nutrition and high temperatures further aggravates the problem. Thus it would appear that tree-dwelling aphids that hatch early are adapted to cool conditions (Dixon, 1973), and are unable to thrive when

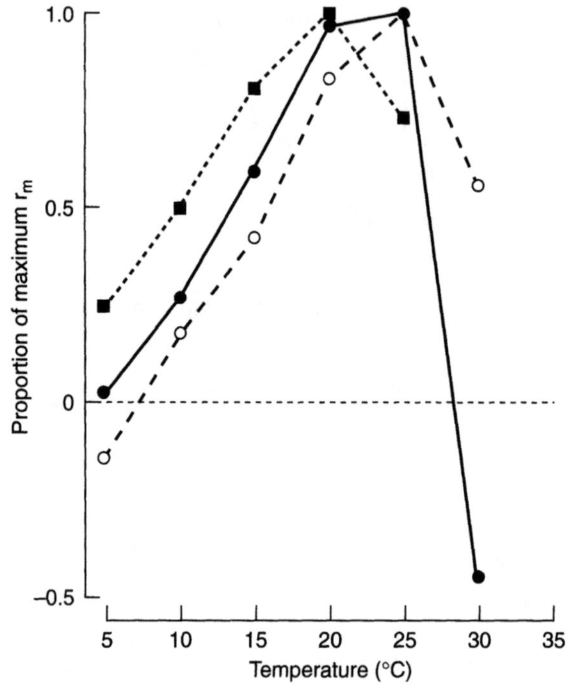

Figure 7.23 The proportion of the maximum natural intrinsic rate of increase achieved in relation to temperature for three aphids. Filled circles, *Myzus persicae*; open circles, *Lipaphis erysimi*; filled squares, *Macrosiphum euphorbiae*. *Macrosiphum euphorbiae* does relatively better at low than high temperatures compared with *Lipaphis erysimi*. (After Barlow, 1962; De Loach, 1970.)

the host quality is poor and temperatures are high, when they either aestivate or host alternate. By colonizing herbaceous plants during summer, aphids that hatch early possibly avoid the very high temperatures that occur in tree canopies, especially those aphids that live on the roots of their secondary hosts, and benefit from the higher nutritional quality of herbaceous plants. In this context it is interesting that several species of host-alternating Pemphigidae spend the summer not on herbaceous plants but on the roots of other species of trees, mainly of Gymnosperms. Several species of the autoecious Lachnidae leave the canopy and live on the roots of their coniferous host trees during summer. Therefore, as is well illustrated by the sycamore aphid, although canopy temperatures during summer do not reach lethal levels, nevertheless the aphid is not adapted to grow and reproduce on the poor-quality nutrition and at the high temperatures that prevail in summer (Figure 7.19). This is surprising as the summer morphs of several species show differences in, for example, the length of their gut or proboscis (Figure 7.24) and fat content that enable them to survive the conditions they are most likely to experience (p. 162).

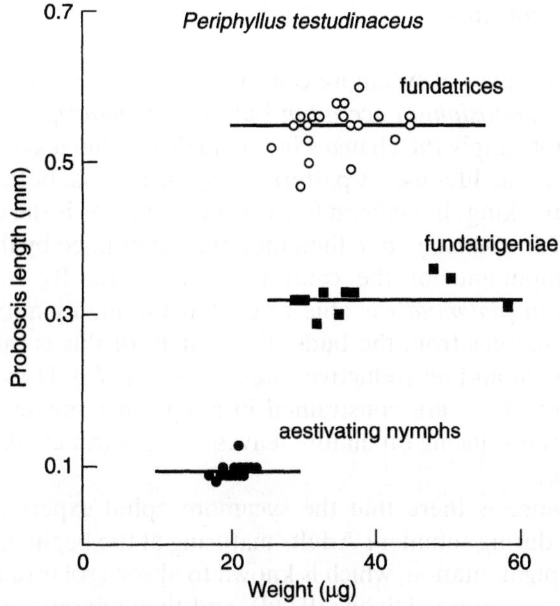

Figure 7.24 Proboscis length in relation to the weight of the nymphs of the fundatrices, fundatrigeniae and aestivating forms of *Periphyllus testudinaceus*.

7.4.3 Resource partitioning

This temporal patterning in aphid life cycles might also be explained in terms of resource partitioning (primary reason) followed by morphological and physiological adaptations to the particular conditions (secondary reason). However, there is no evidence for this. For example, *Drepanosiphum acerinum* is not present throughout most of the range of *Drepanosiphum platanoidis*, therefore it is difficult to understand why the latter has not exploited the summer months, unless it is constrained by its physiology.

7.4.4 Natural enemies

An alternative hypothesis is that the temporal patterning of life cycles is a means by which some aphids escape their natural enemies. *Periphyllus testudinaceus* hatches early and the first generation matures before the buds of sycamore break. That is, the first generation is active long before insect predators like ladybirds and syrphids emerge from hibernation. However, these fundatrices are attacked by an Hymenopterous parasite, *Trioxys falcatus*, which has been recorded killing 52–86% of the fundatrices of *Periphyllus testudinacens* and 36% of *Periphyllus acericola*. Therefore, as specific natural enemies appear to be very good at finding their hosts, whatever the time of the year, it is unlikely that enemy-free time has been a factor in the evolution of temporal patterning.

7.4.5 Resource tracking

In summer the leaves of sycamore can sustain the active growth and reproduction of *Drepanosiphum acerinum* but not *Drepanosiphum platanoidis*. That is, it is not simply the changes in the quality of the leaves of sycamore that determines the life-history pattern. It is possible that both these species are resource tracking. If confined to living on leaves, as is the case for these two species of *Drepanosiphum*, then they are constrained by the spatial and temporal components of the canopy of sycamore. By hatching early *Drepanosiphum platanoidis* is able to exploit the highly nutritious young foliage as it emerges from the buds. The benefit of this is that it achieves very high growth and reproductive rates (Dixon, 1976). However, there is possibly a cost. They are constrained in a way that prevents them from growing and reproducing on mature leaves in the warm conditions prevailing in summer.

What evidence is there that the sycamore aphid experiences environmental stress during summer? Adults maturing at the beginning of summer lack melanic pigmentation, which is known to absorb solar radiation and so raise body temperature (Dixon, 1972d), and they migrate vertically down into the lower canopy of sycamore and aggregate on leaves in shade (Dixon, 1969; Dixon and McKay, 1970). This contrasts with the behaviour of *Drepanosiphum acerinum* and the lime aphid *Eucallipterus tiliae*, which are reproductively active all through summer, do not migrate vertically and retain their melanic pigmentation (Dixon, 1972d). In addition, the sycamore aphid, unlike the lime aphid, shows metabolic acclimatization reducing its metabolic rate when subject to prolonged periods of high temperature (Dixon, 1973).

The implication of the above is that each species is adapted to a limited range of conditions. Species of aphids that are active for a large proportion of a year, e.g. *Periphyllus testudinaceus* (February–November), *Drepanosiphum platanoidis* (March–November), experience a very wide range of temperature and food quality. Is it possible for them to adapt equally well to all these conditions? Zoogeographically there are borealpine species, like *Boernerina depressa* and *Wahlgreniella ossiannilsoni*, which tend to live at high altitudes and latitudes, although their host plants are more widely distributed. Similarly, others are typical of steppe conditions – *Xerophilaphis suaedae* and *Brachyunguis tamaricophilus* – and others of African and Afro-Mediterranean conditions – *Uromelan compositae* and *Dysaphis cynarae*. Even species of the same genus living on the same host plant can show very different distributions, e.g. *Staticobium limonii* occurs on sea lavender, *Limonium vulgare* in north and western Europe and is replaced by *Staticobium latifoliae* in southern Europe and in the steppe regions of central Asia (Barbagallo and Patti, 1985). *Callipterinella calliptera* does better at higher temperatures and occurs more commonly in the south than other species of aphid living on birch (Hopkins, personal communication). It is generally accepted that particular species of insects are adapted to living in particular conditions. However, it is rarely known why they are so restricted. If species are so constrained it is not unreasonable to suggest that those aphids that experience particularly wide ranges of conditions may not be equally adapted to grow and reproduce over the full range of conditions. Species that hatch early may be particularly adapted to exploit the leaves of trees that are actively growing and/or senescing in the relatively cool conditions that prevail in spring and autumn. The cost is that they are then possibly unable to exploit mature leaves in the warm conditions prevailing in summer and enter diapause/aestivation during this period. Those species that hatch late are not so constrained and are able to exploit the foliage of trees during the warm summer months.

Canopy-living species experience a great change in food quality during the course of a season. Some species enhance and extend the period of high food quality in the spring by galling the leaves (p. 146) and others benefit from inducing leaves to senesce prematurely (p. 42). Species like *Periphyllus testudinaceus* extend the period by feeding on twigs just below the terminal buds, the first part of a tree to start growing in the spring, and subsequent generations move onto the leaves after bud burst. For this, however, individuals of the first, second and subsequent generations need a proboscis of a particular length, because the phloem elements are located more deeply in the twigs than in the leaves, i.e. in addition to physiological adaptations they show morphological adaptations (Figure 7.24). Other species, e.g. *Stomaphis quercus* and *Cinara piceae*, are similar to *Periphyllus* in feeding on the twigs below the terminal buds, but instead of subsequent generations moving on to the leaves/and or young growth, they move down the tree, from branches, to trunks and roots following the onset of cambial activity in these structures, which occurs later in the season in the trunk and roots than in the branches. In this way they have access to high-quality food for most of the season.

Life cycles, therefore, can be viewed as the means by which aphids track seasonal changes in resources, and appear to have evolved through the process of optimization.

7.5 Generation-specific strategies

During the course of a life cycle aphids pass through a number of distinct morphs, which in host-alternating aphids can be as many as eight. Some morphs, like the aestivating nymphs of *Periphyllus*, are clearly adapted for a particular function. The majority of aphids, however, are either winged or unwinged and would appear to be adapted mainly for a migratory or reproductive role. The model presented in Chapter 4 simply implies that the development of wings affects the optimal partitioning of resources and results in a reduction in the rate of population increase. This model successfully predicts the trends in size and reproductive behaviour observed in the field. However, the implication is that there is but one strategy for winged and one for unwinged individuals. The striking changes that occur in the relative length of the proboscis in the first three generations of *Periphyllus* indicate the situation is not as simple as the model implies. There is a growing body of evidence that individuals of each of the generations that make up a life cycle have specific reproductive strategies adapted to the conditions they are most likely to encounter: in effect they anticipate the predictable seasonal trends in habitat quality.

The sycamore aphid has only three morphs: alate viviparous females, apterous oviparous females and alate males. The second-generation adults, that develop and are present in summer, enter a reproductive diapause and have a low fecundity thought to be a direct consequence of the poor quality of the food available to the aphid when feeding on mature leaves (Mordvilko, 1908). However, when reared at low temperatures on unfurling leaves, conditions normally experienced by the first generation, the large, second-generation individuals that develop have well-developed fat bodies and poorly developed gonads; even so, as if about to enter aestivation, i.e. they are prepared for the onset of harsh conditions associated with the cessation of growth of the leaves of sycamore (Dixon, 1975b). Second-generation aphids also differ from those of the first generation in having fewer ovarioles (Wellings *et al.*, 1980), larger appendages and a greater number of rhinaria on their antennae (Dixon, 1974b), a longer gut (Dixon, 1975b) and a higher wing-beat frequency, and they fly longer and lower (Mercer, 1979).

In autumn when temperatures are low and nutrients are being recovered from the leaves the quality of the habitat for the sycamore aphid is also good. Thus the quality of the habitat in which the spring (the first-) and autumn (the third-) generation individuals live is good, whereas summer (second-) generation individuals live in a relatively poor habitat.

Summarizing, first- and third-generation individuals have a good habitat, and associated with this, a short adult life, small fat body, weak spacing behaviour and produce many offspring quickly – features characteristic of an *r*-selected species. Second-generation individuals have a harsh habitat, a long adult life, well-developed fat body, marked spacing behaviour and produce fewer offspring – features characteristic of a *K*-selected species. The development of these generation-specific characteristics is not affected by rearing the aphid under constant conditions (Dixon, 1975b; Leather and Wellings, 1981). Interestingly the individuals that hatch from the overwintering eggs weigh 36 mg, whereas those born to these aphids and the second and third generations weigh 80 mg. The first-generation individuals start life feeding on the unfurling leaves in which the phloem elements are likely to be shallowly located. Later generations feed on fully grown leaves in which the phloem elements are deeply located. This association between size at birth and the depth to which aphids have to probe to reach phloem elements is seen as an important constraint on birth size (p. 66).

Life cycles of host-alternating aphids like the black bean aphid are more complicated. *Aphis fabae* overwinters as an egg on spindle (*Euonymus europaeus*), hatching in spring and giving rise to two generations of apterous individuals, followed by a third generation of alatae. The alatae colonize broad beans and other herbaceous plants. The alate exules produced on the secondary hosts then colonize other secondary hosts, and the males and gynoparae that are produced in autumn migrate to spindle. The gynoparae give birth to apterous oviparae that lay the overwintering eggs after mating with the males (Figure 7.2).

The trend in the reproductive traits of the morphs of this aphid tend to be similar on both the primary and secondary hosts. Spindle at bud burst supplies the first-generation aphids with a rich food supply that becomes less suitable as spindle ceases growing. First-generation aphids have relatively large gonads and rapidly produce a large number of offspring, whereas second-generation aphids have relatively small gonads and produce fewer offspring. The emigrants migrate from spindle to actively growing beans, which are a very nutritious host compared to spindle by this time. However, beans soon cease growing and lose quality. Therefore, on both the primary and secondary hosts the later generations experience harsher nutritional conditions than the earlier generations. On both primary and secondary hosts, individuals in the later generations tend to have proportionally more lipid reserves and are better able to survive starvation. Therefore, as in the sycamore aphid, individuals of the first generation put proportionally more of their resources into reproduction, and less into lipid reserves, than do individuals of later generations.

The alate exules that disperse and colonize fresh secondary hosts late in summer also initiate a sequence of morphs whose reproductive traits are similar to those of the sequence of morphs initiated by emigrants. However, it is not done by varying the size of the gonads. The offspring of alate exules grow faster, have more well-developed embryos in their gonads and a higher reproductive rate than aphids of subsequent generations. These characteristics are also adaptive since alate exules are likely to colonize plants that are nutritionally more favourable than the ones on which they developed.

Attempts to change these characteristics by rearing aphids in constant conditions and on high-quality host plants have resulted in a reversal of the downward trend in size from generation to generation, but no change in the generation-specific reproductive characteristics (Figure 4 in Dixon and Dharma, 1980b; Leather and Wellings, 1981).

The generation-specific plasticity in the expression of a genotype may be as important to the success of aphids as the opportunistic multiplicative faculty of parthenogenesis so often cited. The generation-specific strategies evolved in the sycamore and black bean aphids enable them to anticipate seasonal changes in habitat quality, while extrinsic variables tune, through phenotypic plasticity, the reproductive strategy to the actual conditions experienced (Wellings *et al.*, 1980; Dixon and Wellings, 1982). In terms of the aphid model (*cf.* Figure 4.15) this might operate by partitioning energy optimally not only between soma, gonads and flight apparatus, but also to lipid reserves, with the investment in the latter programmed and specific to each generation.

Currently, life-history strategies are classified as species characteristics (Stearns, 1976), and individual differences in life-history strategies within a species are explained on a genetic basis. At present, however, a single genotype, with seasonally expressed adaptations, has no place within the general framework of life-history theory, centred as it is on the ideas of *r*- and *K*-selection.

Life-history theory is largely dependent on comparative studies of closely related species living in different habitats (Bell, 1980). Each species has a set of co-adapted characters appropriate for living in a particular habitat. It has proved difficult to dissociate these characters experimentally and thus determine their relative costs and benefits for a species. However, there are many aphid species that are ideal for testing life-history theory, especially those that show intrinsically programmed and variable reproductive strategies within a morph, i.e. individuals of the same generation and morph show tactical diversity and have different reproductive strategies.

7.6 Habitat predictability and tactical diversity

At least four alternative strategies have evolved in animals to maximize fitness: pessimism, phenotypic plasticity, spreading the risk in space and time, and mixed strategies (the production of phenotypically diverse offspring). Examples of all but the first strategy have been described for aphids.

Let us assume that aphids are food limited. So far no one has been able to demonstrate that food is a limiting resource because it is difficult to measure the quality and quantity of food available to aphids. However, the inverse relationship between mortality and relative growth rate recorded for several species (Figure 7.25) supports this contention, as the highest growth rates are rarely achieved. In addition the large sycamore aphids that develop when reared at 10°C show signs of nutritional stress when transferred to 20°C, compared with the controls (Figure 7.26). Their disproportionately large embryonic mass requires more nutrients than are available to the aphids, and this affects both survival and rate of increase, and the birth weight of their offspring (Wellings and Dixon, 1983).

Theoretically, as there are nutrition-related costs in terms of increased mortality, and benefits, in increased reproduction, if there are more

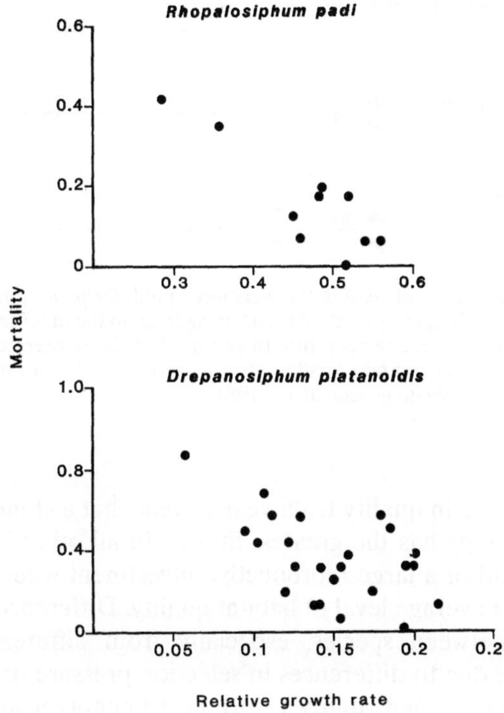

Figure 7.25 Mortality in relation to relative growth rate for *Rhopalosiphum padi* and *Drepanosiphum platanoidis*.

ovarioles in an aphid's reproductive system (Figure 7.27), then seasonal trends in host quality should be reflected in a trend in ovariole number. Fundatrices, present when plants are most nutritious, generally have more ovarioles than individuals of following generations, e.g. fundatrices of *Thecabius affinis* of the subfamily Pemphigidae have 50 ovarioles, in contrast to the 13 recorded for later generations (Wellings *et al.*, 1980; Tashev and Markova, 1982). There are few exceptions: these include the aestivating adults of the sycamore aphid that live for a very long time and span periods of very poor and very rich nutrition but nevertheless develop a relatively large number of ovarioles. During the period of nutrient stress, the reproductive biomass, i.e. the mass of embryonic tissue, which determines the stress-related mortality, is very low. Afterwards, when the host quality has shown a dramatic improvement, the same adults are able to capitalize on the very rich nutrition by virtue of their large number of ovarioles. Thus, in the few species where an individual lives long enough, selection can apparently act separately on reproductive biomass and ovariole number.

The effect of reproductive strategy on overall fitness of aphids in habitats varying in quality between years is given in Table 7.3. It is only in the habitat

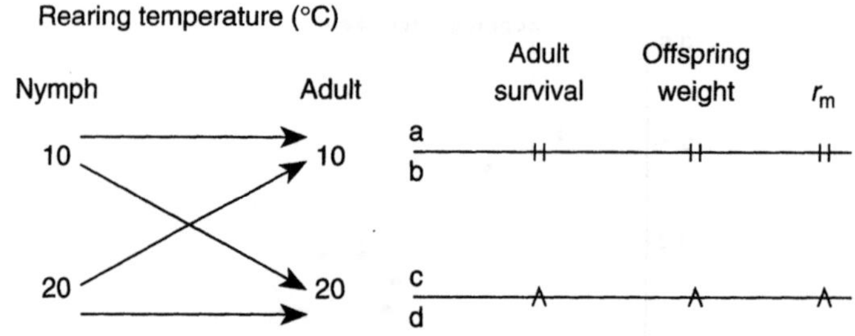

Figure 7.26 The consequence of rearing the sycamore aphid, *Drepanosiphum platanoidis*, at either 10 or 20°C and on becoming adult of transferring them to the other temperature (b and c) or keeping them at the same temperature (a and d). Aphids transferred from 10 to 20°C have a poorer adult survival, produce smaller offspring and have a lower rate of increase than the control aphids. (After Wellings and Dixon, 1983.)

that is most variable in quality from year to year that a clone with a mixed reproductive strategy has the greatest fitness. In all other habitats aphids with either a small or a large reproductive investment would be favoured, dependent on the average level of habitat quality. Differences in reproductive attributes between species, especially from different subfamilies, however, may be due to differences in selection pressure or to differences in the constraints on their ability to produce phenotypically diverse offspring. That the evolution of mixed reproductive strategies has not been found in the Drepanosiphidae argues for a phylogenetic constraint (Ward and Dixon, 1984). However, this is likely to change when the reproductive strategies of more species of the family are known, especially the more polyphenic species.

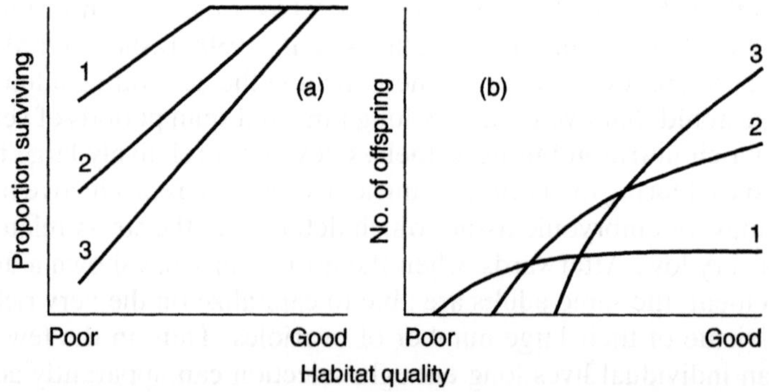

Figure 7.27 The proportion of aphids surviving (a) and the number of offspring they have (b) in relation to habitat quality of aphids with a small (1), medium (2) or large (3) reproductive investment. (After Ward *et al.*, 1983a,b.)

Table 7.3 The effect of reproductive strategy on overall fitness of aphids in habitats varying in quality between years

(a) *Effect of reproductive investment and habitat quality on fitness within a year*

Reproductive investment (strategy)	Habitat quality	
	Poor (P)	Good (G)
High	0	4
Low	1	1
Mixed	0.5	2.5

(b) *Effect of change in habitat quality from year to year on overall fitness*

Reproductive investment	Range in habitat quality within and between years			Overall fitness
	G G G G	⟷	P P P G	
High	4	×	1	4[a]
Low	1 } fitness	×	1	1
Mixed	2.5	×	1	2.5
	G G G P	⟷	P P P G	
High	3	×	1	3[a]
Low	1	×	1	1
Mixed	2	×	1	2
	P P P P	⟷	P P P G	
High	0	×	1	0
Low	1	×	1	1[a]
Mixed	0.5	×	1	0.5
	G G G G	⟷	P P P P	
High	4	×	0	0
Low	1	×	1	1
Mixed	2.5	×	0.5	1.25[a]

[a]Most fit.

7.7 Response to habitat quality

In addition to showing phenotypic plasticity (Chapter 4) some aphids respond to changes in habitat quality by adjusting the proportion of their offspring that is committed to a high or a low level of reproductive effort. The greater the nutrient stress experienced by mothers, the fewer of their offspring are committed to a high reproductive investment and *vice versa* (Walters *et al.*, 1988). Even in a poor-quality habitat, however, some

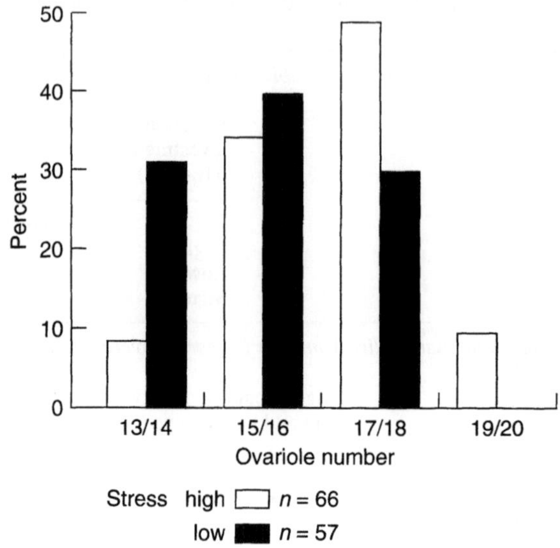

Figure 7.28 Percentage of offspring with 13/14, 15/16, 17/18 and 19/20 ovarioles in relation to the nutritional stress (*cf.* Figure 7.27) experienced by the parents.

offspring will have a large number of ovarioles and *vice versa* (Figure 7.28). Therefore the degree of association between weight, an indicator of habitat quality, and reproductive investment will rarely be good.

The trade-off is in the relative sizes of the fat body and gonads (ovariole number). The unwinged offspring of the vetch aphid, *Megoura viciae*, can have from 14–20 ovarioles in their gonads. Individuals with 14 ovarioles have smaller gonads and a larger fat body than individuals with 20 ovarioles (Figure 7.29) (Brough and Dixon, 1989). Their ability to withstand starvation is negatively correlated with the number of ovarioles (Ward *et al.*, 1983) and their reproductive rate positively correlated with the number of ovarioles/size of gonads (Wellings *et al.*, 1980). Thus, to some extent, certain aphids can tailor the reproductive investment of their offspring to fit trends in habitat quality more closely, maintaining a high reproductive rate in rich habitats and reducing mortality in the poorer habitats. The retention of a range of ovariole numbers in all habitats possibly reflects an advantage of spreading the risk between individuals of a clone, with some allocating more of their resources to dispersal and survival than reproduction and *vice versa*.

Some aphids show flexibility of reproductive capability, in that nutrient stress promotes embryo resorption. Because of its short life an individual aphid on a poor host cannot usually expect any great improvement in habitat quality and it makes the best use of what is currently available: it matures its largest embryos and resorbs the smallest, sacrificing its potential

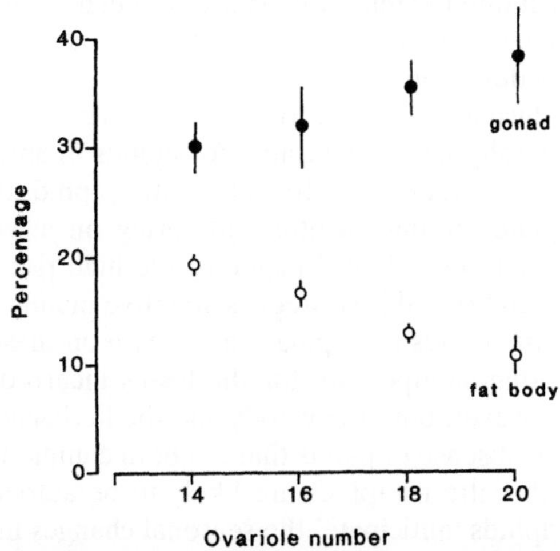

Figure 7.29 The percentage of the sectional area of the body of apterous adults of *Megoura viciae* at the onset of reproduction that is occupied by gonads (filled circles) and fat body (open circles) in relation to the number of ovarioles in their gonads. Bars represent standard errors. (After Brough and Dixon, 1989.)

lifetime fecundity to maintain its immediate reproductive rate (Ward and Dixon, 1982; Leather *et al.*, 1983). In good habitats an aphid may ovulate extra embryos (Figure 7.30) and so increase potential fecundity (Ward *et al.*, 1983).

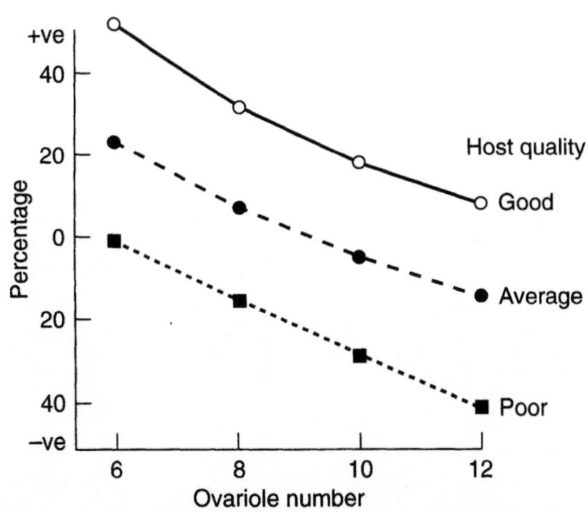

Figure 7.30 Percentage of embryos ovulated after the adult moult in relation to ovariole number and the quality of the habitat in *Aphis fabae*. In poor habitats all but the aphids with lowest reproductive investment resorb embryos, whereas in good habitats they all ovulate additional embryos but this is inversely related to ovariole number.

Thus to a very limited extent, aphids are indirectly able to adjust their reproductive biomass and escape the constraints imposed by decisions made during early development.

Summarizing, during the course of a year plants vary in quality with woody plants generally most favourable for aphids in spring and autumn and herbaceous plants in summer. Host-alternating aphids track this change in quality and spend autumn, winter and spring on a woody plant and summer on an herbaceous plant. In spite of the high risk associated with transferring between hosts this strategy is adaptive mainly because clones, in achieving very high rates of population increase on these two groups of plants, can more than compensate for the losses incurred in transferring between hosts. However, both the woody and the herbaceous host have to be common. In the absence of plants that are both common and show complementary growth patterns aphids are likely to be autoecious. All holocyclic species of aphids 'anticipate' the seasonal changes in habitat quality by producing a sequence of short-lived morphs that are each adapted to the conditions they are likely to experience, i.e. they show generation-specific strategies. In addition, some aphids show intraclonal tactical diversity in reproductive investment, an adaptive response to the variability in the seasonal trend in habitat quality. By responding to current conditions, aphids can also adjust the reproductive commitment of their offspring, and by ovulation or resorption of embryos during adult life, alter their reproductive capability and so modify decisions made earlier in their development. It is likely that the success of aphids is to a very large extent due to their programmed but flexible life cycles, which enable them to track very closely the seasonal changes in the quality of plants.

8 Resource tracking in space

Most species of aphids, at all stages of development, move about over the surface of their host plants and even between adjacent plants. These local movements result in slow diffusive dispersal. In contrast, aphids also show persistent 'straightened out' movements, during which their vegetative responses are depressed, which are a means of transport over greater distances (Kennedy, 1975), i.e. changes occur in an aphid's responsiveness to host plant stimuli. This was first described by Moericke (1955) who noted that after flying for a period aphids switch from migrating flight (Distanzflug) to alighting flight (Befallsflug). The latter has also been called targeted flight (Nottingham and Hardie, 1989). In physiological terms migration occurs mainly when vegetative responses or targeted locomotion is inhibited (Kennedy, 1975; Nottingham and Hardie, 1989). This latter behaviour is characteristic of the winged stages, but is also shown by nymphs. This is well illustrated by *Pemphigus trehernei* that lives on the roots of the sea aster that grows on mud banks in tidal salt marshes. First instar nymphs, on standing erect on the water surface, are caught by the wind which sends them scudding across open water to be deposited by the falling tide on another mud bank. Thirty minutes of this waterborne experience is enough to reverse the reaction of these nymphs to light and send them down a new crevice where aster roots may be found and colonized. This reversal to photo-negative behaviour provides some indirect control over the distance the nymphs travel (Foster, 1978; Foster and Treherne, 1978). Thus local (trivial), movements and distant (migratory), movements can be recognized, and it is possible that they are the extremes of a continuum of movement by which aphids track resources in space. The movement which brings about a redistribution in space, an ecological concept, is migration (Taylor and Taylor, 1983).

8.1 Mechanisms

8.1.1 Take-off

Accidental dispersal is probably relatively rare in aphids. Polyphenic species have to change from apterous to alate development, and this

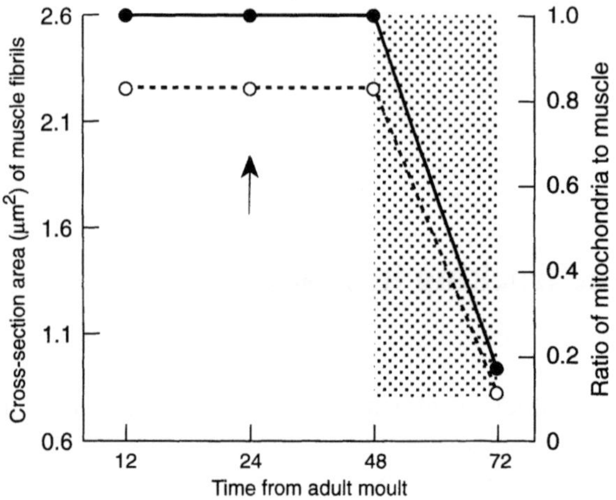

Figure 8.1 Changes in the cross-sectional area of the muscle fibrils (filled circles) and the ratio of mitochondria to muscle area (open circles) in transverse sections through the dorso-ventral indirect flight muscles of *Megoura viciae* in relation to the time from adult moult (the arrow indicates that flight usually occurs at this time and the hatched area indicates wing muscle autolysis). (After Burns, 1971.)

ontogenetic switch takes time and is usually in response to overcrowding or a deterioration in host quality, or both (p. 104). Once alates have settled on a host plant they autolyse their wing muscles (Figure 8.1), which is probably triggered by a rise in the juvenile hormone titre in the aphid (Johnson, 1957, 1959; Kobayashi and Ishikawa, 1993, 1994a,b). The viviparae of some tree-dwelling aphids, however, are all winged and can fly throughout adult life, but even in these species accidental dispersal is unlikely. When feeding on a good-quality host, sycamore aphids increase in weight and *vice versa* (Figure 8.2), and can only fly from a good-quality host after reducing their weight. This is achieved by excreting several drops of honeydew.

Crowding markedly influences the readiness with which aphids fly (Figure 8.3). At low levels of crowding, take-off is delayed and offspring are produced before an aphid flies (Figure 8.4b and c). On a good host some species of aphid will not take-off unless crowded, whereas in others a crowding stimulus is not required. Host quality also affects the readiness to take-off and aphids generally depart sooner from poor-quality hosts (Figure 8.3a), and the shorter the stay the fewer offspring they are likely to give birth to before leaving (Figure 8.3c).

Thus, although capable of flight, alate aphids, respond to current cues of host quality (Dixon *et al.*, 1968). They leave quickly if it is poor, but if it remains above the average expectation they either deposit young before flying or stay. This facultative dispersal is characteristic of aphids living on plants that at any particular time can vary in quality. However, when the seasonal trend in quality is predictable obligate migrants are produced. The

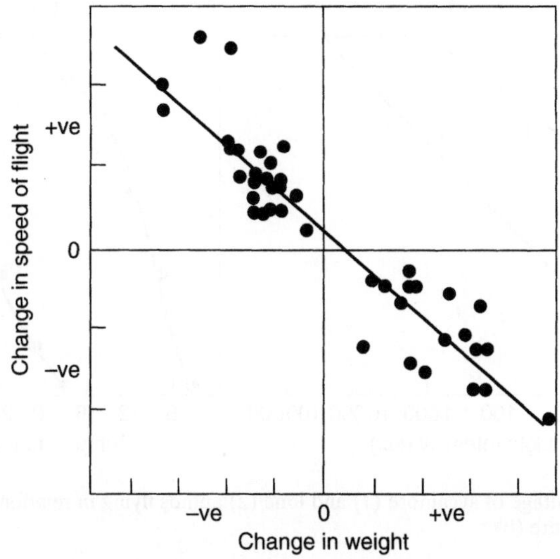

Figure 8.2 Changes in flight speed in the sycamore aphid, *Drepanosiphum platanoidis*, in relation to changes in weight achieved by feeding the aphid on good- or poor-quality hosts. (After Mercer, 1979.)

emigrants and sexuals of host-alternating aphids (p. 128) are produced in anticipation of seasonal changes in host quality. These morphs are obligate migrants (Dixon, 1971b).

Physical factors also affect take-off. Aphids are more likely to leave at high than low light intensities and temperatures, and at low rather than high wind speeds (Figure 8.4). Species differ in their response to these factors. The sycamore aphid flies at lower light intensities and temperatures than the lime aphid. This partly determines their different daily flight

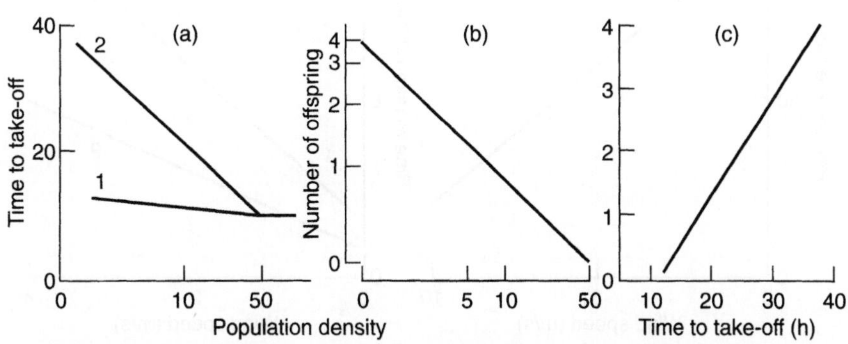

Figure 8.3 Time to take-off in hours (a) and the number of offspring produced before flying (b and c) by *Sitobion avenae* in relation to population density (a and b) and time to take-off (c). 1, Poor-quality host; 2, high-quality host. (After Walters and Dixon, 1984.)

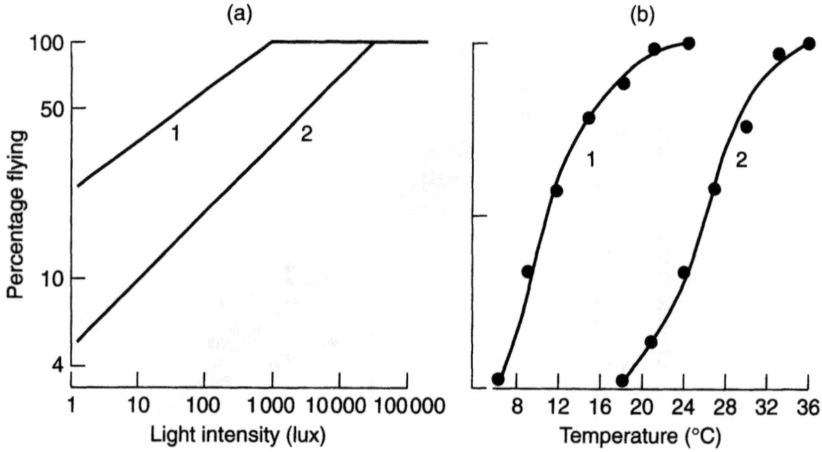

Figure 8.4 Percentage of sycamore (1) and lime (2) aphids flying in relation to light intensity (a) and temperature (b).

activity curves (*cf.* Figure 8.12). Take-off in tree-dwelling aphids, such as the sycamore aphid, is more markedly inhibited by high wind speeds than is the take-off of aphids that live on short-lived host plants (Figure 8.5) (Haine, 1955; Dixon and Mercer, 1983; Walters and Dixon, 1984). Delay is more critical when living on short-lived hosts that are likely to deteriorate rapidly in quality. Even in these species high winds may inhibit take-off for a day, but not longer, indicating a change in threshold. However, although it is not known whether other tree-dwelling aphids share the

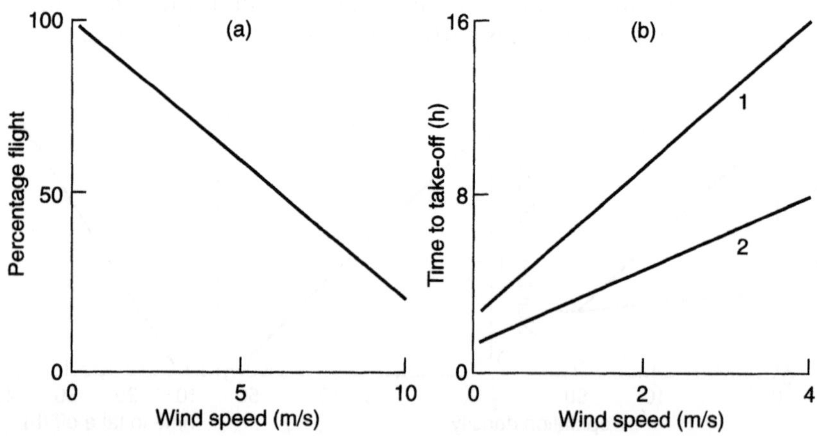

Figure 8.5 Percentage of sycamore aphids flying in 24 h (a) and time to take-off of the cereal aphid, *Sitobion avenae* (b), from good- (1) and poor-quality hosts (2) in relation to wind speed. (After Mercer, 1979; Walters and Dixon, 1984.)

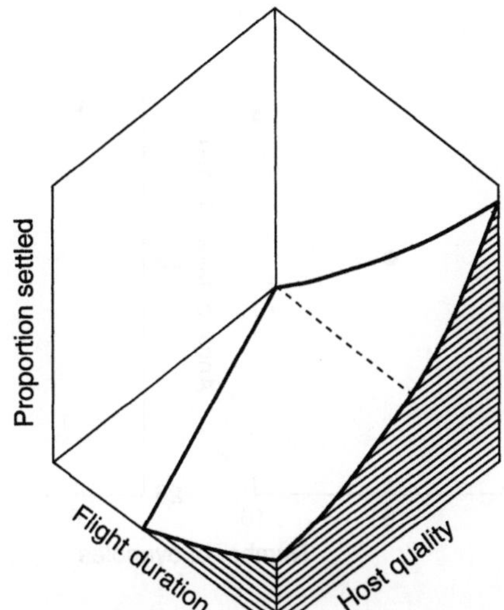

Figure 8.6 Proportion of aphids settled in relation to flight duration and host quality. (After Johnson, 1958; Kennedy and Booth, 1963a.)

sycamore aphids' reluctance to fly on windy days, it is possible that this behaviour is associated with tree dwelling, i.e. living on a relatively permanent host. In general then, adverse weather conditions are only likely to delay rather than prevent take-off (Johnson, 1969; Walters and Dixon, 1984).

8.1.2 Settling

Aphids are more likely to settle on a good- than on a poor-quality host (Walters and Dixon, 1982). If an aphid settles on a non-host plant or poor-quality host the settling response is quickly inhibited and flight is resumed. Repeated flights have a cumulative positive effect on the settling response (Figure 8.6) (Johnson, 1958; Kennedy and Booth, 1963b). The length of time for which an aphid has flown and the number of plants it has sampled possibly give some indication of the abundance of potential hosts. The optimal strategy after a long and unsuccessful search would be to settle and feed on the next host plant encountered because of the low likelihood of finding a better host.

8.1.3 Migratory urge

At the population level, the tendency to disperse has been measured in terms of the proportion of a population that develops into alatae (Lamb and Mackay, 1979) and, at the individual level, in terms of whether an aphid flies

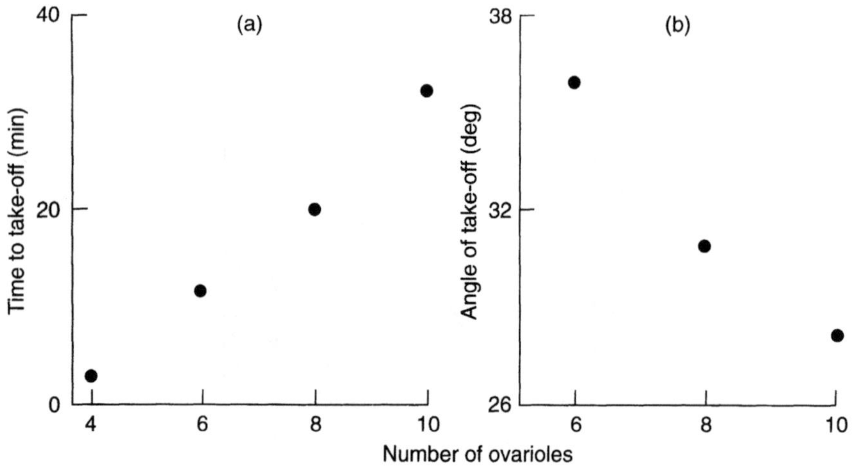

Figure 8.7 Time to take-off (a) and angle of take-off (b) in relation to the number of ovarioles (reproductive investment) in *Sitobion avenae*. (After Walters and Dixon, 1983.)

before or after giving birth to offspring (Shaw, 1970b). Crowding during nymphal life affects the development of wings, flight musculature, fuel reserves and gonads. Consequently, migrants have a well-developed flight apparatus but poorly developed gonads – the oogenesis flight syndrome (Johnson, 1969; Shaw, 1970c). Flight is partly dependent on crowding experience during nymphal (Shaw, 1970b,c) and early adult life (Dixon *et al.*, 1968; Walters and Dixon, 1982).

Variability in the tendency of individuals to disperse is also associated with programmed differences in the level of reproductive investment (number of ovarioles in their gonads; p. 113) between individuals within a clone and this is usually independent of their size (p. 168), those with few ovarioles take-off more readily and at a steeper angle (Figure 8.7), delay wing muscle autolysis longer (Figure 8.8), are more resistant to starvation and have relatively more olfactory sense endings on their antennae than aphids with many ovarioles (Walters and Dixon, 1983). In *Aphis fabae* the proportions flying for increasing lengths of time (Kennedy and Booth, 1963a) are similar to the proportions that have decreasing numbers of ovarioles (Figure 8.9). Similarly, the proportions of alates of *Aphis fabae* that take-off before giving birth, after giving birth or that do not fly at all (Shaw, 1970b) may also be associated with reproductive investment in terms of ovariole number. In those species or clones of species in which the winged individuals do not show variation in reproductive investment the migratory urge appears to be more related to size; with the large individuals more reluctant to take-off and flying for shorter periods than small individuals (Kidd and Cleaver, 1984, 1986).

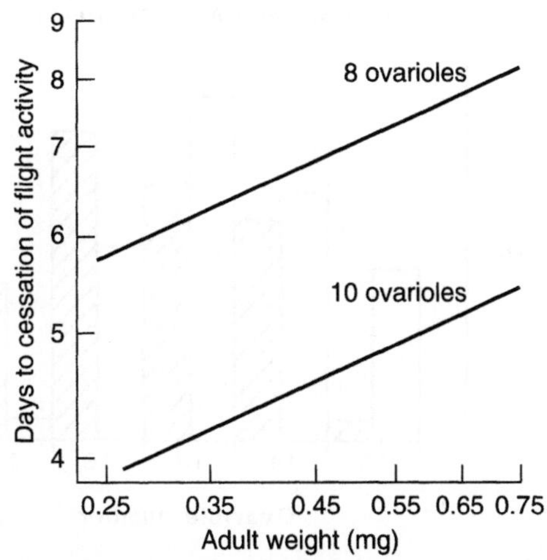

Figure 8.8 Days to the cessation of flight activity (wing muscle autolysis) in relation to adult weight in *Sitobion avenae* with eight and 10 ovarioles. (After Walters and Dixon, 1983.)

8.1.4 Differential migration

Within a species, aphids with relatively small gonads (low ovariole number) have larger lipid reserves (p. 168), a greater urge to take-off and do so at a steeper angle, may fly for longer and more frequently, and are better able to survive starvation (p. 168) than aphids with large gonads. The result should be differential migration, with the long-distance migrants

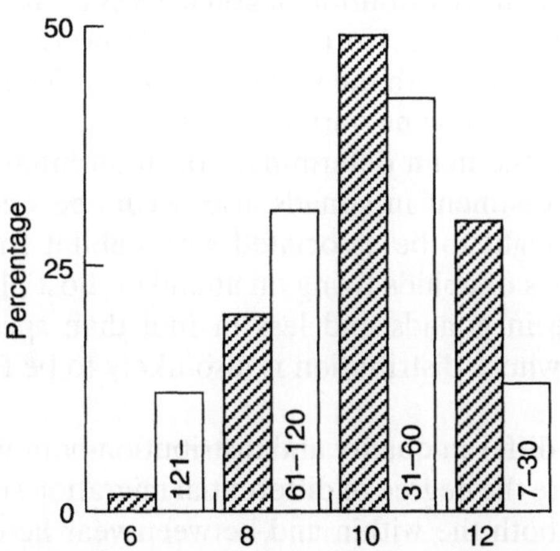

Figure 8.9 The percentage of *Aphis fabae* that flew for decreasing periods of time in minutes (open bars) compared with the percentage that have six, eight, 10 and 12 ovarioles (hatched bars). (After Kennedy and Booth, 1963.)

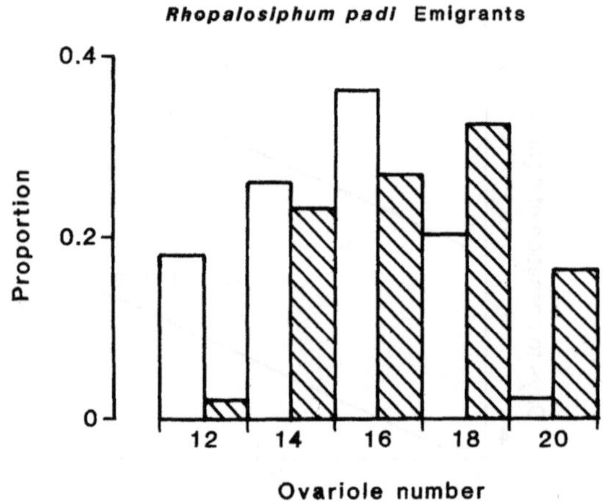

Figure 8.10 The proportion of spring migrants of the bird cherry-oat aphid of ovariole classes 12–20 leaving bird cherry (open bars) and colonizing oats (hatched bars) 1 m away.

consisting mainly of individuals with small gonads and the flitters of individuals with large gonads. There is some support for this from a greenhouse experiment using emigrants of the bird cherry-oat aphid (*Rhopalosiphum padi*), which are obligate migrants (Walters and Dixon, 1983). Amongst the individuals colonizing secondary host plants located close to the primary host were proportionately more individuals with large gonads than one would expect from the distribution of gonad sizes of the individuals leaving the primary host (Figure 8.10) (Dixon and Howard, 1986). Variation in body size might interact with the variation in reproductive investment, and in those species that show no variation in the relative size of their gonads body size may be the main determinant of differential migration. A very wide range of investment in gonads also occurs between species (Figure 8.11). This is thought to be associated with habitat fragmentation/abundance, with species of aphids living on abundant host plants investing proportionally more in gonads and less in fuel than species living on less abundant hosts, whose distribution is also likely to be fragmented (Dixon *et al.*, 1993).

That is, plants differ in quality and distribution both within and between seasons. Therefore the degree of differential migration shown by aphids will probably reflect both the within and between year heterogeneity of their environments. In favourable environments individuals with a high reproductive investment, the 'risk takers', will be fitter than those with a low reproductive investment, the 'risk averse' and *vice versa* (p. 165).

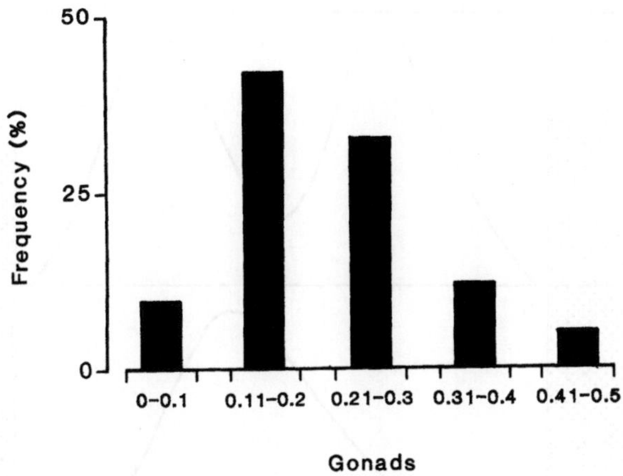

Figure 8.11 Frequency distribution of the proportion of the dry weight made up of gonads for alatae of 43 species of aphids.

8.2 Timing

8.2.1 Daily patterns

By using suction traps, whose sampling efficiency is independent of wind speed and which segregate the aphids into hourly catches, Johnson and his colleagues (1957) revealed the factors that determine the number of *Aphis fabae* flying over a bean crop. It mainly depends on the rate of eclosion of the winged adults and the time it takes them to complete their teneral development. Flight is inhibited at night by low light intensity and sometimes during the day by low temperatures. The high rate of moulting in the early morning, followed by a lower rate during the rest of the day, sometimes rising again in the evening and falling almost to zero during the night, is possibly controlled by temperature. The teneral period between final ecdysis and first flight is temperature dependent: aphids fly much sooner at higher temperatures.

Aphids become flight-mature all through the day and night, temperature permitting. When light intensity and temperature rise above the thresholds for take-off, the aphids that mature overnight, but are prevented from flying by cold and darkness, take-off and give rise to the first flight peak of the day (Figure 8.12a). The second flight peak later in the day represents the peak in number of newly moulted alatae that appear in the morning and complete their teneral development by the afternoon. The second flight peak declines in the evening because the temperature drops and light fades to intensities below the thresholds necessary for take-off.

The decline in aerial numbers at mid-day is due to a lack of flight-mature alatae and to the fact that most of the aphids in the air are newly

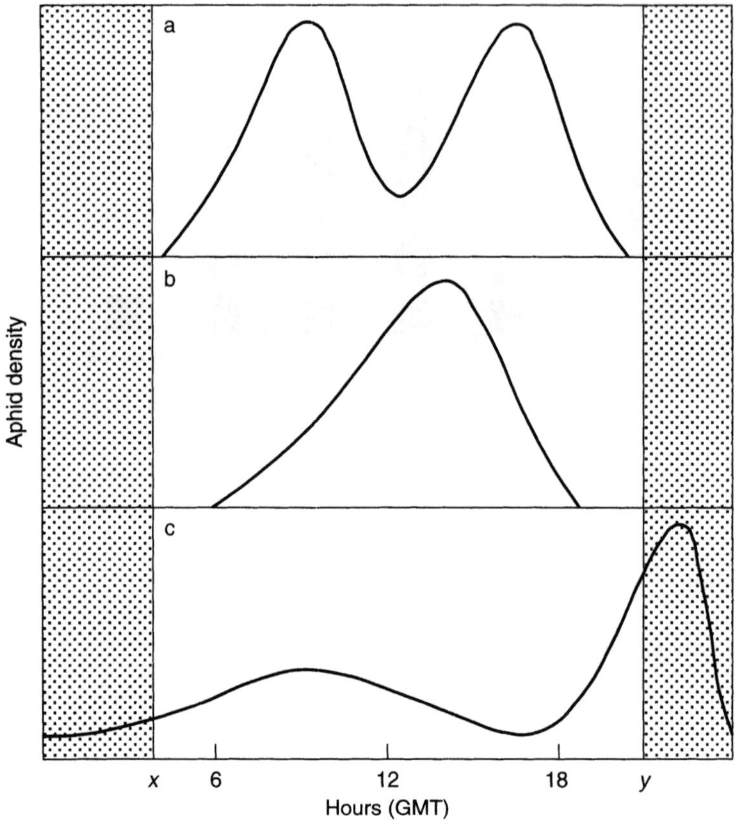

Figure 8.12 Daily flight activity of *Aphis fabae* (a), *Eucallipterus tiliae* (b) and *Drepanosiphum platanoidis* (c) (*x* and *y* time of sunrise and sunset, respectively).

flight-mature and fly at most for 2 h. The model developed by Johnson *et al.* (1957) accurately predicts the daily flight curves of *Aphis fabae*. Wind and humidity, which had previously been regarded as important factors in determining aphid flight, can be ignored when considering flight periodicity in the black bean aphid. In fact Johnson (1969) presents convincing evidence that most aphid migration occurs in windy weather. However, this is for aphids living on annual or short-lived plants, which are often widely scattered and less apparent than trees.

Many species of aphids show the same bimodal flight periodicity as the black bean aphid, and the model developed by Johnson *et al.* (1957) also predicts the daily flight activity of the emigrant alatae of *Phorodon humuli* leaving hop (Taimr *et al.*, 1978). There would not be a bimodal flight periodicity if, after completing the teneral period, most alatae were able to fly for several days (Taylor, 1965). For example, the lime aphid does not autolyse its wing muscles, is capable of flight throughout its adult life and has a single peak of flight activity just after midday (Figure 8.12b). The timing of greatest activity in this case is possibly determined by the aphid's high light intensity and temperature thresholds for flight (Figure 8.4).

In contrast, although capable of flight throughout adult life, the second-generation sycamore aphids present in summer nevertheless show a bimodal flight periodicity with a peak early in the morning and another after sunset (Figure 8.12c). The low light intensity and temperature thresholds for flight in this species (Figure 8.4) enable it to fly early and late in the day, and the decline in aerial numbers at mid-day (Figure 8.12c) is associated with high wind speeds, which inhibit take-off (Haine, 1955; Dixon and Mercer, 1983). The dramatic increase in flight activity around sunset, although dependent on the calm conditions that tend to prevail then, is heightened by an intrinsic urge to take-off at this time of the day. However, the flight activity of first-generation sycamore aphids in still conditions is strikingly different, as in this case most take-off in the early afternoon when, in the field, wind speeds are likely to be high (Dixon and Mercer, 1983).

In still conditions second-generation sycamore aphids, possibly in common with other tree-dwelling aphids, can retain control of their flight paths, and remain within sight and reach of trees. Flight in still conditions may be associated with the permanence and apparency of trees. Directed local flight may be the most effective means of dispersal from tree to tree within a habitat, referred to as trivial flight activity.

Thus the form of the daily flight curves is dependent on the availability of flight-mature aphids, their urge to fly and their light intensity, temperature and wind-speed thresholds for take-off. However, only the study of other species will decide the extent to which one can generalize about the mechanisms underlying the different daily flight activity curves, their association with particular habitats and the evolutionary significance of the timing.

8.2.2 Seasonal patterns

Seasonal periodicity in the flight activity of aphids is also well marked. The green spruce aphid (*Elatobium abietinum*) has a single flight peak in May just as its host is breaking its buds; the host-alternating hop aphid (*Phorodon humuli*) has two peaks and the bird cherry-oat aphid (*Rhopalosiphum padi*) three (Figure 8.13). The spring and autumn peaks of flight activity in the two host-alternating aphids represent the populations leaving and returning to the primary hosts, respectively. In the bird cherry-oat aphid there is an additional period of flight activity during the summer when the aphid redistributes itself between its secondary host plants. The flight behaviour of the aphids that make up the spring, summer and autumn flights differ. In a flight chamber the autumnal migrants have greater initial rates of climb and spend longer in the migratory phase than the spring and

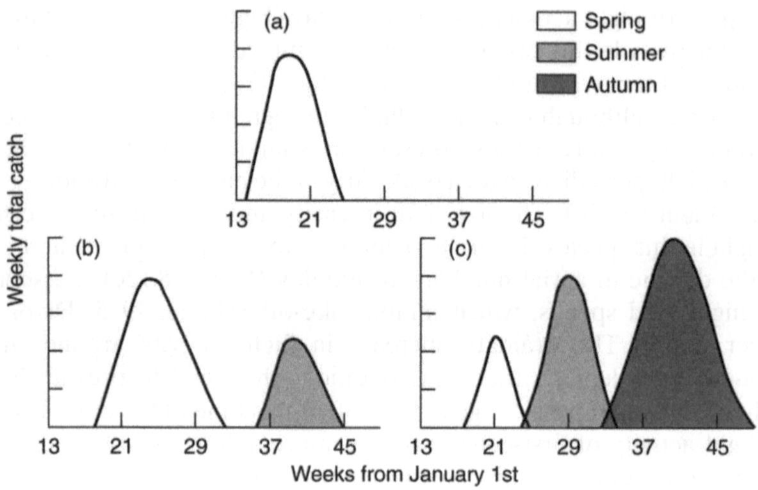

Figure 8.13 Seasonal flight activity of *Elatobium abietinum* (a), *Phorodon humuli* (b) and *Rhopalosiphum padi* (c).

summer migrants (David and Hardie, 1988; Nottingham and Hardie, 1989; Nottingham *et al.*, 1991b). These different morphs also differ in their reproductive behaviour and fuel reserves (Dixon, 1975b). It is tempting to think that this could be associated with the size of the plant target to be located by each morph, with the secondary hosts generally being more abundant and presenting a larger target than the primary host. These physiological/behavioural responses might result in the aphids staying in or moving out of the boundary layer and so determine the distance travelled. However, it remains to be shown just how these responses affect the fitness of clones.

In host-alternating species the seasonal pattern of flight activity reflects the optimum time for host transfer between primary and secondary host plants. In polyphenic autoecious species, winged forms develop in the second/third generations so flight in these species tends to occur just before summer and when each clone consists of many individuals, which increases the probability that a clone will successfully locate another host plant. That this flight mainly occurs in early summer, just before host quality generally starts to deteriorate, could serve to spread clones across several host plants and so reduce the risk of extinction.

8.3 Distance travelled

Atmospheric circulation dominates the aerial movement of aphids because of their low flight speed, 1.6–3.2 km/h, relative to the speed of movement of air. In the very still conditions close to vegetation some aphids fly within sight of vegetation, alight at short intervals and do not fly very far. Taylor (1965, 1974) calls this 'boundary layer migration'.

Aphids that fly upwards and out of the boundary layer enter air masses that are likely to be moving. Up to 5 km/h, air flow is laminar and under such conditions an aphid flies upwards until there is a balance in the tendency to fly upwards in response to light from the open sky, and to fly downwards in response to light reflected from vegetation (Kennedy, 1976). This height is a few metres, probably less than 8 m, from the ground (Taylor, 1965). The aphid is then borne downwind adding its own air speed to that of the air mass. After a period of flight aphids actively fly downwards, a consequence of the downward phototactic response increasing in strength relative to the upward one, rather than a simple switch over from one to the other (Kennedy, 1976). This kind of migration is called 'stratiform' (Taylor, 1965).

On most days the average wind speed exceeds 5 km/h, and flow is not laminar, because of either convection or turbulence produced by surface roughness. The eddies produced vary in size from 15 to 460–600 m in diameter, taking up to 1 h to circulate. Aphids entering these air masses are borne along in a rolling motion until they actively fly down out of the air mass and into the boundary layer.

Support for the view that most aphid migration occurs by riding turbulent convection comes from their aerial vertical density profile. It is similar in form to that for dust and plant spores:

$$f(z)=c(z + ze)^{-\lambda} \tag{8.1}$$

where $f(z)$ is the number of aphids per unit volume (density) at height z, c is a scale factor related to the population size of the source, λ is the gradient of density and ze is a measure of the departure from linearity on a double logarithmic scale. In the lower strata of the air at least, the gradient for dust (–0.28) is remarkably similar to that for aphids (–0.35). This led to the suggestion that the relationships for the deposition of dust and plant spores in relation to distance from source could be adapted to describe aphid dispersal.

This proved difficult to test because the aerial populations of most species of aphid are made up of individuals from many contiguous sources. However, in Britain the gynoparae and males of the host-alternating hop aphid come mainly from two discrete sources, the hop gardens in Hereford and in Kent, which are 180 km apart (Figure 8.14). Using 7 years of census data collected by the Rothamsted Insect Survey, Taylor *et al.* (1979) produced density profiles radiating out from each of the sources, showing that the migration is not directional. As the sources reached similar densities the two relationships between density and distance were combined (Figure 8.15). This elegantly revealed that the aphid aerial distribution curve is similar to that expected if the aphids were simply transported on the wind

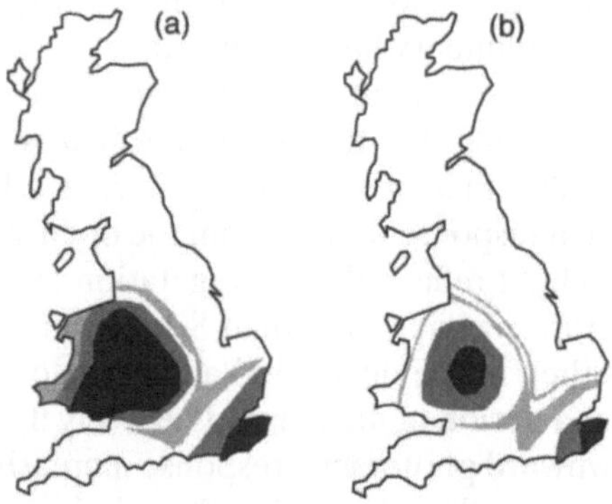

Figure 8.14 Mean aerial distribution of sexuals of the hop aphid, *Phorodon humuli*, for 1971–1976. (a) Gynoparae and (b) males. The concentrations are over the Hereford and Kent hop gardens. (After Taylor *et al.*, 1979.)

with the gradient dependent on the degree of turbulence rather than by random diffusion.

An aphid's migratory tendency, defined as its tendency to fly upwards, is also likely to determine the form of the density distance function. However, it is unknown what selection pressures have shaped the frequency distribution of the distances travelled by individuals, in terms of short and long flights, i.e. the migratory ambit of a species, although the actual distances travelled are determined by wind speed (Taylor *et al.*, 1979). It is likely that

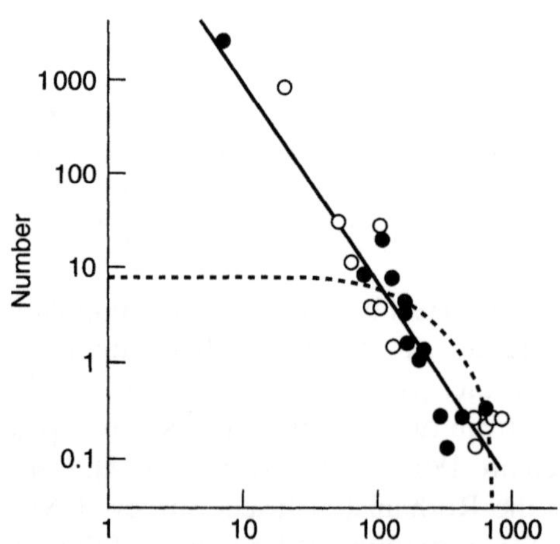

Figure 8.15 The average number of gynoparae and males of the hop aphid caught from 1971 to 1977 in suction traps at varying distances from the Hereford (filled circles) and Kent (open circles) hop gardens (*cf.* Figure 8.14). Solid line, deposition of wind-borne fine soil particles; dotted line, expected from random diffusion. (After Taylor *et al.*, 1979.)

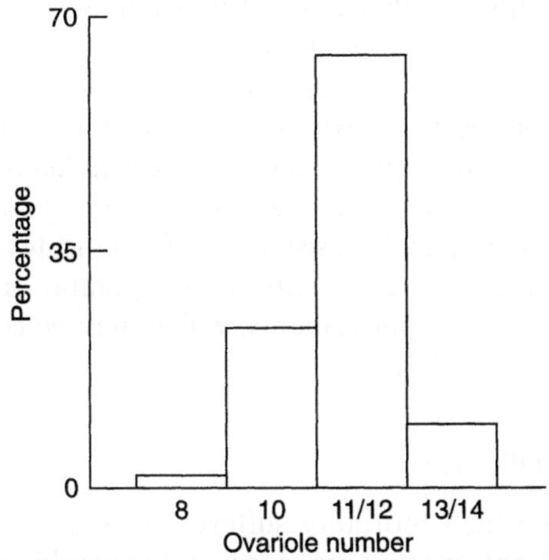

Figure 8.16 Percentage of emigrant hop aphids that have eight 10, 11/12 and 13/14 ovarioles ($n = 50$).

the migratory ambit is linked with reproductive investment and/or size, with the smallest individuals and/or those with the fewest ovarioles showing the greatest migratory urge and travelling furthest. That the hop aphid, and many others, show differential reproductive investment and all show a wide range in size lends support to this idea (Figure 8.16).

The median distance travelled by hop aphid sexuales is about 15–20 km, with a 95% limit of 100–150 km. The direct interchange of sexuales between the two hop centres is therefore minimal in a season, but inevitable over a series of years, as the aphid can survive on the widespread winter host and on wild hops in surrounding areas, albeit at a low density (Taylor *et al.*, 1979).

Aphids lifted by convection high enough to reach the persistent high-altitude (300–900 m) air streams, called low-level jet streams by meteorologists, could be transported very rapidly over great distances. However, an aphid would have to keep flying in order to stay in the jet stream as it would experience little convective lift. It is likely that the living spruce aphids (*Cinara piceae*) Elton (1925) found on a glacier in Spitzbergen had made the 1300 km journey from the Kola peninsula in this way, flying for up to 24 h in the long summer days of the northern latitudes. As Wellington (1983) stressed, although they are weak fliers, aphids and other insects potentially can disperse over a wide area by exploiting the ephemeral but very structured nature of air movement.

Elton's finding spruce aphids on a glacier in Spitzbergen, Captain W. Parry's finding of the same species of aphid 100 years earlier on ice flows at 82 N in the polar sea, 160 km from the nearest land (Ross, 1828), and a few other similar observations, are striking and have tended to divert attention from the norm. Although these long-distance movements are very impressive for such small insects it is likely to be disadvantageous for aphids to move great distances. It could result in their moving into areas where the host plant is rare or absent, or exhibit a phenology and/or live in an environment to which the aphid is poorly adapted. A detailed review based mainly on indirect evidence from plant virus epidemiology and the spread of new genotypes, such as insecticide-resistant clones, indicates that short-distance movements have a much greater impact on population and genotype distribution than long-distance movements, which may be comparatively infrequent (Loxdale *et al.*, 1993).

8.4 Adaptive significance

In addition to the high mortality suffered during dispersal, alate aphids incur other disadvantages that are more easily quantified. The development of a flight apparatus and/or its maintenance is costly in terms of time and fecundity (p. 109). If aphids fly then they may incur an additional cost in that their potential fecundity is further reduced (Burns, 1971) and there is a further delay in the onset of reproduction. The combined effect is a marked reduction in reproductive potential. However, although costly for the individual, dispersal is likely to increase the overall fitness of a clone.

Although dispersal results in the colonization of suitable plants (Dixon, 1969) it is not always clear what short-term advantages there are in dispersing, except from annual plants that are about to die. The tendency of aphids to migrate appears to be density dependent (Dixon, 1969; Kidd and Cleaver, 1984, 1986). This is viewed as the means by which aphids maximize the production of migrants. On short-lived plants the density-dependent production of winged individuals and flight results in the production of 50% more migrants than a strategy in which the winged individuals make up a fixed proportion of the offspring in each generation (Kidd, 1990). As it is likely that all the aphids on short-lived plants have originated from a single colonist and therefore belong to the same clone, the production of migrants is cheap as it relieves the density-dependent pressure on resources experienced by the rest of the colony. However, the situation is more complicated on long-lived plants, which are likely to have been colonized by many clones of varying degrees of relatedness.

As, on long-lived plants like trees, aphids also produce either sedentary or migrant offspring, it is of interest to speculate on the relative benefits of the two options. The variable incidence of the two options in aphids living on trees gives some clues as to their adaptive significance. The sycamore aphid (*Drepanosiphum platanoidis*) declines by up to 63% per week during late summer mainly as a result of migration (Dixon, 1969); populations of the Turkey-oak aphid (*Myzocallis boerneri*) may lose over 90% to migration (Dixon, 1990c; Kindlmann and Dixon, 1996), whereas other species, *Kaltenbachiella japonica*, appear to be more or less immortal, with low reproduction and dispersal rates (Komatsu and Akimoto, 1995).

Egg hatch in the sycamore aphid is synchronized with the time of average bud burst of sycamore but not necessarily with the time of bud burst of a particular tree (Dixon, 1976a). The high levels of migration observed in this species, which is likely to result in considerable between tree movement, rules out synchronization of egg hatch with bud burst on individual trees. Poor synchronization is costly in terms of survival and fecundity (Dixon, 1976) as is the between tree movement (Ward *et al.*, 1998), which prevents adaptation to bud burst of particular trees. If synchronization is very uncertain next year even though the aphid was successful this year, then the optimum strategy might be to spread the risk in space and for clones to colonize several trees. Sycamores in common with other trees consistently break their buds either early or late from year to year. However, temperature in spring and aphid infestation levels the previous autumn can affect the relative time of bud burst (Chambers, 1979). Possibly the better strategy in this case is to track the variability in phenology by spreading the risk in space.

Alternatively aphids can adapt to individual trees, a phenomenon which has been reported in scale insects (Edmunds and Alstad, 1978). This can only evolve if there is very little movement between trees. In the case of the monoecious galling aphid, *Kaltenbachiella japonica*, living on elm, the egg hatching time is genetically determined and closely correlated with the bud burst time of the tree on which the eggs were laid, and although winged forms are produced they have rarely been observed to fly. The fine-scale adaptation observed in this species appears to be due to the variation in bud burst phenology between trees acting as an intense selection pressure (Komatsu and Akimoto, 1995). The striking difference between the sycamore aphid and *Kaltenbachiella japonica* is that the latter is a galling aphid. For the successful induction of a gall the aphid must hatch and start feeding on the leaves just as they start growing (p. 141). In the host-alternating and gall-forming aphid, *Smynthurodes betae*, egg hatch does not coincide with bud burst (Burstein and Wool, 1993). In this case the regular seasonal host transfer between hosts is likely to result in clonal mixing and prevent fine-scale adaptation to individual host plants. The implication is that the benefits of host alternation (p. 147) offset the cost of poor synchronization with bud burst, and the fine-scale adaptation seen in aphids like *Kaltenbachiella japonica* is most likely to occur only in monoecious galling aphids.

Summarizing, although costly for an individual, dispersal enables clones to spread the chance of survival in space and locate plants that are of above average quality. Although the speed and direction of flight are usually governed by the wind, aphids nevertheless terminate a flight by actively flying downwards and thus have some control over the distance that they travel, which in most cases is relatively short. The infrequent but impressive long-distance migrations are achieved by the accidental riding of high-altitude jet streams. The patterns of daily and seasonal flight activity are shaped by the availability of flight-mature aphids and their light intensity, temperature and wind-speed thresholds for take-off. However, except for the spring and autumn flights of host-alternating aphids the adaptive significance of the timing of the flights is poorly understood. Similarly, why some species of tree-dwelling aphids show little dispersal and closely track the phenology of individual trees, whereas other species show high levels of dispersal and appear to track the average phenology of their host tree, remains to be resolved.

9 Population dynamics

Aphids can become very abundant, noticeably so on agricultural crops. A hectare of field beans can produce 4000 million alatae of the black bean aphid (*Aphis fabae*) (Way and Banks, 1967) and the cereal aphid, *Metopolophium dirhodum*, can achieve population densities of 220 per tiller or 1000 million per hectare of wheat. Although aphids are small these numbers are equivalent in mass to an elephant and a cow respectively. Insect predators feeding on these aphids can also reach impressive numbers. In England in 1976, 280 000 seven-spot coccinellid adults, *Coccinella septempunctata*, matured per hectare of wheat after feeding on the cereal aphid *Sitobion avenae*, which nevertheless achieved a peak population of 200 million per hectare. The estimated 24 000 million seven-spot ladybird beetles produced in the county of Norfolk combined with some produced in other counties in 1976 gave rise to a ladybird 'plague' that drove people off the beaches in parts of East Anglia.

Populations have either been studied on a few plants in a relatively small area or, uniquely, in the air over Europe. The often detailed studies of aphids on plants only cover a minute fraction of the population and range of these highly mobile insects. However, by combining the results of the two approaches it is possible to extrapolate from the understanding of what determines the numbers of aphids on a few plants to understanding the population dynamics of the species over a wide area and so consider their spatial dynamics (Dixon, 1979).

9.1 Dynamics of aerial populations

Suction traps, positioned 20 m above the ground, have been collecting flying aphids for 24 h a day in some locations for periods in excess of 30 years. These traps were set up at sites throughout the UK and Western Europe with the objective of monitoring and forecasting the abundance of pest species of aphids (Taylor, 1973a). The traps are sited in agricultural areas and positioned well off the ground so that the catch is more representative in terms of species and abundance of the aphids in the area. The very large data sets that are now available have not only proved useful for monitoring pest aphids but have also been used by ecologists interested in population regulation and life-history patterns.

Over the last few years this long-term census data has been analysed with the objective of identifying deterministic dynamical behaviour in nature. The impetus for these analyses was the counter-intuitive conclusion from life table analyses that density-dependent regulation of abundance was infrequent in insects (e.g. Dempster, 1983; Stiling, 1987, 1988). An important step in the resolution of this apparent conflict between theoretical expectation and empirical results was Turchin's (1990) study showing that the detection of density dependence in abundance data was dependent upon the nature of the density-dependent processes involved and analytical methodology used. He showed that, in general, direct (non-delayed) density dependence would not be detected in data on populations that are characterized by time-lagged effects and complex dynamical behaviour.

Because of their very high reproductive rates and overlapping generations, aphids often become very abundant. The overlapping of generations and the pronounced seasonal fluctuations in abundance have proved to be major obstacles in the analysis of aphid population dynamics (Dixon, 1990c). In an attempt to avoid these obstacles some authors have resorted to using annual suction-trap catches. Turchin and Taylor's (1992) analysis of annual abundance data includes two aphid species, the sycamore aphid, *Drepanosiphum platanoidis*, and the beech aphid, *Phyllaphis fagi* (Figure 9.1). The former was categorized as a case of 2-year limit cycles and the latter as a case of chaotic dynamics. A subsequent analysis of a longer data set for the beech aphid, however, revealed that this species has stable, rather than chaotic endogenous dynamics (Perry *et al.*, 1993).

Woiwod and Hanski (1992) analysed annual trap catch data for 557 species of moths and aphids, and concluded that some form of density-dependent regulation is typical for these insects, with aphids showing a higher incidence of density dependence than the moths. This difference between aphids and other insects has also been revealed by the analysis of another data set consisting of nearly 300 species of insects spanning nearly 16 000 years (Mackenzie *et al.*, 1997). Two-thirds of the aphids showed direct density dependence but less than a quarter of the non-aphids. One striking difference between these two groups of insects is their potential rate of population increase, which is much greater in aphids than in the moths that make up most of the non-aphid group. Indirect evidence of this comes from the relationship between the incidence of direct density dependence and the number of generations per year in moths. A significantly higher proportion of the species that have more than two generations per year showed density dependence compared with those that are mono or bivoltine, and aphids achieve even more generations and showed an even greater incidence of density dependence (Figure 9.2). In addition, a higher percentage of the censuses of monoecious tree-dwelling aphids (81%) show density

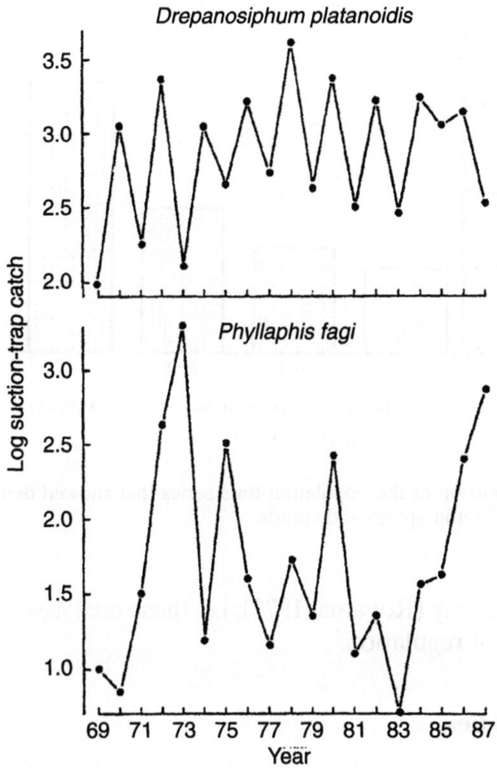

Figure 9.1 The total numbers of sycamore aphids, *Drepanosiphum platanoidis*, and beech aphids, *Phyllaphis fagi*, caught each year in the suction trap at Dundee, Scotland, 1969–87.

dependence than those that are restricted to herbaceous host plants or host alternate (64%). This may be because the percentage cover of many trees tends to be greater than that of herbaceous plants and subdominant trees, and the proportional cover of the host plant could be important in determining the realized rate of increase and abundance of these aphids (p. 206). Thus it appears that the dynamics shown by the aerial populations of aphids may be a consequence of their very high potential rate of increase, which can result in them rapidly achieving very high levels of abundance.

In using yearly totals of aerial abundance the above studies promote, albeit unintentionally, the notion that regulation occurs between years. This runs contrary to the conclusions of the studies carried out on aphids on plants (p. 192), which stimulates a closer look at the nature of the suction-trap data. What do yearly aerial population totals represent when dealing with such short-lived insects as aphids? It is questionable whether the analysis of this yearly data reveals anything more than can be gleaned from the observation that the populations persist for many years – the numbers are

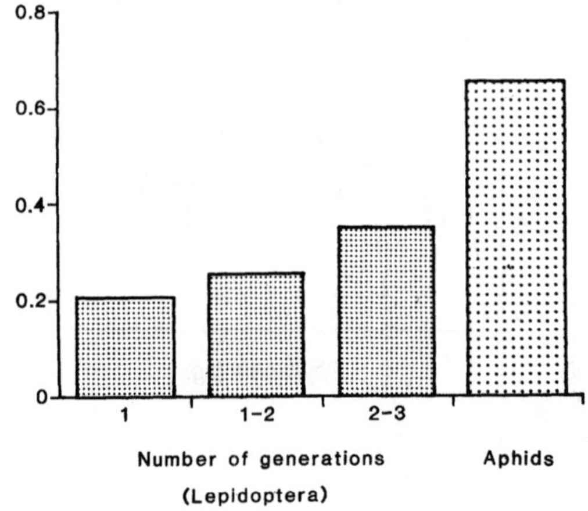

Figure 9.2 The proportion of the population time series that showed density dependence in relation to voltinism in non-aphids and aphids.

regulated in some way (Royama, 1977), i.e. these analyses reveal nothing of the mechanisms of regulation.

9.2 Intensive studies

The dynamics of aphid populations, with their overlapping generations and unstable age structures, are not easily studied using explicit formulae as in analytical models. Most analyses of aphid population dynamics on plants have involved detailed, often large simulation models consisting of a specified routine of arithmetic operations. Although few aphid populations have been studied long enough for time-series analysis, population studies of aphids on both herbaceous and woody hosts are beginning to reveal patterns in the changes in abundance from year to year and we have some understanding of the underlying processes. However, for a very good practical reason, continuity in space, the longest runs of population data are for tree-dwelling aphids.

9.3 Patterns

In the lime aphid (*Eucallipterus tiliae*) and sycamore aphid (*Drepanosiphum platanoidis*) there is an inverse relationship between spring and autumn numbers, and a positive relationship between autumn and spring numbers (Figures 9.3 and 9.4). Although there are several parthenogenetic generations between egg hatch in spring and egg laying in autumn, the ratio of the densities of the aphids at these two stages gives a measure of the overall rate of increase of the population. Thus, in the analysis of changes

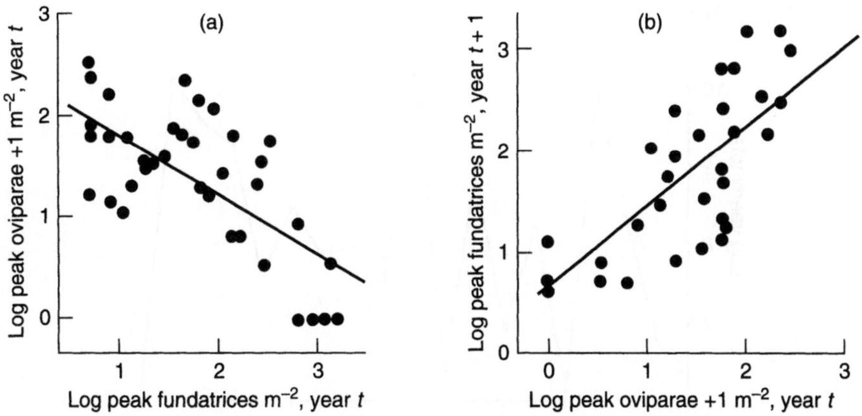

Figure 9.3 The peak number of oviparae of the lime aphid (*Eucallipterus tiliae*) in relation to the peak number of fundatrices each year (a) and the peak number of fundatrices in relation to the peak number of oviparae the previous year (b). Each point represents one tree in 1 year.

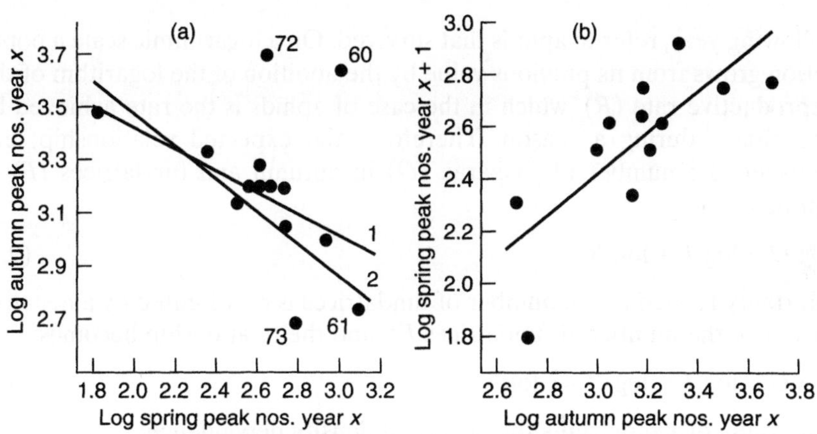

Figure 9.4 The peak number of sycamore aphids (*Drepanosiphum platanoidis*) in autumn in relation to the peak number present in spring (a) and the peak number present in spring in relation to the peak number present the previous autumn (b) for each year from 1960 to 1973. (Regression line 1 in (a), $y = 4.65 - 0.59x$, $r = 0.81$, $P < 0.01$, does not include 1960 and 1972; 2, $y = 4.25 - 0.41x$, $r = 0.91$, $P < 0.01$, does not include 1960, 1961, 1972 and 1973. (After Dixon, 1979.)

in lime and sycamore aphid numbers a meaningful first step is to determine the relationship between the number of aphids present after egg hatch in spring and those present in autumn prior to egg laying.

The equation expressing the relationships between the number of fundatrices in spring and the number of oviparae in autumn of the same year, and the number of oviparae in autumn and the number of fundatrices in the

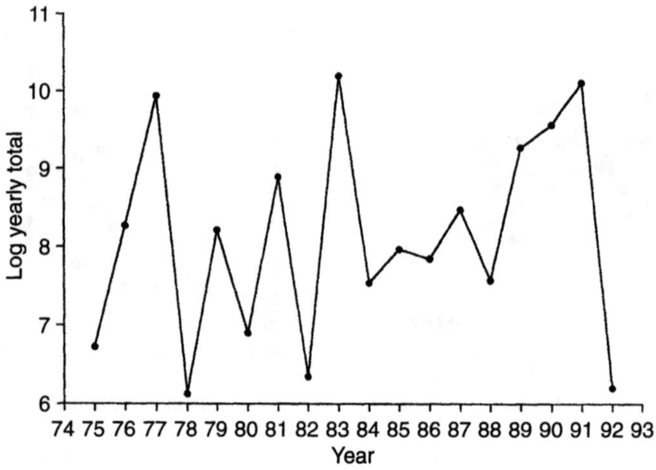

Figure 9.5 Annual abundance (totals) of Turkey-oak aphids on one tree from 1975 to 1992.

following year, refer to aphids that survived. On a logarithmic scale a population grows from its previous value by the addition of the logarithm of the reproductive rate (R), which in the case of aphids is the rate achieved by the clones during a season. Therefore, the expected relationship, e.g. between the number of oviparae (O) in autumn and fundatrices (F) in spring is:

$$\log O = \log F + \log R \quad (9.1)$$

Mortality related to the number of fundatrices is represented by a negative power of the number of fundatrices F^{-x} and the relationship becomes:

$$\log O = \log F + \log R - x \log F \quad (9.2)$$

The value of x represents the degree of density dependence.

In both the lime and sycamore aphids the value of x is approximately 1.5, which indicates that there is an overcompensated density-dependent factor acting within years. Between autumn and the following spring (equation 9.1) the value of x is less than 1, which indicates that the mortality over this period is inversely density-dependent (Dixon, 1970b, 1971d, 1979).

A time-series analysis of 18 years of census data of the Turkey-oak aphid, *Myzocallis boerneri* (Figure 9.5), which is not known to have specific natural enemies such as parasitoids, similarly shows that regulation is mainly by the operation of processes acting within years. Each year aphid abundance follows the same regular seasonal pattern of density changes (Figure 9.6). In most years, density grows exponentially until the middle of June, followed by a steep decline until the middle of September after which populations increase slightly in abundance. Population density fluctuates around

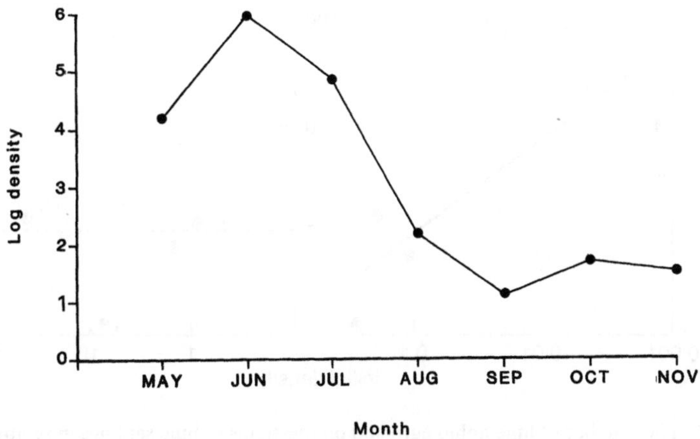

Figure 9.6 The average seasonal pattern of changes in the numbers of Turkey-oak aphids on a tree, from May to November, for the period 1975–92.

this seasonal mean or 'equilibrium' value. In the field populations are subjected to frequent perturbations. When growth conditions are highly favourable as in spring, density may overshoot the local mean for this time of year. Alternatively, weather and natural enemies may depress density below the local seasonal equilibrium. In these situations density-dependent processes come into operation and bring density back to the seasonal equilibrium, i.e. there is no evidence of time-delayed effects of density in the data adjusted for the seasonal trend in density (Sequeira and Dixon, 1997).

As in the lime and sycamore aphids there is some evidence for the seesaw effect in the Turkey-oak aphid, i.e. a negative correlation between the number of first-generation aphids and the number of sexuales produced in autumn (Dixon *et al.*, 1996). However, it is statistically weak possibly because of stochasticity during the summer decline, when density is low.

9.4 Processes

In natural populations where density may be influenced by environmental perturbations, persistence over long periods of time implies some form of regulation (Royama, 1977, 1981). Theoretically, density dependence is the only form of regulation that can explain stable long-term patterns of density fluctuations (Nicholson, 1933, 1954; Varley *et al.*, 1973; Berryman, 1991; Royama, 1992). Deterministic density-dependent relationships in census data are not, in themselves, evidence of regulation in nature. The ultimate evidence of regulation must come from experiments.

The regulatory mechanisms operating in the lime, sycamore and Turkey-oak aphids have been determined by studying laboratory populations, by

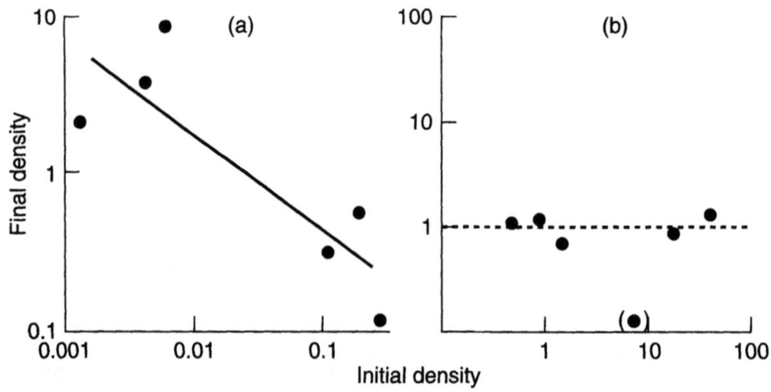

Figure 9.7 The number of lime aphid eggs laid on the stems of lime saplings in relation to the number of aphids initiating the population at the beginning of the year (a) and peak number of sycamore aphids on sycamore saplings in relation to the number of aphids per unit area on the buds at the beginning of the year (b) (lime aphid numbers are per square centimetre of leaf, sycamore aphid numbers per square centimetre of bud surface and leaf, respectively, the point in parentheses indicates a sapling that died at the end of the year). ((a) after Brown, 1975; (b) after Chambers, 1979.)

field experiments and simulation modelling. Seasonal changes in the abundance of these three species have been followed for 9, 15 and 18 years, respectively.

Even in the absence of natural enemies laboratory populations of the lime aphid suffer a sudden decline in numbers in June or July if the numbers are high at the beginning of the season (Dixon, 1971d). This gives an inverse relationship between initial and final numbers within a year, similar to that seen in the field (Figure 9.7a). Although overcompensation occurs in cage populations it is less marked than in the field. This reduced overcompensation is most likely mainly due to reduced losses from migration in these caged populations, i.e. the migrants cannot leave the system.

A simulation model indicates that weather is important as a disturbing factor and as a major determinant of peak numbers, and numbers of fundatrices in any given year. The ultimate limit to growth in any given year does not appear to be available space or death of the host through overexploitation, but rather to intraspecific competition for food and an aphid-induced decline in its quality. At high population densities the intense competition results in the development of relatively small adults, proportionally more of which migrate, which depends on the densities experienced during nymphal development, and those that remain have very low reproductive rates (Barlow and Dixon, 1980).

This interpretation rests on the existence of a cumulative or integrated density-dependent effect. The weights of individual aphids developing either in the field or laboratory are more closely associated with the

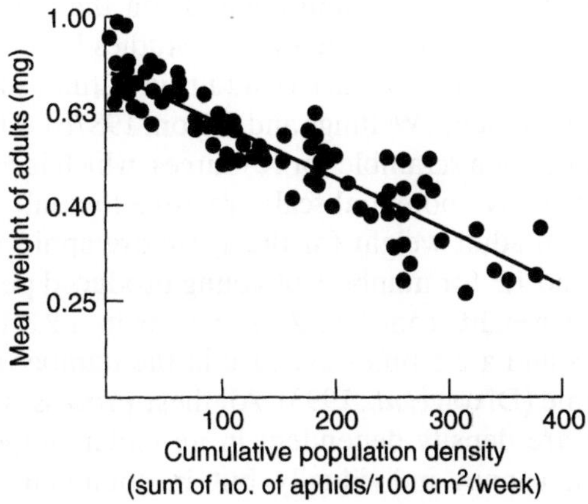

Figure 9.8 Mean weight of adult lime aphids from the field in relation to the average cumulative population density experienced by the leaves on which the aphids developed.

cumulative density of aphids that have fed on the leaves up to that time (Figure 9.8), than the density experienced by an aphid during its development. In addition, in the field the mortality of adults and incidence of flight increases strikingly after prolonged heavy infestation of the trees, corresponding to a cumulative density of about 250 aphid-weeks/100 cm^2 of leaf surface. This could be due to the transmission of crowding effects from generation to generation, possibly through birth weight or from changes in the quality of the host plant in response to aphid attack, or both. Aphid-induced changes in the host plant have been shown experimentally to affect the time aphids take to reach maturity and adult weight, but do not affect aphid mortality or flight activity. Thus intraspecific competition and the adverse affect of aphids on host plant quality appear to regulate the numbers of the lime aphid.

The maximum numbers reached in autumn in cage populations of the sycamore aphid are similar, irrespective of the number of fundatrices initiating the populations in the spring. Overall the populations behave in a compensatory density-dependent manner, attaining a maximum autumnal population of 1 aphid/cm^2 (Figure 9.7b). This is at variance with the over-compensatory relationship observed in the field (Dixon, 1970b; Wellings *et al.*, 1985). As with the lime aphid this difference may be a consequence of the cages preventing the migrants leaving the system. This is particularly the case for the sycamore aphid where the cages were relatively very small.

In the cage populations, two density-dependent processes operate: (1) flight of first-generation aphids and (2) a reduction in the fecundity of second-generation aphids as a result of crowding experienced during their development. Analysis of field data has identified the duration of aestivation of the second generation as being density dependent, and has also shown that reproduction in autumn depends on the cumulative effect of densities in spring and summer. Laboratory studies have failed to reveal a plant affect in the poor performance of adults in autumns following springs when aphids are abundant (Wellings and Dixon, 1987b). The effect appears to be a consequence of a scramble for resources, which is mediated through body size. Reproductive potential scales relative to adult weight, w, as w^b. The exponent b on adult weight for the sycamore aphid is approximately 1.5 for biomass and 1.3 for numbers of young produced per unit time, i.e. a doubling of adult weight from 1 to 2 mg results in a 2.8 times increase in terms of biomass and a 2.5 times increase in the number of offspring produced per unit time (Dixon *et al.*, 1993). All these processes except the effect on adult weight are density dependent in an undercompensatory fashion and, as a result, cannot individually but in combination may possibly account for the population behaviour observed in the field.

However, weather, in particular wind, can have a dramatic effect on sycamore aphid population dynamics (Dixon, 1979). Wind causes leaves to brush together, which dislodges the aphids, most of which are then lost to the population (Dixon and McKay, 1970). Thus the amount of space available for sycamore aphids may be limited to a small fraction of the total leaf area, especially during periods of high winds. The leaves in fact can act as devices converting weather into aphid mortality. In exceptionally still autumns, aphids are able to achieve most of their potential rate of increase, and however few there are at the beginning of autumn, there is sufficient time to reach saturation levels of 1 aphid/cm^2 before the end of the season. In the more usual windy autumns the aphid only achieves a fraction of its potential rate of increase, and the abundance achieved will reflect the potential rate of increase: low in years when the aphid is abundant in spring and summer and *vice versa* (Figure 9.9) (Chambers *et al.*, 1985).

Although not as consistent as in the previous two species, high populations of Turkey-oak aphid in spring are likely to be followed by low populations in autumn and *vice versa* (Figure 9.10). Predators undoubtedly play a role in determining the magnitude of the dramatic crash in numbers that occurs in early summer in those years when aphids are very abundant in spring. However, they do not account for the lack of recovery of the population in autumn. For example, from week 13 onwards the autumnal populations of the Turkey-oak aphid in 1982 and 1983 (Figure 9.10) follow markedly very different trends but both experienced similar slight predation from week 19 onwards. That a similar response is observed in the within-year population dynamics of caged populations also argues against natural enemies having an important role in determining the response.

In the Turkey-oak aphid poor performance in autumns of those years when the aphid is abundant in spring is associated with the small size of the aphid, e.g. in the autumn of 1983 this aphid was less than half the size, and therefore less fecund, than those present in 1982. As in the other species small size

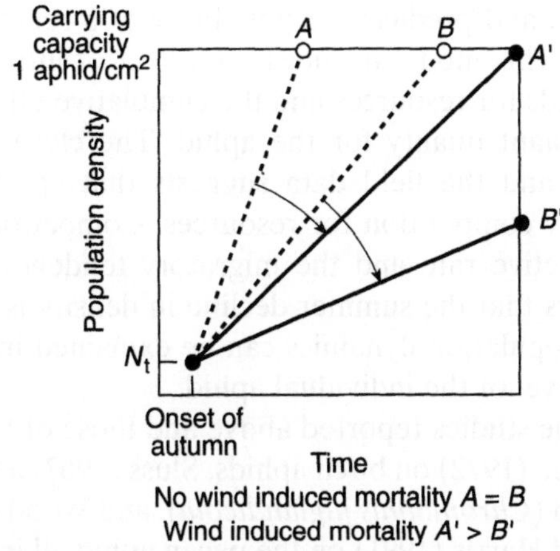

Figure 9.9 The effect of wind-induced mortality (→) on the peak autumnal numbers achieved by sycamore aphid populations with a high (*A*) and low (*B*) potential rate of increase. Slopes of the lines represent the realized rates of increase.

is mainly a consequence of competition for resources and even in the absence of competition it takes several generations for this aphid to achieve the size of aphids that have not experienced crowding (Sequeira and Dixon, 1996).

An individual-based simulation model of the Turkey-oak aphid system (Kindlmann and Dixon, 1996; Dixon, *et al.*, 1996) shows that regulation

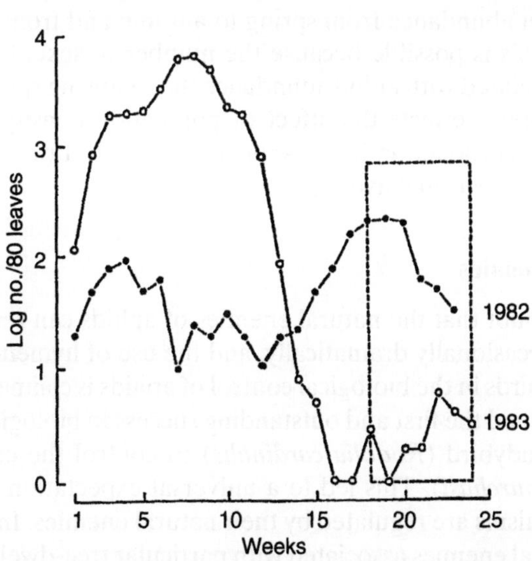

Figure 9.10 The Turkey-oak aphid population trends observed in 1982 and 1983. The dashed box marks the period of sexual production, with eight times more sexual females produced in 1982 than in 1983.

occurs within years, and predicts the possibility of a see-saw effect under certain conditions. This model includes the effect of intraspecific competition between aphids for resources and the cumulative effect this drain on resources has on plant quality for the aphid. The close match between model predictions and the field data suggests that aphid abundance is mainly regulated by competition for resources. Competition affects adult body size, reproductive rate and the migratory tendency of adults. The model clearly shows that the summer decline in density is driven by adult dispersal, i.e. the population dynamics can be explained in terms of selection acting at the level of the individual aphid.

To summarize, the studies reported above and those of Gange (1985) on the alder aphid, Heie (1972) on birch aphids, Sluss (1967) and Dixon (1977) on the walnut aphid (*Chromaphis juglandicola*), and Wood *et al.* (1985) and Bumroongsook and Harris (1991) on the pecan aphid all indicate that high abundance in spring is usually followed by low abundance in autumn. This see-saw effect is driven by a cumulative effect of density operating in one of several ways: through its effect on host plant quality at the end of the year, through the aphid or both. High numbers of pecan aphids early in the year adversely affect the quality of Pecan as food for the aphid later in the year. In the case of the sycamore aphid the effect is pronounced with high aphid abundance early in a year resulting in very small aphids in autumn, which have lower reproductive rates. In the lime and Turkey-oak aphids it appears to operate through its affect on both the quality of the host plant and the aphid. Thus, a mechanism exists whereby the system can transfer information on abundance from spring to autumn and from autumn to the next spring. This is possible because the number of sexuales in autumn is positively correlated with aphid abundance the following spring. In addition to these long-term effects the affect of population density on migration, body size and reproduction in the short term tend to bring the population back to the local seasonal average.

9.5 Natural enemies

There is no doubt that the natural enemies of aphids can reduce their rate of increase, occasionally dramatically, and the use of hymenopterous parasites and ladybirds in the biological control of aphids is claimed to have been successful. Indeed the first and outstanding success in biological control was the use of a ladybird (*Rodolia cardinalis*) to control the cottony cushion scale (*Icerya purchasi*). This led to a universal expectation that the abundance of organisms are regulated by their natural enemies. Indeed the array of insect natural enemies associated with particular tree-dwelling aphids can be awesome (Figure 9.11).

In spite of the early success with *Rodolia*, ladybirds have not proved effective biological control agents of aphids (Hodek, 1973). The reason for this

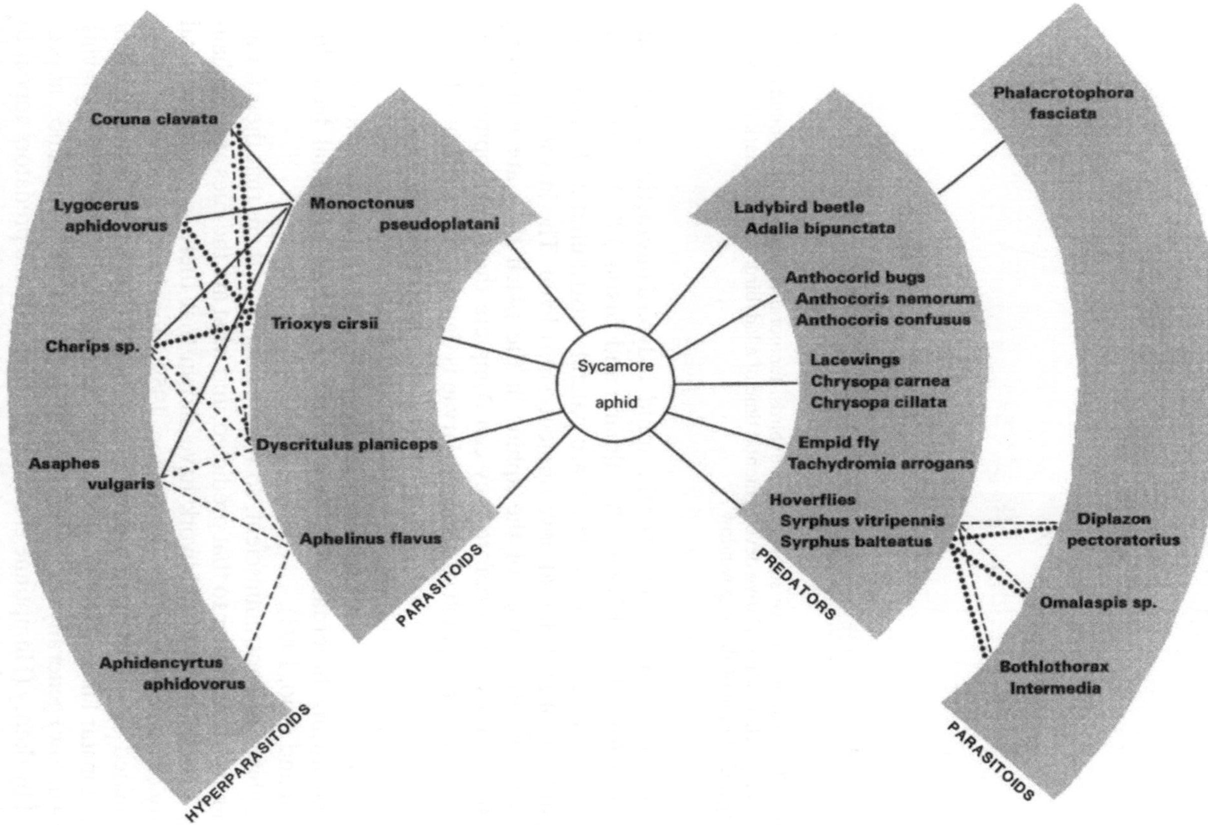

Figure 9.11 The predators, parasitoids and hyperparasitoids that attack the sycamore aphid.

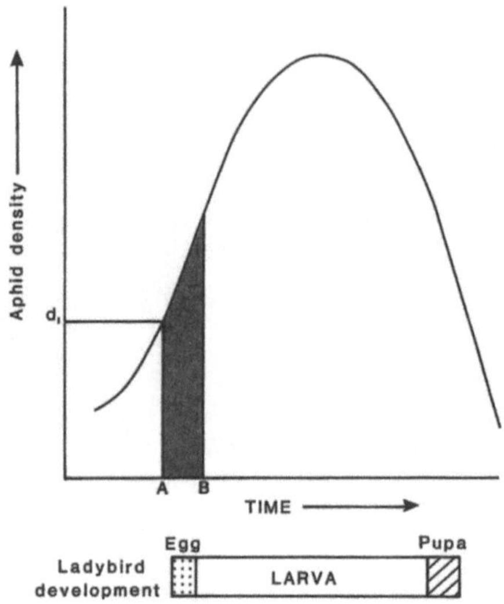

Figure 9.12 The reproductive strategy of ladybirds: temporal changes in the abundance of aphids in a patch and the relative developmental time of a ladybird. A and B mark the time of opening and closing of the 'egg window'.

is to be found in the dynamics of the aphid–ladybird interaction and its consequences for ladybird fitness (Kindlmann and Dixon, 1993). These predators exploit aphid colonies, which are patchily distributed and, relative to the duration of their larval development, short lived. The larvae risk starvation if the abundance of the aphids in the patch they are exploiting declines to a low level before they can complete their development. The period for which a patch can sustain larvae is dependent on the number of larvae present. That is, larval survival is low if too many eggs are laid in a patch or they are laid too late, i.e. when aphids are about to disperse, and the ladybirds that survive are likely to be small, which adversely affects their survival during hibernation (Zhou *et al.*, 1995) and potential fecundity (Dixon and Guo, 1993).

Ladybirds are not omniscient, so what evidence is there that their foraging behaviour is similar to that predicted, and what are the cues and mechanisms? Laboratory and fieldwork have shown that there is a lower critical density of aphids for egg laying (Dixon, 1959; Honek, 1978) and that this is determined by the minimum density of aphids required for the survival of the first instar larvae (Dixon, 1959). Field data also indicates that egg laying (egg window) generally occurs prior to the peak in aphid abundance, as predicted by theory (Hemptinne *et al.*, 1992). That is, these predators appear to be able to assess the age of a patch. Given that aphid abundance, through its effect on early larval survival, determines the lower threshold (d_1) for oviposition and the opening of the egg window (A in Figure 9.12), why do they cease laying although aphid numbers are still increasing (B in Figure 9.12)? Experimental evidence indicates that the closing of the egg window is also associated with survival, but in this case the mechanism is egg cannibalism. If adults oviposit in a patch over a prolonged period of time the survival of the last laid eggs is likely to be very low because they stand a high risk of being eaten by the larvae that emerged from the first laid eggs. Ladybirds appear to be inhibited from ovipositing by chemical cues associated with conspecific larvae (Doumbia *et al.*, 1997), i.e. foraging ladybirds respond to cues that affect their fitness. Feeding rates, aphid skins and honeydew (Dixon, 1959; Carter and Dixon, 1982; Evans and Dixon, 1986) indicate the abundance of aphids and other chemical cues the presence of larvae, and the relative strengths of these cues possibly determine both thresholds for egg laying; the egg window. These thresholds are also likely to be affected by the cost incurred in moving between patches, i.e. the adults respond to cues that determine their fitness mainly in terms of the survival of the early stages of their offspring. An additional benefit of restricting the number of eggs laid in a patch is that the survival of the later stage larvae and their quality (size) is greatly enhanced. However, the mechanism determining the overall interaction is mainly the risk of egg and early larval cannibalism.

It is likely that aphidophagous predators generally show the same patch dynamics. The elegant studies of Kan (Kan and Sasakawa, 1986; Kan, 1988a,b, 1989) indicate that some syrphids avoid laying eggs in patches that are just about to be vacated by the aphids. Others avoid laying eggs in patches which already contain syrphid larvae (Hemptinne *et al.*, 1993). In the case of lacewings, as in ladybirds, adults are reluctant to oviposit in areas contaminated with chemical cues indicating the presence of larvae (Ruzicka, 1994).

Hymenopterous parasitoids can mature on one aphid and therefore would appear to be potentially more likely to regulate aphid abundance. Their effectiveness, however, is reduced by their longer developmental times relative to their hosts and by the action of their hyperparasitoids and predators, which in many cases are less specific than the primary parasites (Hamilton, 1973, 1974; Holler *et al.*, 1993; Mackauer and Volkl, 1993; Figure 9.11). In addition, because of the risk from hyperparasitism, primary parasites are likely to cease ovipositing in a patch of aphids that are already parasitized. High levels of primary parasitism makes the patch attractive to hyperparasites. By continuing to oviposit in patches of aphids where the incidence of parasitism is high, a primary parasite may reduce its potential fitness (Ayal and Green, 1993).

Summarizing, there is good and increasing experimental evidence in support of the theoretical prediction that aphidophagous predators should lay a few eggs early in the development of a patch, which would account for why they are not important in aphid population regulation. Similarly, the ineffectiveness of primary hymenopterous parasitoids might be the effectiveness

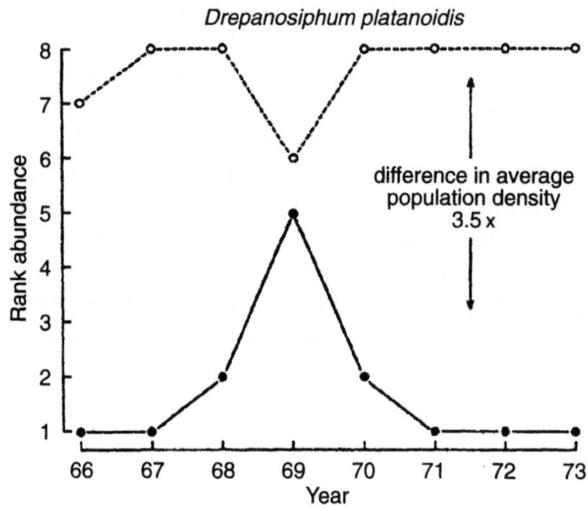

Figure 9.13 The rank abundance of sycamore aphids on two trees in Glasgow, Scotland, 1966–73. (From Wellings *et al.* 1985.)

of their hyperparasitoids, and the selection pressure it generates favouring those primary parasitoids that avoid ovipositing in patches of aphids where the incidence of parasitism makes the patch attractive to hyperparasitoids.

9.6 Intraspecific differences in abundance

Some trees are regularly heavily infested with aphids and others lightly infested, from year to year (Figure 9.13). There has been a tendency to attribute this to either plant resistance or exposure to the elements. For both the sycamore and Turkey-oak aphids, the within-year population dynamics on trees most heavily and lightly infested, is similar (Figures 9.14 and 9.15). However, the elevation of the relationships is significantly higher for the heavily infested trees. How can we account for this? In the case of the Turkey-oak aphid (Figure 9.15), the two trees are similar in terms of the success of the aphids in switching from asexual to sexual reproduction and survival of the overwintering eggs. However, the leaves of the more heavily infested tree start to senesce earlier and are shed some 2 weeks earlier than those of the other tree. The consequence of this is that although the reproductive rate in the first half of the year is the same on both trees, the reproductive rate is very much higher in autumn on the heavily infested tree. The difference in average abundance on the two trees (1.4 times) is similar to the difference in the average fecundity achieved on the two trees (1.5 times); therefore, the difference in the elevations of the relationships between $\log(A_n/S_n)$ and $\log(S_n)$ can be accounted for in terms of the recruitment achieved on these two trees.

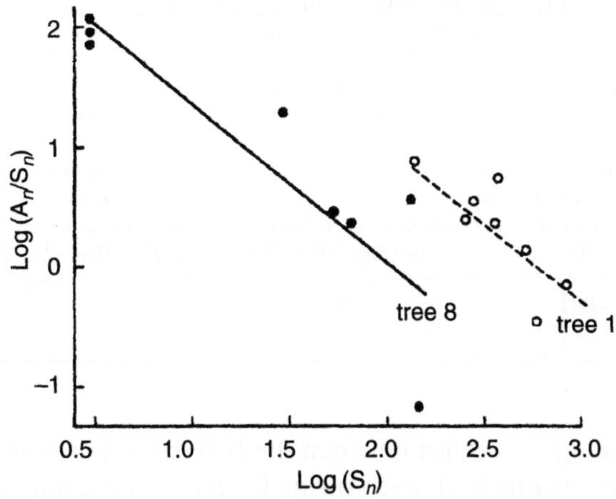

Figure 9.14 The relationships between the logarithm of the average numbers of sycamore aphids present in autumn (A_n) divided by the average numbers present in spring (S_n) and the logarithm of the average number present in spring for two trees (1 + 8) in Glasgow, Scotland, 1966–73.

The difference in average abundance of the sycamore aphid on the most heavily and least heavily infested trees is 3.5-fold (Figure 9.13 and Table 9.1). This cannot be attributed to the reproductive rate achieved on the two trees or the success in switching from asexual to the sexual egg-laying forms.

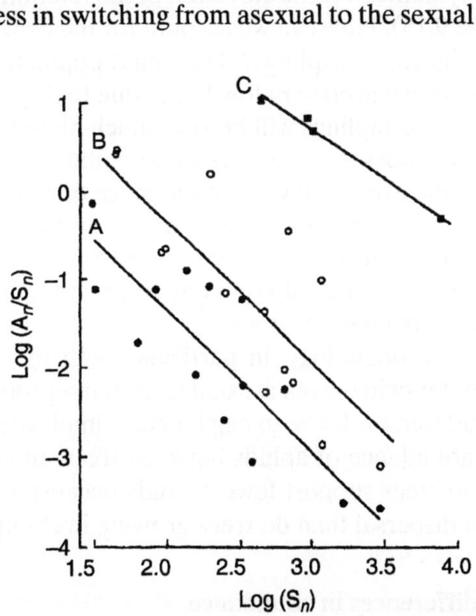

Figure 9.15 The relationship between the logarithm of the average numbers of Turkey-oak aphids present in autumn divided by the average numbers present in spring and the logarithm of the average number present in spring for two adjacent Turkey-oak trees (A and B), 1975–88, and for four caged saplings (C).

Table 9.1 The average abundance (*A*), recruitment (*B*), success in switching from asexual to sexual reproduction in autumn (*C*) and in recruiting aphids (F_{n+1}) to the population in spring relative to the numbers (A_n) present the previous autumn (*D*) for the aphids on the two sycamore trees referred to in Figures 9.13 and 9.14

Tree	*A* Average logarithm of the total numbers recorded each year 1966–73	*B* Average recruitment/ adult,day 1968–73	*C* Average success in switching from asexual to sexual reproduction 1966–73	*D* Average log (F_{n+1}/A_n) 1966–73
1	4.09]	0.45	0.66	−0.35]
8	3.52] ***	0.44	0.67	−0.98] *

The most striking difference between the two trees is in how successful the aphid is in establishing itself each spring from overwintering eggs. Bud burst on the lightly infested tree occurs very late and few of the aphids that hatch survive to colonize the unfurling leaves. Although this is undoubtedly a major factor keeping the average level of abundance low on this tree, it does not account for the significant difference in the elevations of the relationships between $\log(A_n/S_n)$ and $\log(S_n)$ for the two trees (Figure 9.14). This indicates that there is another important factor.

The population trends of Turkey-oak aphids on caged saplings show the same within year dynamics but the elevation of the relationship is very much higher (C in Figure 9.15). How can we account for the generally higher level of abundance on the caged saplings? The aphid populations on the caged saplings will have experienced very low losses due to dispersal and the realized recruitment on the saplings will be very much higher than on the trees sampled in the field. Losses due to dispersal could therefore be a major factor determining the average level of abundance on a tree. Sycamore tree 8 (Table 9.1) is different from all the other trees sampled in being some distance from another sycamore tree and completely surrounded by oaks. Thus losses of adults by dispersal are likely to greatly outnumber gains by recruitment of adults from other trees.

To summarize, tree phenology, in particular both the time of leaf senescence relative to the switch from asexual to sexual reproduction in autumn and the time of bud burst relative to egg hatch, is implicated in determining the difference in abundance of aphids between trees. In addition, it is suggested that isolated trees support fewer aphids because they sustain much higher losses from dispersal than do trees growing in clumps.

9.7 Interspecific differences in abundance

The equilibrium population density of a species is seen as the outcome of the interaction between its rate of increase and the strength of the density-dependent factor acting upon it. Natural enemies do not appear to have a great role in regulating the abundance of deciduous tree-dwelling aphids.

Moreover, there is no evidence that the efficiency of the natural enemies of the various aphid species differ; therefore, differences between aphid species in r_m or, more particularly, realized rate of increase, R, appear to be the most likely cause of differences in abundance between species. The theoretical grounds for this is presented in Dixon and Kindlmann (1990).

Analysis of the empirical data for tree-dwelling aphids presented above has given the following difference equation for the between-year dynamics:

$$\log X_{t+1} = \log X_t + \log R - M \log X_t \qquad (9.3)$$

where X_t and X_{t+1} are the peak numbers in spring of year t and $t + 1$, R is the realized rate of increase and M is the density-dependent factor. After antilogging, equation (9.1) gives:

$$X_{t+1} = Rx_t^{1-M} \qquad (9.4)$$

The equilibrium density (X^*) of which is:

$$X^* = R^{1/M} \qquad (9.5)$$

What factors are likely to affect the degree to which r_m is realized? Dixon *et al.* (1987) argued that one such factor is the probability of finding a host plant [$P(C)$] assuming aphids disperse regularly. Thus, the realized rate of increase, R, which includes the losses incurred in dispersal, is:

$$R = r_m P(C) \qquad (9.6)$$

Given that the probability of finding a host plant [$P(C)$] after D trials is (Dixon *et al.*, 1987)

$$P(C) = 1 - (1 - C)^D \qquad (9.7)$$

then the equilibrium density is given by:

$$X^* = \{r_m(1 - [1 - C^D])\}^{1/M} \qquad (9.8)$$

This indicates that, all other things being equal, the proportional cover of the host plant through its effect on realized r_m can markedly affect the abundance of an aphid.

The indigenous deciduous tree-dwelling aphids of Britain all belong to the same family, the Drepanosiphidae. They are either highly host specific or live on at most two species of a particular genus of tree. The positive relationship between the qualitative estimates of the ranked relative abundances of these aphids and their host trees (Figure 9.16) lends support to the idea that plant abundance is a major factor determining aphid abundance.

9.8 Abundance and global warming

The projected increases in global temperatures are likely to have a positive effect on the developmental and reproductive rates of poikilothermic organisms like aphids. Field cages have been used in the Arctic to raise the

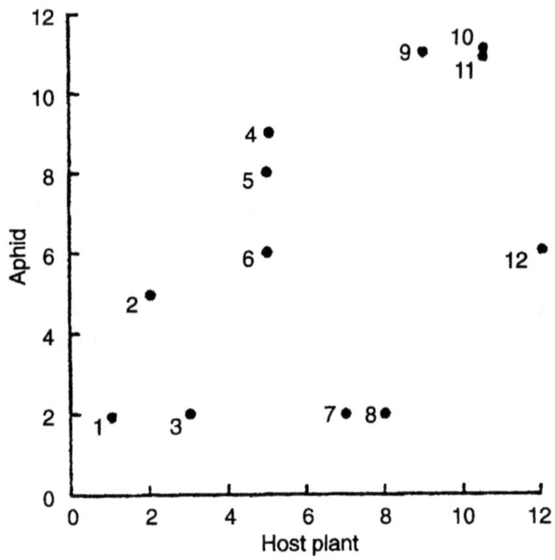

Figure 9.16 The relationship between the rank order of abundance of 12 species of deciduous tree-dwelling aphids and the rank order of abundance of their host plants ($r_s = 0.52$, $P < 0.05$). (1) Oak aphid, *Tuberculoides annulatus*; (2) beech aphid, *Phyllaphis fagi*; (3) birch aphid, *Euceraphis punctipennis*; (4) alder aphid, *Pterocallis alni*; (5) sweet chestnut aphid, *Myzocallis castanicola*; (6) hornbeam aphid, *Myzocallis carpini*; (7) sycamore aphid, *Drepanosiphum platanoidis*; (8) lime aphid, *Eucallipterus tiliae*; (9) field maple aphid, *Drepanosiphum aceris*; (10) Turkey-oak aphid, *Myzocallis boerneri*; (11) holm-oak aphid, *Myzocallis schreiberi*; (12) walnut aphid, *Chromaphis juglandicola*.

ambient temperature experienced by aphids (Strathdee *et al.*, 1995b). The elevated temperatures resulted in higher peak abundances of *Acyrthosiphon svalbardicum* and *Acyrthosiphon brevicorne* feeding on *Dryas actopetala*. However, these studies ignore all the other factors that have been shown to affect aphid abundance. If temperature is the sole determining factor then it is puzzling why the abundance of *Acyrthosiphon brevicorne* outside the cages was highest at the site in Swedish Lapland where the average temperature was the lowest.

In those species that overwinter, not as eggs, but as living aphids (anholocyclically) winter temperatures can have a very marked effect on survival and cause populations to fluctuate in abundance dramatically from year to year. In the green spruce aphid, *Elatobium abietinum*, it has been shown that an increase in the winter temperatures at sites in the UK where temperature does not currently limit the population size will have no effect. The average population density at these sites is regulated by density dependence. However, at sites where mortality caused by winter temperatures is at least occasionally severe, there is still capacity for the mean population size to increase. The increase will probably be small, as many sites have begun to reach a plateau of density with respect to the effect of temperature.

The absence of cold winters could remove the violent oscillations seen in density between years at these sites, because warmer sites generally seem to have more stable populations. In addition, and contrary to expectation, populations at warmer sites have a relatively lower mean population density. Therefore, it is unlikely that global warming will result in a greatly increased population density of *Elatobium* in the UK (Thacker, 1995).

Summarizing, global warming is unlikely to result in a general increase in aphid abundance, however, it could result in a change in the distribution of species and the appearance of major pest aphids in countries where they have not previously been recorded.

9.9 Forecasting of pest aphid outbreaks

In the late 1960s and early 1970s concern over the prophyllactic application of pesticides resulted in the recommendation that governments in Europe should establish pest forecasting systems in their respective countries (Taylor, 1973a). Thirty years on, prophyllactic spraying of pesticides is still widespread and disappointingly few forecasting systems are in place. In fact the long-term collection of data necessary for developing accurate forecasting systems is not being encouraged, and many of the suction traps of the Rothamsted Insect Survey are no longer operational. Therefore, although it is possible to develop accurate forecasting systems it is currently not politically or economically desirable. However, if the cost of pesticide reflected not only the cost of production but the damage they do to the environment and human health then attitudes would change. Below are examples of forecasting systems and one case that is currently being applied.

9.9.1 Green spruce aphid

Outbreaks of the spruce aphid (*Elatobium abietinum*) are associated with mild winters (Bejer-Petersen, 1962; Carter, 1972; Powell and Parry, 1976). When temperatures fall to –7°C or below, ice forming inside the needles of the host seeds ice crystals in the attached aphid and kills it (Powell, 1974). However, the waxy bloom of this aphid that serves as an anti-wetting layer may hold off ice crystals and, in preventing them from contacting the aphid's body, delay fatal inoculative freezing (Bevan and Carter, 1980). Although mortality is catastrophic as a result of freezing, the aphid also steadily decreases in numbers once the average weekly temperature falls below 4°C (Figure 9.17a). In other species of aphid this pre-freeze mortality and reduced longevity and fecundity in the survivors (Knight *et al.*, 1986; Bale *et al.*, 1988) is associated with damage to the aphid's symbionts (Parrish and Bale, 1991). Thus, in springs following mild winters there are often large numbers of overwintered aphids that have survived on spruce. Just prior to bud burst the aphids multiply rapidly and, if numerous to begin with, will achieve outbreak levels (Figure 9.17b) and cause defoliation.

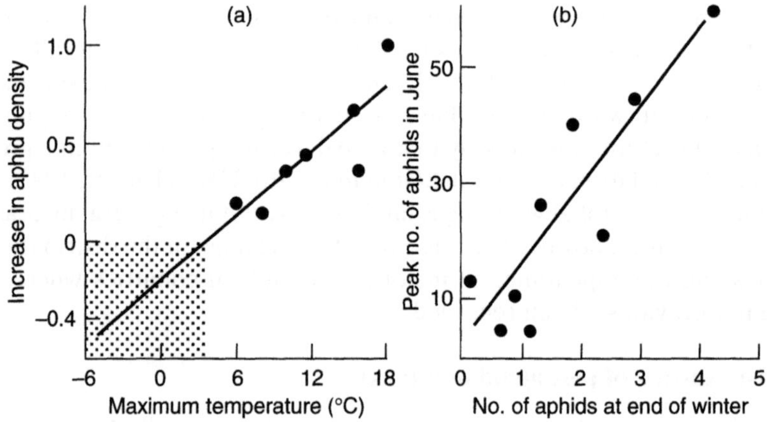

Figure 9.17 Factors determining the abundance of the green spruce aphid. (a) The increase in density per week, measured as the logarithm of the number of aphids per gram dry weight of needles, in relation to the maximum temperature experienced each week. (b) Peak number of aphids per 100 needles in June in relation to the number of aphids per 100 needles present at the end of winter. Stippling, negative rates of increase. (After Powell and Parry, 1976.)

9.9.2 Cereal aphids

The number of cereal aphids (*Sitobion avenae*) flying in East Anglia, UK, in early summer is also correlated with the severity of the previous winter (Figure 9.18). It is likely that the aphid survives well and possibly even increases in abundance in mild winters (Figure 9.19). Further increase in numbers in spring give rise to numerous alatae, many of which colonize cereal crops. Winter weather also affects the growth of winter wheat, the major cereal crop, so that the number of day-degrees remaining at the beginning of summer before wheat matures, and becomes unsuitable for aphids, varies from year to year. Therefore the abundance of aphids in summer is not mainly due, as previously thought, to the number of aphids that successfully overwinter (Carter *et al.*, 1982; Dewar and Carter, 1984) but to the time remaining before wheat matures. In the county of Norfolk, UK, aphids on wheat are more likely to reach outbreak levels following severe than mild winters (Figure 9.20). This negative relationship with winter day-degrees is greatly improved (27–62% of the variation accounted for) by including the accumulated winter temperatures (day-degrees above 1°C) from October to March for the four preceding years (Maudsley, 1993), i.e. severe winters delay the development of wheat and increase the likelihood of an aphid outbreak, but only if several previous winters have been mild, which probably results in a large proportion of the aphids overwintering anholocyclically.

This long-term forecast can be used to identify high-risk years, which need to be confirmed by means of short-term forecasts. Systems that use

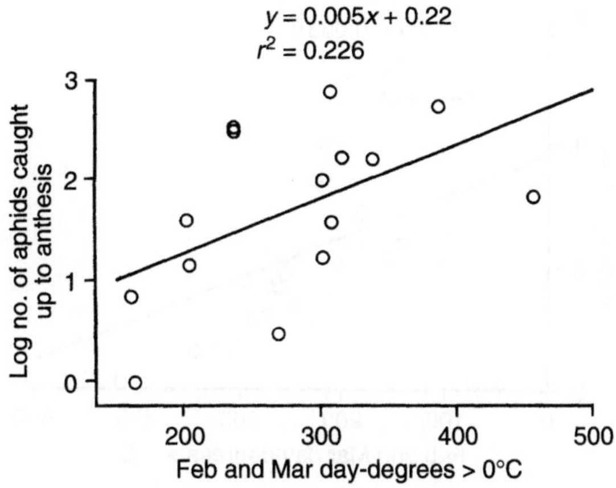

Figure 9.18 The number of alates of *Sitobion avenae* caught in the Broom's Barn suction trap before the end of flowering of wheat in the field, from 1976 to 1983 in relation to the number of day-degrees above 0°C in the period February to March of the previous winter.

field estimates of the numbers of aphids present on wheat in early summer along with their maximum potential rate of increase tend to overestimate peak numbers, because the aphids rarely achieve maximum rates of increase. The accuracy can be increased by taking not one field census but several and using these to obtain an estimate of the realized rate of increase. This can be used to obtain a more accurate prediction of the peak numbers (Entwistle and Dixon, 1986, 1987).

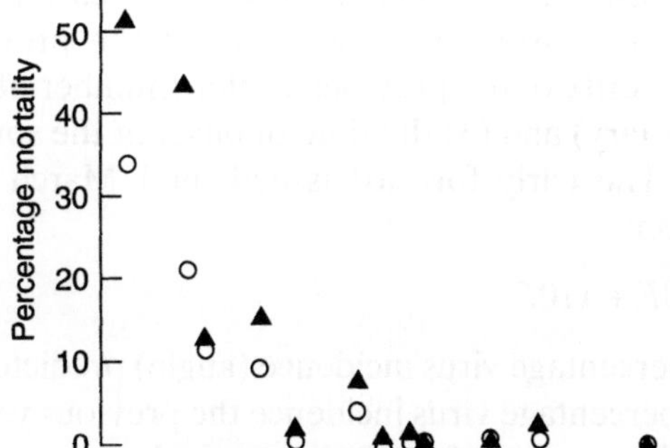

Figure 9.19 The percentage mortality over 4 days of two cereal aphids (filled triangles, *Metopolophium dirhodum*; open circles, *Sitobion avenae*) in relation to the minimum temperature experienced during each period. (After Williams, 1980.)

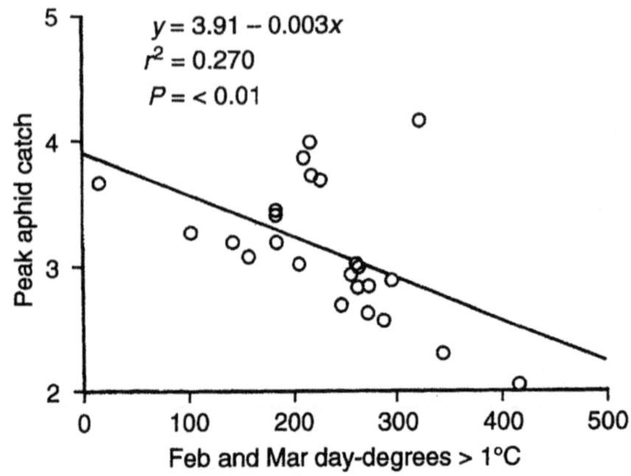

Figure 9.20 The relationship between peak catch of *Sitobion avenae* and day-degrees above 1°C in February and March, 1968–92.

9.9.3 Green peach aphid and sugar beet yellows

The green peach aphid (*Myzus persicae*) is an important vector of beet yellows viruses and is mainly anholocyclic in England. The incidence of beet yellows virus in sugar beet is closely associated with (1) the number of days in the first 3 months of the year when the temperature falls below 0°C, (2) the mean weekly temperature in April, and (3) the numbers of *Myzus persicae* flying in May and June (Figure 9.21). The temperature in late winter determines how many aphids survive and the rate of increase of the survivors is determined by the temperature in April. When the aphid is abundant in spring sugar beet crops are at risk (Watson *et al.*, 1975). This forecast has been revised and extended (Harrington *et al.*, 1989). The equation used to forecast the incidence of virus in the eastern counties of the UK includes (1) a measure of virus infection the previous year, (2) a measure of the severity of the previous winter (number of ground frosts in January and February) and (3) the time of onset of the spring migration of *Myzus persicae*. The early forecast issued on 1 March is based on the following equation:

$$V_E = 0.2V_p - 67.9F + 110.7 \qquad (9.9)$$

where V_E is the percentage virus incidence (angle) predicted for the eastern region, V_p is the percentage virus incidence the previous year, and $F = \log(n + 1)$, where n is the number of ground frosts in January and February.

This relationship accounts for 63% of the variance in the incidence of virus from year to year. A later forecast based on:

$$V_E = 0.37V_p - 25.7F + 0.0092A^2 + 303 \qquad (9.10)$$

where A is the Julian date of the first suction-trap catch of *Myzus persicae*,

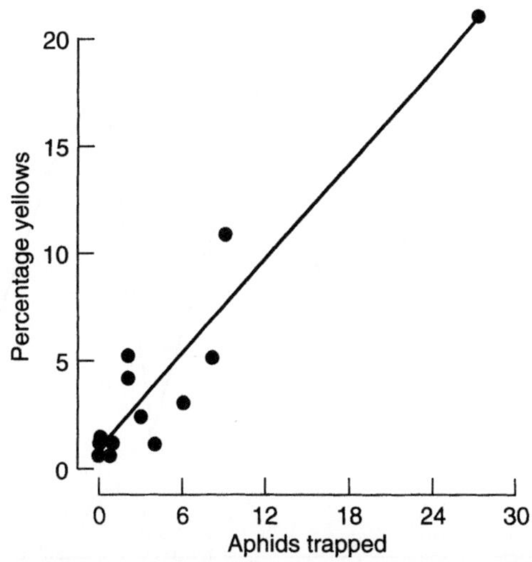

Figure 9.21 Percentage of sugar-beet plants with yellows in relation to the number of *Myzus persicae* caught on sticky traps in May and June. (After Heathcote, 1974.)

accounts for 87% of the variance but is unlikely to be made before most sugar beet has been sown, which rules out the possibility of varying the seed treatment with pesticides from year to year, i.e. winter and spring temperatures through their effect on the abundance and timing of the spring flight of *Myzus persicae* greatly affect the year to year incidence of yellow viruses in sugar beet.

9.9.4 Black bean aphid

This aphid can inflict severe losses on broad bean crops of up to 46% (Way and Heathcote, 1966). Initially in the UK the number of overwintering eggs on spindle, the winter host of this aphid, was used to forecast subsequent infestation of beans (Way and Banks, 1968). The reliability of this forecast was considerably improved by taking into consideration the suction-trap catches of the numbers of *Aphis fabae* migrating to spindle the previous autumn and those migrating from spindle to beans in spring, and the peak number of aphids on spindle in spring (Way *et al.*, 1981). The shorter the interval between the estimate of population size and time of crop infestation the more accurate the forecast, e.g. the correlation with the number migrating to spindle the previous autumn accounts for 28% of the variance, whereas the number migrating in spring accounts for 64%. If the distribution and local abundance of spindle bushes are taken into consideration then the accuracy of the forecast can be further increased (Way and Cammell, 1982). An earlier study in Germany by Behrendt (1966, 1969) also

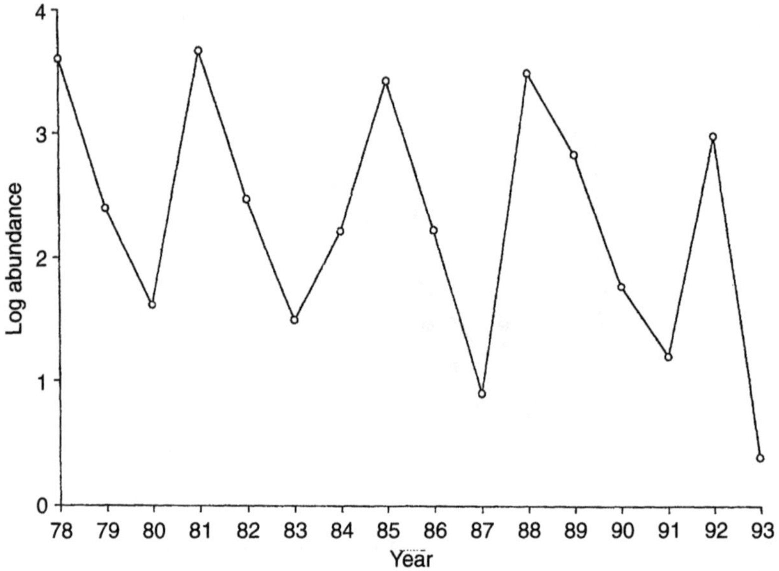

Figure 9.22 The observed time series of *Aphis fabae* at Rostock, 1978–93, showing apparent cycling with a 3–4 year period.

indicated that egg counts were useful for forecasting, as the greater the number hatching in spring the greater the number leaving bean crops in summer.

The plant protection service based in Rostock, Germany, has been collecting and identifying aphids caught in yellow water traps over a considerable number of years. Field counts of egg density were also recorded. From this time-series (Figure 9.22) it appears, rather surprisingly, that the numbers of *Aphis fabae* migrating from bean crops are cyclical with a period of 3–4 years. Of note is that egg densities are *not* correlated with aphid numbers in the preceding summer, nor can they be used to predict the population size in the following summer, as they could by Way *et al.* (1981) in the UK and by Behrendt (1966a,b, 1969, 1971) for another region of Germany.

The initial regression analysis of this data set (1978–90) indicated that the pattern is driven by May temperatures, perhaps acting on population growth rates. Subsequently, upon the addition of an extra 2 years of data (1991 and 1992) this assessment changed, with the densities of the aphid in previous years (i.e. the cyclical behaviour) added to May temperatures as important factors. Finally, March rainfall replaced the other factors when the 1993 data was taken into account. This continually changing inclusion of different variables emphasizes the need for long runs of data if we are to develop efficient forecasting systems; years outside the previous experience of the forecasting system may undermine its otherwise accurate predictions. Multiple regression models contain an inherent flaw in that each variable is included each year; no new variables can be considered; and if, for example all the host plants die in a drought and rainfall has not hitherto been considered important, the forecast fails. Similarly, the role of, for example, density dependence might be significant if weather is 'normal', but if conditions are particularly harsh or clement the effect of previous densities might be overridden, despite being included in the current model, which is based on 'normal' years. Thus, even if a forecasting system based on weather is an accurate one, an unusual year will still be damaging to the system; also, if climate change occurs, the frequency with which 'unusual' years occurs will increase. If climate continues to change it may be impossible to arrive at a precise forecasting system using weather; there is no substitute in terms of accuracy for direct estimates of population size. However, these are costly and the alternative of prophyllaxis regardless of damage predictions may be preferable (at least in monetary terms). If prophyllaxis occurs and upon successful validation of a weather-based forecasting system is stopped, we must be certain that there is no density-dependent element in the year to year dynamics or else a population held at continually low levels could escape suddenly from its confinement to outbreak levels even in a year with 'normal' weather, when only a moderate abundance of the pest is expected. In summary, the cycles in *Aphis fabae* abundance in the Rostock area appear to be weather-driven; we should not assume, as Haukioja (1991) did, that density cycles cannot be caused by weather; a lot of data is needed to ensure that the true driving factors have been defined if we are to accurately forecast abundance, i.e. apparently simple systems can be anything but (Thacker *et al.*, 1997).

9.10 Spatial dynamics

Species like the elderberry aphid (*Aphis sambuci*) have an unstable spatial aerial pattern of distribution, apparently occupying a dramatically varying fraction of the potential distribution in any one year (Figure 9.23). Elderberry, the winter host of this holocyclic species, is ubiquitous throughout Britain and so too are the summer hosts, various species of dock (*Rumex* spp.). In contrast the whole of Britain is almost uniformly occupied by species such as *Rhopalosiphum insertum* and *Cryptomyzus galeopsidis*, but *Anoecia corni*, not surprisingly, is restricted to parts of England and Wales where its host, dogwood (*Cornus sanguinea*), occurs and to south-west Scotland where dogwood is grown as a garden shrub (Figure 9.24). Although these maps do not provide direct evidence of movement between areas, they have been offered, along with others, in support of the contention that these changing distributions involve movement as well as mortality (Taylor and Taylor, 1979; Taylor, 1979).

It is now well established that the spatial variability in population abundance and average population density of a species are related in both space

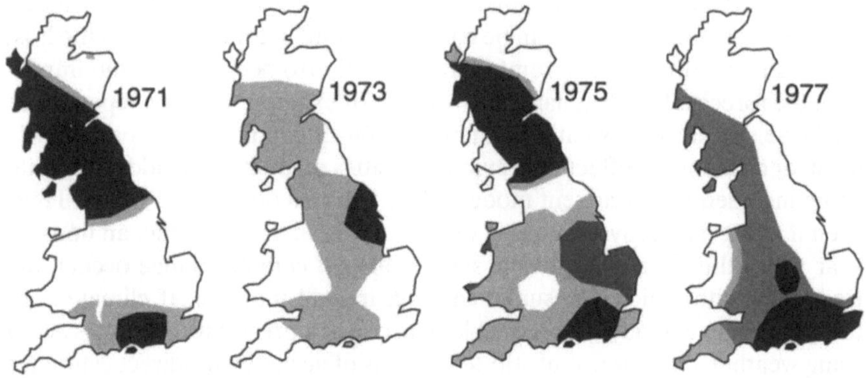

Figure 9.23 Maps showing changes in the aerial distribution of the elder aphid, *Aphis sambuci*. (After Taylor and Taylor, 1979.)

and time. For a wide range of plants and animals the relationship between variance (V) and mean (M) conforms to a simple power law:

$$V = aM^b \quad (9.11)$$

where a and b are constants (Figure 9.25) (Taylor, 1961). The slope of the relationship (b) generally falls within the range 1–2, is a species-specific characteristic and is thought to reflect the balance (Δ) between the opposing

Figure 9.24 Maps showing aerial distribution of *Rhopalosiphum insertum* (a), *Cryptomyzus galeopsidis* (b) and *Anoecia corni* (c). (After Taylor, 1973b.)

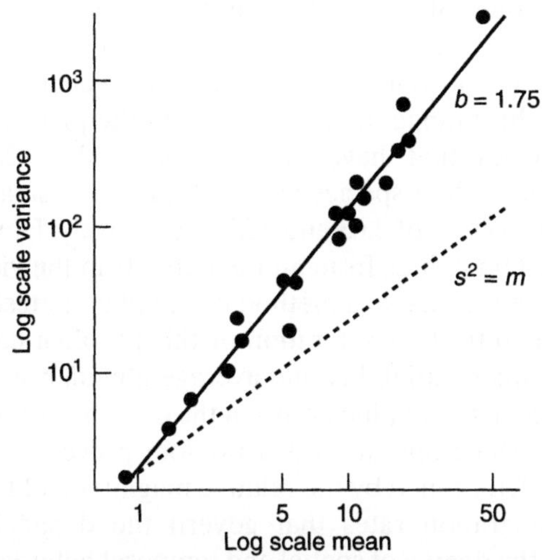

Figure 9.25 Logarithm of the variance of the distribution of sycamore aphids between leaves in relation to the logarithm of the mean number per leaf for each sample taken at weekly intervals throughout a year.

behavioural tendencies to aggregate within ($H\rho^q$) and migrate from ($G\rho^p$), centres of population density:

$$\Delta = G\rho^p - H\rho^q \qquad (9.12)$$

where the exponents p and q are rate constants for density-dependent migration and aggregation, respectively, G and H are constants of proportion and ρ is population density (Taylor and Taylor, 1977, 1978). Simulation studies based on this model yield censuses that satisfy $V = aM^b$. That migration between patches is rarely random is deduced from the slope (b) of the relationship between the variance and the sample mean, which is greater than 1, and non-random dispersal is also proposed on evolutionary grounds (Taylor *et al.*, 1978, 1983).

This interpretation has been challenged by Hanski (1980, 1982) on the grounds that it is unreasonable to expect insects, especially aphids, to move between population sites in order to maximize individual fitness, particularly in an area as large as Britain. Hanski also re-analysed the results of Taylor *et al.* (1980) to indicate that a value of b greater than unity does not necessarily imply density-dependent movement. Like Taylor and Taylor (1977), Hanski also argues that the value of b may reflect the ecology of species, but in a different sense. A specialist is only present in some habitats, even when very abundant, whereas a generalist possibly increases faster in the poorer habitats because the preferred habitats are fully occupied. Thus the slopes of the relationships between variance and sample mean for different habitats for specialists should have values of $b > 2$ and for generalists values of $b < 2$. This is not supported by the values of b for 97 species of aphids cited by Taylor *et al.* (1980). Specialist species of aphids, that live mainly on only one species of plant, and generalist species are very similar in both their range in values of b and the proportion of species in the two categories that have values greater than 2. The two most polyphagous (generalist) species cited, *Myzus persicae* and *Macrosiphum euphorbiae*, have values of 2.63 and 2.25, respectively. However, as suction-trap catches contain aphids from all the habitats in the vicinity of the trap, they may provide too coarse a method of sampling to test Hanski's idea.

A further theoretical consideration of this problem has shown that the relationship between variability and average abundance could be a consequence of chance demographic events in the dynamics of population growth and decline, in which migration is a random process. The form of these relationships is determined by the relative magnitude of birth, death, immigration and emigration rates that govern the dynamics of population change, and by the degree of spatial and temporal heterogeneity. If density-dependent factors are of limited significance, as is suggested for species that are r-strategists, chance variations in birth and death rates alone are sufficient to account for the relationship between variance and mean abundance and for the slope lying on average between 1 and 2. For K-selected species, in which strong density-dependent factors are assumed to be acting, a high degree of environmental heterogeneity additionally ensures that such relationships remain approximately linear. Therefore there would appear to be no need to invoke any complex behavioural mechanisms to explain the observed patterns (Anderson *et al.*, 1982). However, Taylor *et al.* (1983) justifiably argue that the simulations of Anderson *et al.* fit field results less well than simulations based on the behavioural Δ model.

Support for the idea that the patchy distribution of aphids reflects more the environmental heterogeneity rather than metapopulation dynamics comes from a study of the mechanisms that govern the aggregated distribution of sycamore aphids between leaves (Figure 9.25). Periodically aphids cease feeding, move off and colonize another leaf at random. The length of time for which an aphid stays on a leaf depends on its temperature, exposure to the sun, frequency with which its underside is brushed by other leaves (Figure 9.26), its nutritive status and the presence of other aphids (Dixon, 1970a; Dixon and McKay, 1970; Dixon and Mercer, 1983). Thus kinetic movements in response to the variable quality of the environment result in this aphid aggregating and reproducing mainly on the relatively few leaves that have a favourable combination of microclimate, degree of infestation and nutritive status.

A calm autumn results in a dramatic increase in the abundance of sycamore aphids, which is reflected in suction-trap catches (Dixon, 1979). In addition, the numbers of sycamore aphids on trees in the Glasgow area and those caught in suction traps throughout Britain are correlated with one

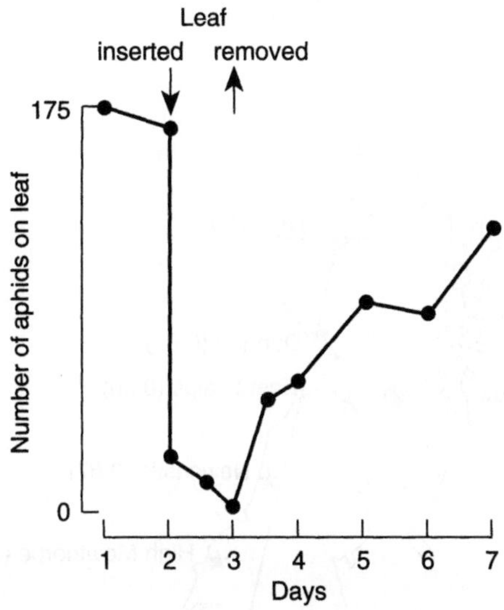

Figure 9.26 The effect on the number of aphids on a leaf of inserting another leaf immediately below it.

another, which indicates that good and bad years for this aphid are general throughout Britain. However, the further apart the traps the weaker the correlation, possibly reflecting regional differences in weather from year to year (Figure 9.27). Thus it is possible, as with the distribution of sycamore aphids between leaves on a tree, that the dramatically changing spatial and temporal mosaic of aerial population densities mainly just reflects a similar changing mosaic of habitat quality at ground level. This may not be true but only empirical studies can settle the issue.

Summarizing, some aphids can become very abundant. The analyses of long runs of suction-trap catches indicate that aerial populations of aphids are regulated. Intensive studies of aphids living on trees indicate that it is mainly intraspecific competition for resources through its direct affect on the incidence of migration, body size and reproduction that regulates aphid abundance. Although natural enemies reduce the rate of increase of aphids, their effectiveness in regulating aphid abundance is constrained by their natural enemies in the case of parasitoids and by the need to forage optimally in order to maximize their fitness in the case of parasitoids and particularly aphid-specific predators. That some trees are regularly heavily infested with aphids and others of the same species only lightly infested is largely due to differences in tree phenology and degree of isolation from other trees of the same species. The difference in the average abundance of the different species of tree-dwelling aphids can largely be accounted for in terms of the abundance of the host trees. Although aphids generally do better at higher temperatures it is unlikely that global warming will result

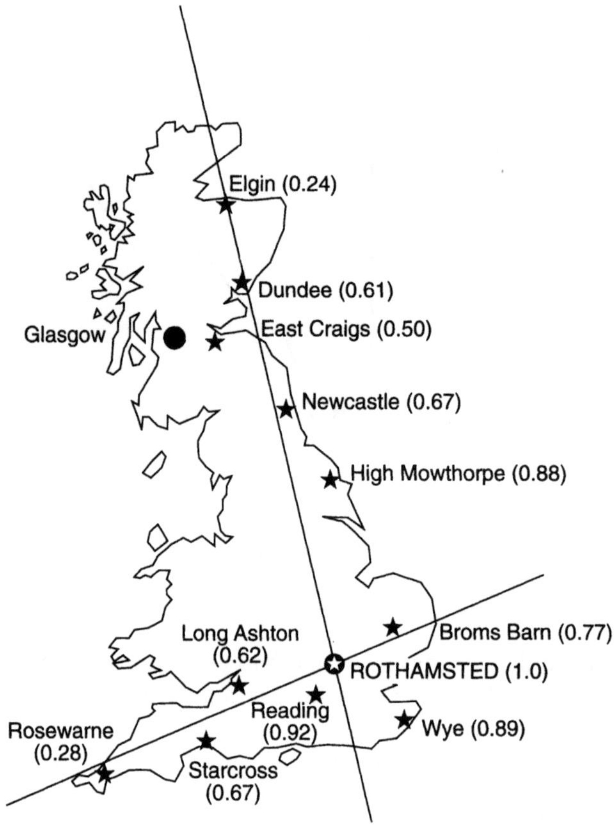

Figure 9.27 The location of some of the Rothamsted Insect Survey suction traps in the UK and the correlation coefficients between the number of sycamore aphids caught by these traps and the Rothamsted trap.

in a general increase in aphid abundance; however, it could result in a change in the distribution of some species. It is possible to develop accurate forecasting systems for pest aphids, but with a few notable exceptions, this endeavour is currently receiving little economic or political encouragement. The dramatic spatial and temporal changes in the aerial abundance of aphids over wide areas, so strikingly portrayed in the maps published by Taylor and his colleagues, are very suggestive of large-scale population movements, i.e. metapopulation dynamics. However, only if these changes in spatial distribution can be related to what is occurring on plants, and then subjected to experimental analysis, can we determine the extent to which movement shapes the dynamics of aphid populations.

10 Community structure and species diversity

Although there are many studies of aphids at the individual and population levels there have been few studies of the interactions between natural enemy, aphid and plant populations or even interactions between aphid species, especially in the field. This is partly due to the difficulty of identifying aphids, especially the highly polyphenic species, and the tendency to regard plants solely as a source of food. However, as keys for the identification of aphids become available and the role of plants in aphid biology is more fully appreciated (Dixon, 1977), there is likely to be an increasing interest in the role of aphids in community ecology. It is an aspect of ecology that is poorly understood with great uncertainty whether there are rules governing the structure and functioning of communities.

10.1 Mutualism between plants and aphids

Large quantities of honeydew are produced by aphids, which in some species contains a high proportion of the trisaccharide melezitose. This led Owen and Wiegert in a series of papers (Owen, 1978, 1980a; Owen and Wiegert, 1976, 1981) to propose that plants benefit from being eaten and, in particular, plants release surplus sugars by enlisting the help of aphids. The sugar in the form of honeydew is utilized by free-living nitrogen fixing bacteria in the soil, which increase in number beneath aphid-infested plants and make more nitrogen available to these plants. Melezitose, or a particular mixture of sugars in honeydew, are thought to have an optimal affect on nitrogen fixation. Thus these authors view aphids as a necessary 'part' of a plant, releasing surplus sugars that promote a better supply of nitrogen.

The addition of the four sugars commonly found in honeydew – fructose, glucose, melezitose and sucrose – to soil at rates equivalent to those recorded beneath trees (Llewellyn, 1972) causes an increase in the abundance of microbial populations, at least in woodland soils (Dighton, 1978a,b), and fructose is more effective at promoting nitrogen fixation than melezitose (Petelle, 1980). However, as single sugars rather than a mixture were used, Petelle's results do not refute Owen's hypothesis (Owen, 1980b). A more rigorous test of the mutualism hypothesis, in which alder aphids

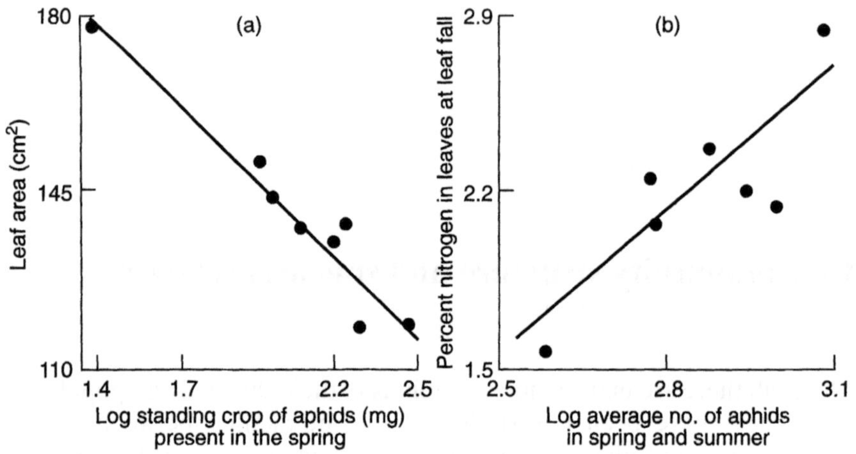

Figure 10.1 The average area of sycamore leaves in relation to the standing crop of sycamore aphids on the leaves in spring (a), and the percentage of nitrogen in the leaves at leaf fall in relation to the average number of sycamore aphids on the leaves in spring and summer (b.)

(*Pterocallis alni*) were removed from a plot of red alder (*Alnus rubra*) by spraying with malathion, has revealed that aphid infestation results in a decrease in ammonification and nitrification in the soil and a decrease in above ground primary production (Grier and Vogt, 1990). Although their experiment does not rule out the possibility of melezitose stimulating nitrogen fixation by free-living nitrogen-fixing bacteria, nevertheless, contrary to the prediction of the hypothesis, nitrogen availability in the soil is markedly reduced by large quantities of aphid honeydew. Whatever the positive effect, if any, of honeydew in promoting the long-term availability of nitrogen for plant growth, however, one cannot ignore the negative effect aphids have on plant growth and seeding.

10.2 Aphids and plant growth

As relatively few aphids gall or distort leaves their presence often goes unnoticed. However, they can become very abundant (p. 189), e.g. 116 000 leaves of a 20 m sycamore tree may be infested with 2.25×10^6 aphids, equivalent in mass to a large rabbit. It is not surprising, therefore, that infested trees achieve less growth than uninfested trees.

Trees vary in their response to aphid infestation. Although aphids do not affect the number of leaves borne by lime, oak or sycamore, sycamore produces smaller leaves, which contain more nitrogen, when heavily infested in spring (Figure 10.1). However, the leaf area equivalent to the energy removed from a leaf by sycamore aphids only accounts for a small proportion of the observed diminution in leaf area. If the drain imposed is

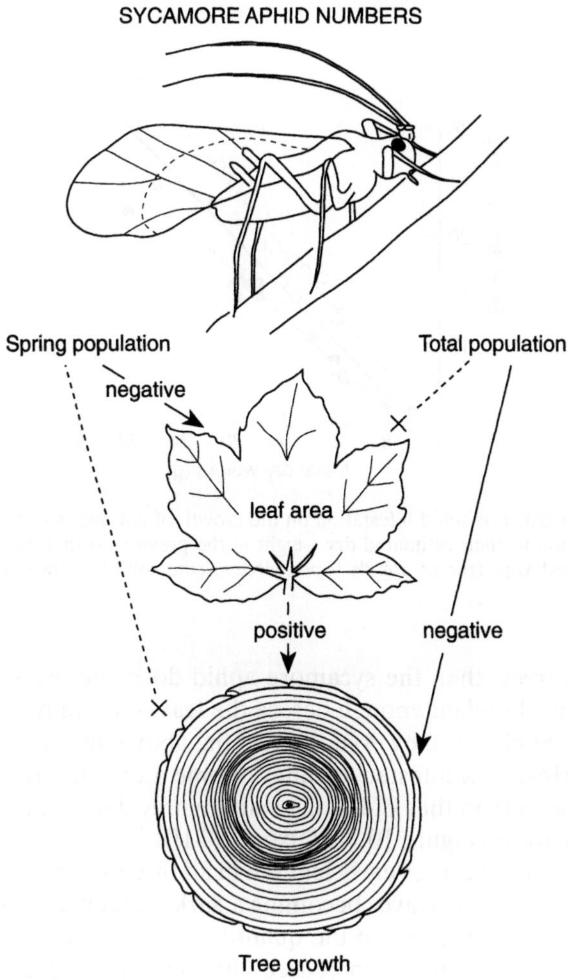

Figure 10.2 Diagrammatic representation of the effect of sycamore aphid numbers on leaf area and width of the annual rings laid down by sycamore.

expressed in terms of nitrogen rather than energy then aphids again remove far less nitrogen than expected from the reduced size of the leaves. This implies that the effect aphids have on plant growth is not just a direct consequence of energy and nutrient drain. Other factors, e.g. the saliva aphids inject into plants (p. 13), might also adversely affect plant growth.

The width of the annual rings of sycamore is positively correlated with the average size of the leaves, and negatively with the numbers of aphids on the tree throughout a year (Figure 10.2). In the absence of aphids some sycamore trees could produce as much as 280% more stem wood (Dixon, 1971e). Lime and oak aphids hatch later, relative to the time of bud burst

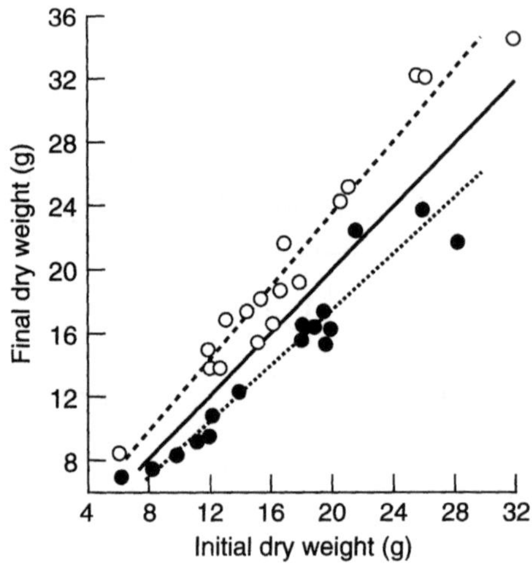

Figure 10.3 The effect of aphid infestation on the growth of oak saplings: their dry weight in autumn in relation to their estimated dry weight in the previous spring for saplings infested (solid circles) and kept free of aphids (open circles) (the solid line indicates no change in weight).

of their host trees, than the sycamore aphid does and, as a consequence, rarely become abundant enough before the leaves are fully grown to affect their growth. Similarly, they do not affect the above-ground growth in girth and length. However, infested saplings of lime and oak often weigh less at the end of a year than they did at the start, mainly due to a reduction in the mass of their roots (Figure 10.3) (Dixon, 1971f).

Aphid infestation causes early leaf fall in all three species and, in oak and sycamore, results in the leaves becoming a darker green. In oak this is a consequence of a 25% increase in the quantity of both chlorophylls A and B. Associated with this is an increased dry matter production, which in sycamore can be 1.7 times greater in infested than in uninfested leaves. Following years of heavy aphid infestation lime and sycamore break their buds later than usual, and in the case of lime the leaves are smaller and a darker green, and have a net production 1.6 times greater than the leaves of previously uninfested saplings (Dixon, 1971e,f).

Although saplings and mature trees may differ in their reaction to aphid infestation (Llewellyn, 1975), there is no doubt that aphids have a pronounced adverse effect on the growth of trees and could be a major factor in determining the outcome of intra- and interspecific competition between trees, and the establishment of seedings as well, as they often become heavily infested with aphids that fall from the parent trees. Some trees can compensate in part at least for the nutrient drain imposed by aphids and it remains to be shown that aphids affect the genetic fitness of mature deciduous trees. However, high numbers of the spruce aphid (*Elatobium*

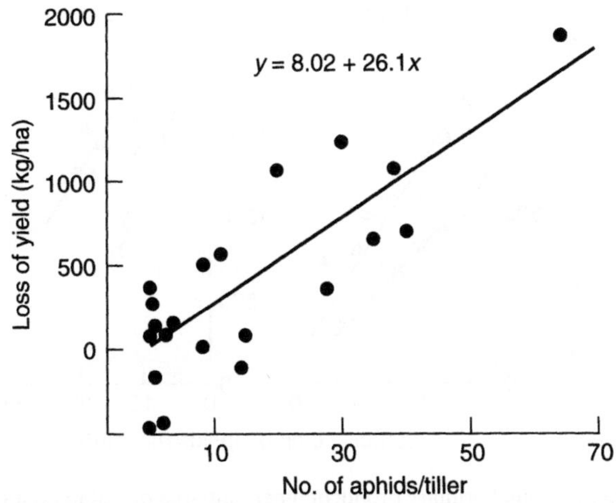

Figure 10.4 Loss of yield of wheat in relation to the peak number of *Sitobion avenae* per tiller. (After Vereijken. 1979.)

abietinum) can cause the defoliation of spruce, which, if recurrent, can cause the death of the tree.

10.3 Aphids and the seeding of plants

The number and quality of seeds produced by plants have been used as a measure of genetic fitness. As many trees do not fruit and seed equally every year but at irregular intervals, they are not convenient for studying the effect of aphids on seed production. For example, the seed production (weight) in sycamore is not correlated with either aphid numbers in the current or previous years, but is strikingly dominated by the very infrequent good year when seed production can be as much as 4 times the long-term average. However, annual, biennial and short-lived perennial plants are more convenient for measuring both the total seed production and any beneficial effect of the proposed honeydew-increased nitrogen fixation.

Aphids affect the yield and quality of the seed of several crop plants. For example, *Aphis fabae* reduces the number of seeds produced by bean plants (*Vicea faba*) by up to 86% and the average weight of a bean by 45% (Banks and Macaulay, 1967). Similarly, aphids reduce the yield of cereals (Figure 10.4) (Vereijken, 1979). In natural habitats aphids also reduce the fitness of their host plants. In a study of a saltmarsh in the UK, Foster (1984) showed that in peak aphid years virtually no seed is produced by most sea lavender plants. This is due to the aphid (*Staticobium staticis*) causing the plants to abort their flowers. Similarly, heavy aphid infestation of the herb, *Senecio sylvaticus*, growing in pine forests in the Netherlands, causes a marked reduction in the number and quality of the seed it produces (Ernst, 1987).

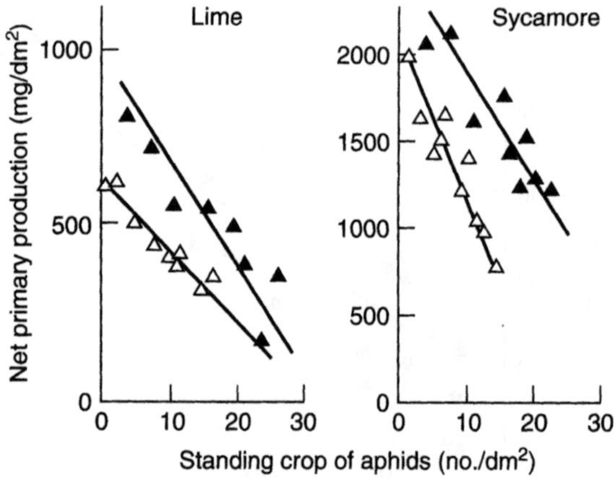

Figure 10.5 The net annual primary production per unit area of leaf in relation to the average number of aphids on the leaves during the year for lime and sycamore saplings watered with either full- (solid triangles) or half-strength (open triangles) nutrient solutions. (After White, 1970.)

Thus aphids do affect the reproductive potential of plants, even when, as in the case of cereals, they are given high levels of nitrogen fertilizer. As in a nutrient-rich environment aphid-infested plants still grow more slowly than uninfested plants, they are likely to suffer even more markedly from aphid infestation in nutrient-limited environments (Figure 10.5).

As well as imposing a severe nutrient drain on plants, aphids transmit viruses, which can also adversely affect the fitness of plants. However, some genera and groups of plants are notable for not having aphid-transmitted viruses, e.g. Coniferae and Rosaceous trees and shrubs such as *Rosa*, *Malus* and *Pyrus*. During their long life trees must be exposed to millions of aphids and therefore subject to strong selection for tolerance or resistance to aphid-transmitted viruses (Eastop, 1991). Depending on the type of virus, stylet borne or circulative, there are costs and benefits of virus transmission for aphids. In transmitting viruses to non-host plants growing in the immediate vicinity of their specific host plant, it has been suggested that aphids could indirectly improve the competitive status of their host plant (Eastop, 1991). However the mechanism by which this is likely to have evolved is unclear.

In summary, there is no evidence to support the contention that plants infested with aphids are fitter than those that are free of aphids. Quite the contrary, aphids generally reduce the fitness of their host plants.

10.4 Ants and plant fitness

In certain years only those plants of ragwort (*Senecio jacobaea*) that are protected by ants (*Lasius alienus* and *Lasius niger*) escape defoliation by larvae of the cinnabar moth (*Tyria jacobaeae*) and produce seed. These aphid-infested and ant-attended plants have a lower pyrrolizidine alkaloid content than uninfested plants. These observations led to the suggestion that the variation in the concentration of the alkaloid between plants of ragwort is maintained by the above tritrophic interaction. In years when the cinnabar moth is abundant plants that have low levels of the alkaloid and are infested with aphids and ant-attended have a higher fitness because they are the only seed-producing individuals. In contrast, in years when the cinnabar moth is uncommon, these plants might have a lower fitness because of reduced seed output due to feeding by the aphids (Vrieling *et al.*, 1991), i.e. by using aphids to attract ants plants may benefit more in terms of their fitness from the negative effect the ants have on other phytophagous insects than it costs to host an enhanced number of aphids. However, it is a dynamic process with the selective pressure imposed by the two phytophagous insects varying from year to year.

An analysis of the incidence of ant attendance of the aphids of the Rocky Mountain region of the USA has revealed that:

1. Some genera of plants host more ant-attended species of aphid than others.
2. Some species of aphid are ant-attended on some of their host plants but not on others and the permissive plants host proportionally more species of ant-attended aphids.
3. Seasonally the attractiveness of a host-specific aphid for ants can vary.

That is, plants may play a critical role in allowing aphids to attract ant associates – the permissive host plant hypothesis of Bristow (1991). The assumption underlying this hypothesis is that the removal of defoliating insects by ants more than offsets the energy and nutrient drain imposed by the aphids. It is thought that the greater attractiveness of aphids feeding on permissive plants is because they produce a higher quality honeydew than those feeding on non-permissive plants. However, other studies have not revealed any evidence for host quality mediating an aphid–ant mutualism (Breton and Addicott, 1992b).

The permissive plant or ant plant mutualism hypothesis has not met with universal acceptance. For example, in south-western Montana there is a negative correlation between the number of ant colonies (*Formica altipetens*) and the abundance of herbaceous plants and sage. It is suggested that the spatial patchiness of the plants is because the ant-attended aphids markedly reduce plant fitness. Similarly, a growth ring analysis of Scots pine most, least and not visited by the ant *Formica aquilonia* in southern Finland, revealed that the trees most frequently visited by ants attending the aphid *Cinara pini* showed the poorest growth (Rosengren and Sundström, 1991).

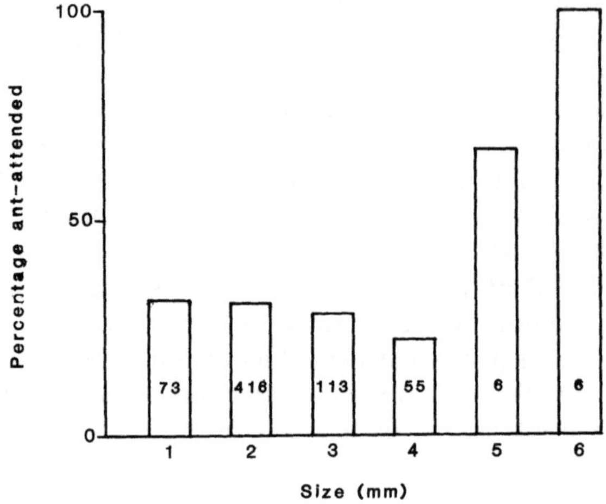

Figure 10.6 Percentage of the species of aphids in the different size classes that are ant-attended.

10.5 Mutualism between aphids and ants

Not all aphids are ant-attended nor do all ants attend aphids, and myrmecophilous species of aphids differ in their dependence on ants. Myrmecophilous species of aphids that live above ground appear to be adapted to ant attendance as they are often conspicuously coloured, gregarious and do not avoid an approaching ant, predator or parasite (Dixon, 1958) and myrmecophilous species generally have poorly developed structural adaptations for defence against natural enemies (Way, 1963). If it is advantageous to be attended by ants then it is surprising that only about a quarter of all aphids are ant-attended (Bristow, 1991).

To account for why so few species of aphids are ant-attended, Bristow (1991) proposed the permissive host hypothesis (p. 227). However, the fact that only 37% of the species of aphids on permissive plants are ant-attended indicates that other factors play a role. What factors are likely to make it advantageous for an aphid to be attractive to ants? As habitats are very heterogenous in terms of their occupancy by ants there are risks for aphids associated with being dependent on ants. The preferred feeding location of an aphid might, however, affect the balance in terms of costs and benefits in favour of ant attendance.

Species of aphids feeding on the branches and trunks of trees need to be large and have a long proboscis in order to feed on the deeply located phloem elements. These species are particularly vulnerable to natural enemies because rapid escape is impeded by the time it takes them to withdraw their stylets. Not surprisingly, therefore, most of the largest species of aphids are ant-attended (Figure 10.6), with some like *Stomaphis*

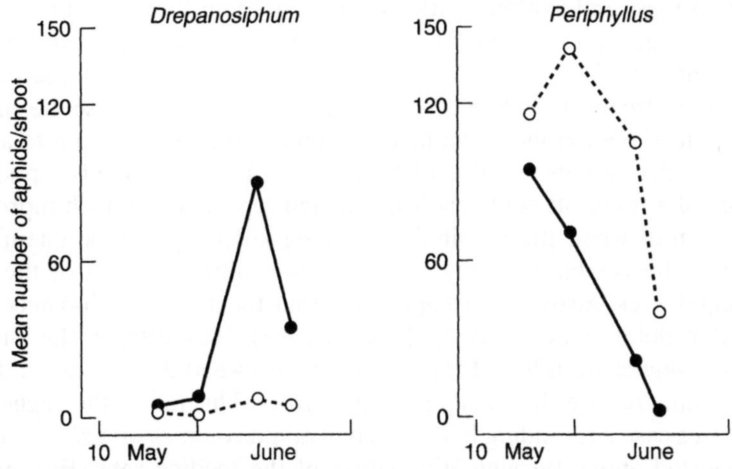

Figure 10.7 Mean numbers of *Drepanosiphum platanoidis* and *Periphyllus testudinaceus* on shoots of sycamore that are ant-attended (open circles) and from which ants are excluded (filled circles). (After Skinner and Whittaker, 1981.)

quercus (p. 65) only found associated with a certain species of ant, *Lasius fuliginosus*. Another potentially high-risk feeding location for aphids is the roots of plants, where they are exposed to attack by many insect ground predators in addition to aphid-specific natural enemies. Of the aphids that live on the roots of plants in Scandinavia a greater proportion (64%) than one would expect by chance are ant-attended ($\chi^2 = 4.4$; $P < 0.05$). Interestingly, of those that are ant-attended, proportionally more (41%) of the small species (3.5 mm or less) live on the roots of plants as opposed to above ground, than for the large species (larger than 3.5 mm, 0%; $\chi^2 = 14.0$; $P < 0.01$), i.e. although feeding location appears to be a major determinant of the likelihood of ant attendance, it is associated with small size on the roots and large size on the branches and trunks of trees.

Ants can have a pronounced effect on the abundance of aphids. For example, numbers of the ant-attended *Periphyllus testudinaceus* on sycamore are higher on branches foraged by the ant *Formica rufa* than on adjacent branches from which they are excluded. However, the other species of aphid on sycamore, *Drepanosiphum platanoidis*, which is not ant-attended, does better in the absence of ants (Figure 10.7) (Skinner and Whittaker, 1981; Whittaker, 1991). Similarly, of the four aphids on fireweed three do better, and one worse, when on plants that are ant-attended (Addicott, 1979).

When ant-attended, *Aphis fabae* increases uptake of phloem sap and so increases production of honeydew. The greater growth and fecundity rates of these aphids are attributed to the improved nutrition resulting from increased feeding (El-Ziady, 1960; Takeda *et al.*, 1982) and to their tendency to disperse less and, consequently, for most of them to aggregate and feed on the youngest parts of their host plant (Banks, 1958; Banks and Nixon, 1958; Banks and Macaulay, 1967). It is possible that in the absence of ants such aphids use more of their resources for dispersal to increase the chance of survival of their clones. The higher proportion that develop into alatae in the absence of ants (Banks, 1958; El-Ziady, 1960) also lends support to the idea of a trade-off between dispersal and reproduction, with reproduction favoured when the possibility of being killed by natural enemies is reduced. The mechanism may be by a 'tranquillizing' effect of the ants reducing the restlessness of the aphid and thus tactile stimulation necessary for alatae determination (p. 112) (Way, 1963). However, in the case of *Formica cinerea* attending *Aphis varians* on fireweed there is no evidence that the ants reduce the numbers of predators. This led to the suggestion that in this case ant tending benefits small aphid colonies *directly*, probably as suggested above through stimulation of the feeding rate (Breton and Addicott, 1992a), but it is unclear why unattended aphids do not feed at the higher rate.

Newly founded colonies of *Aphis varians* and *Aphis helianthi* on fireweed, and of *Pterocomma populifoliae* on aspen, are more likely to persist if they are found and tended by ants within the first week. Exclusion of ants causes such colonies to go extinct faster than ant-attended colonies. This is due to the ants attacking and killing aphid predators like coccinellid beetles (Sanders and Knight, 1968; Addicott, 1979). Similarly, the ant *Lasius niger* affords small colonies of *Aphis fabae* protection against a range of natural enemies (Banks, 1962).

Thus the dynamics of colonies of ant-attended species of aphid are very dependent on the availability of ants (Bradley and Hinks, 1968). Unattended small colonies are more likely to decline than attended colonies. However, in large colonies, the effect is reversed, with ant-attended colonies more likely to decline than unattended colonies, i.e. ant attendance is having a stabilizing affect on aphid abundance (Addicott, 1979). As the relative number of ants tending colonies of *Aphis varians* decreases as aphid numbers increase and ant tending did not reduce the number of predators, it has been suggested that in this case the cause of the density-dependent mutualism is probably a decline in the direct affect of the ants on the aphids (Breton and Addicott, 1992a). However, this does not account for why there are proportionally fewer ants tending large than tending small colonies of aphids. However, as ant attendance can result in a rapid increase in the size of the aphid colonies, as well as a possible increase in the rate of feeding of the individual aphids, the combined effect could be a more rapid deterioration in the quality of the host plant for the aphid and in the quality of honeydew produced by the aphid. It seems likely that this could also account for the decline of the larger colonies and in the relative numbers of ants tending these colonies. In those species of aphids that are only found in

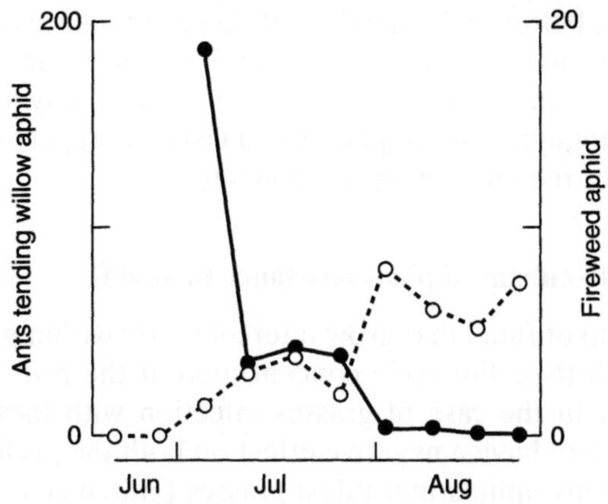

Figure 10.8 Numbers of ants attending colonies of the willow aphid, *Aphis farinosa* (solid circles), and the fireweed aphid, *Aphis varians* (open circles), in relation to time of year. (After Addicott, 1978b.)

association with ants it is possible that the ants keep the aphids at densities below that which is harmful to their host plants and may even keep the total numbers at levels that satisfy the food requirements of the ant colony (Way, 1963) or more likely the ants are a limited and limiting resource for certain species of aphids (Cushman and Addicott, 1989). A possible mechanism is that increase in aphid abundance relative to that of the ants tending them is known to result in an increase in the incidence of the number of aphids being taken by the ants as food (Sakata, 1994). This may be because when the aphids are abundant they provide honeydew in excess of the ants requirements and they turn to aphid predation to maintain the balance of sugar to protein in the ant colony's diet (Pontin, 1978; Sakata, 1994).

In foraging for honeydew, ants are likely to be attracted to the richer sources, only exploiting poor sources when honeydew is in short supply. Thus there could be competition between aphids for ants. There is evidence that aphids compete with intra- and interspecific neighbours for the service of ants (Cushman and Addicott, 1989). In addition, there is evidence that ants prefer to attend certain species of aphids, possibly because of the higher quality of their honeydew (Cushman and Addicott, 1989, 1991) and that ants switch their attention to other species of aphids as a source of honeydew when the numbers of the favoured species of aphid decline (Figure 10.8). Ants are also known to preferentially prey on species of aphid that produce little honeydew. When the honeydew production of their favoured species of aphid is experimentally reduced by girdling its host plant, the ant switches from tending to preying on the species (Edinger, 1985).

The studies on ant/plant and ant/aphid mutualisms have been done on many systems, which differ greatly in generation times and population growth rates. A better understanding of these mutualisms is dependent on knowing what factors influence the strength, symmetry and outcome of the interactions (Cushman and Addicott, 1991). This is likely to reveal that the outcome – an ant/aphid or ant/plant mutualism – is dependent on the functional biology of the interacting components.

10.6 Endophyte-enhanced plant resistance to aphids

There is a group of fungi that grow intercellularly within plants and spend all or nearly all their life cycle concentrated at the root collar or in the seed of plants. In the case of grasses infection with these fungal endophytes is known to have a negative effect on both the preference and performance of many aphids that infest grasses (Hatch *et al.*, 1985; Johnson *et al.*, 1985; Breen, 1993). It is unknown what the relative benefits and costs are to grasses of an endophyte infection and reduced infestation by aphids and *vice versa*. However, this tritophic interaction indicates that fungal endophytes could be important in determining the distribution and abundance of aphids and other phytophagous insects, and it is likely that such interactions could in some cases have positive rather than negative effects.

10.7 Energy and nutrient flow

A large amount of energy is used annually by aphid populations. Lime aphids consume 1.5×10^7 J/m^2/year (Llewellyn, 1972), compared with 3×10^6 J/m^2/year for grazing bullocks (Macfadyen, 1964) and 6.4×10^5 J/m^2/year for browsing oak tree caterpillars (Varley, 1967). Thus aphids may extract much more energy per unit area than grazers and browsers, and do so without consuming any of the plant structure. However, of the food energy ingested by aphids, little is used for growth and as much as 90% is excreted. Therefore, although relatively little of the food energy ingested is available to the natural enemies of aphids, a large proportion passes to the decomposers. The energy content of the honeydew falling from a lime tree in the course of a year is equivalent to that in the leaves at leaf fall (Llewellyn, 1970). In addition the percentage nitrogen content of the leaves at leaf fall is much higher (*cf.* Figure 10.1) in years when aphids are abundant. Thus, fluctuations in aphid numbers have a dramatic effect on the quantity and quality of the food available to the decomposers. Although relatively little of the total energy ingested by aphid populations is available to their natural enemies, nevertheless some aphids can become so abundant that they can give rise to 'plagues' of natural enemies (p. 189), which could have a very marked affect on the abundance of other generally less common species of aphid. This phenomenon has been termed apparent competition (Holt and Lawton, 1994).

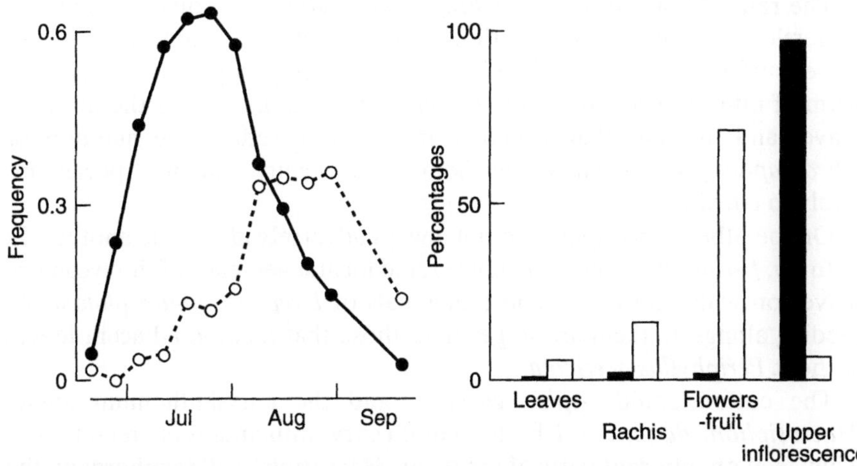

Figure 10.9 Frequency of infestation of fireweed plants by *Aphis varians* (solid circles and bars) and *Macrosiphum valerianae* (open circles and bars) in relation to time of year, and the distribution of each aphid on the plants.

10.8 Competition

Aphid species may feed on different parts of a plant at different times of the year (Figure 10.9). The spatial separation of species may enable them to occur at the same time on the same plant, but does not prevent aphid species and other phloem feeding homoptera from competing. In the absence of a species its niche is likely to be occupied by another species of aphid, providing its morphology, physiology or behaviour does not constrain it. *Aphis fabae*, living on bean, changes its feeding site from the stem to the leaves, and from the upper to the lower surface of leaves, when *Acyrthosiphon pisum* is present (Salyk and Sullivan, 1982). However, in other coexisting species there is no indication that the presence of one species alters the feeding position of the other (Addicott, 1978a; Chongrattanameteekul *et al.*, 1991).

Although the feeding positions on a plant may not overlap, nevertheless the aphids are feeding on a common resource – phloem sap. Intraspecific competition for this resource can be intense and result in longer developmental times, smaller aphids or a switch to the development of another morph better adapted for dispersal. Similarly, *Macrosiphum valerianae* adversely affects the weight of *Aphis varians*, and *Aphis varians* the survival and reproduction of *Macrosiphum valerianae* when both are feeding on the same plant (Addicott, 1978a; Antolin and Addicott, 1988); and in the presence of the sycamore aphid, *Drepanosiphum platanoidis*, proportionately more of the other aphid on sycamore, *Periphyllus testudinaceus*, develop into alatae (Shearer, 1976). The interspecific effect on alate production is less, however, than the intraspecific effect.

The rain of honeydew from *Chromaphis juglandicola* that feeds on the undersides of walnut leaves is thought to prevent the colonization of the upper surface of the leaves by *Calaphis juglandis* (Stroyan, 1977); this is a form of interference. However, as *Calaphis juglandis* prefers the terminal leaves and the ants that attend it are likely to reduce the numbers of *Chromaphis juglandicola* in the immediate vicinity, the two species are likely to coexist.

On the other hand, aphids do not always adversely affect one another. In autumn, *Periphyllus acericola* induce rapid local senescence of the sycamore leaves on which they feed and individuals of *Drepanosiphum platanoidis* feeding alongside them are larger than those that feed on adjacent leaves without *Periphyllus acericola*.

The cotton aphid, *Aphis gossypii*, and the chrysanthemum aphid, *Macrosiphoniella sanborni*, both occur on chrysanthemums in greenhouses. They feed on different parts of the plant, *Macrosiphoniella sanborni* on the growing stem and *Aphis gossypii* on the young leaves. This habitat segregation is less marked, however, when large numbers of aphids cause the plant to stop growing, and no young growth is available. The presence of one species inhibits the growth and population density achieved by the other species (Figure 10.10). Also, when in a mixed population *Aphis gossypii* only produces 55% and *Macrosiphoniella sanborni* 78% of the number of alatae they each produced when not competing (Figure 10.11). It is also interesting to note that the equilibrium populations achieved in the mixed- and single-species populations are similar. This suggests that both species affect the plant similarly and do not affect the plant's carrying capacity. However, on the same plant *Aphis gossypii* does not appear to affect the population growth rate of *Myzus persicae* and *vice versa* (Vehrs *et al.*, 1992). It is reported that the spirea aphid, *Aphis spiraecola*, which has a wide range of hosts, has displaced the apple aphid, *Aphis pomi*, which is limited to rosaceous hosts, from its dominant pest status on apple in Israel (Zehavi and Rosen, 1987) and eastern North America (Pfeiffer *et al.*, 1989). Support for this claim comes from a combined field census of apple orchards and a simulation study by Brown *et al.* (1995), which indicates that it is likely that the spirea aphid has displaced the apple aphid from apple. The two species do not affect one another's distributions on apple but the spirea aphid is the more fecund, especially at higher temperatures. The latter is thought to enable the spirea aphid to dominate on apple. The reduced tendency of the apple aphid to emigrate from apple at high aphid densities appears to account for its persistence at very low population densities on apple despite its reproductive disadvantages.

The few studies that have been done on interspecific competition in natural habitats are consistent in reporting the coexistence of aphids. The component species of each aphid guild appears to partition the resource by plant part and this is associated with differences in their morphology (stylet

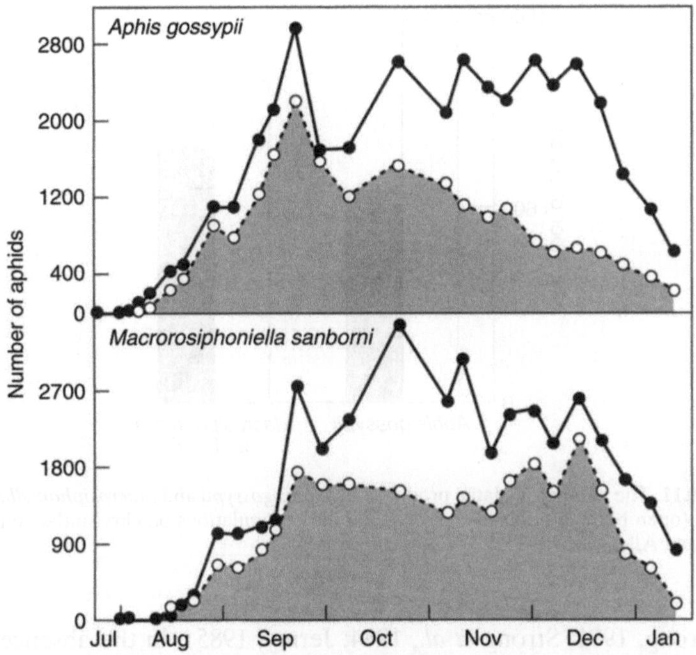

Figure 10.10 The numbers of *Aphis gossypii* and *Macrosiphoniella sanborni* on chrysanthemums in single species (solid circles) and mixed species populations (open circles). (After Tamaki and Allen, 1969.)

length/body size), physiology (tolerance of secondary plant chemicals) and behaviour (hatching times) (Hajek and Dahlsten, 1986; Akimoto, 1988; Völkl, 1989; Inbar and Wool, 1995; Jackson and Dixon, 1996). Natural enemies, in particular predators, and the spatial distribution of the resource, i.e. whether it is in the form of solitary plants or clumps of plants, appear to determine whether the aphid species are segregated. That is, predation and interplant movement by aphids are thought to influence community structure (Edson, 1985). The short survival times of aphid colonies is also thought to prevent aphid numbers increasing to levels when interspecific competition is likely to become important (Antolin and Addicott, 1988). Interestingly, gall parasitism by the aphid *Eriosoma yangi* does not appear to influence niche divergence within the *Eriosoma* gall-forming community on elm as the species have the same niche relationships in areas of Japan where *Eriosoma yangi* is absent (Akimoto, 1988).

The role of interspecific competition in structuring community organization is an important issue in ecology. A large body of positive evidence exists for vertebrates but studies on phytophagous insects have generally cast doubt on its importance (Hairston *et al.*, 1960; Connell, 1980; Lawton

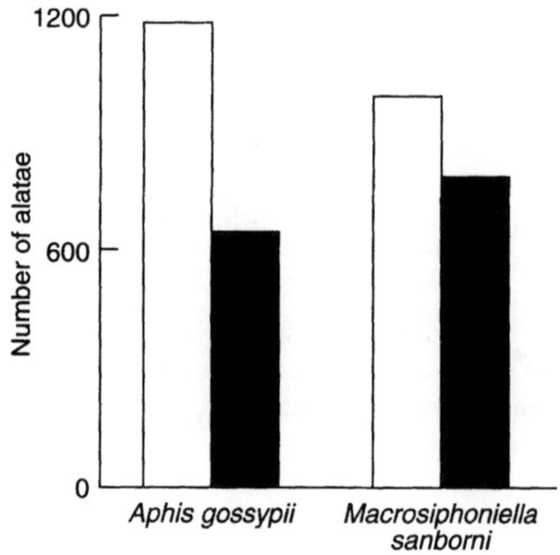

Figure 10.11 The number of alatae produced by *Aphis gossypii* and *Macrosiphoniella sanborni* in single (open bars) and mixed species (solid bars) populations on chrysanthemums. (After Tamaki and Allen, 1969.)

and Strong, 1981; Strong *et al.*, 1984; Jermy, 1985). In the absence of evidence for interspecific competition in natural conditions this doubt is justified. However, for each species other species are an important component of the environment. If an aphid is closely adapted to a particular niche, then the effects of interspecific competition on its fitness could prevent a niche shift (Akimoto, 1988). This is particularly so where the resource is a tree, which because of its long life and abundance is likely to be consistently infested with several aphid species.

10.9 Species diversity and plant architectural complexity

Generally, the more structurally complex plants, like trees, are regularly infested with more species of aphids than herbaceous plants (Dixon *et al.*, 1995). Although grasses can have one aphid species on the flower and different species on the stem, leaves and roots, this is rarely achieved and then only for short periods. However, on trees, several species of autoecious aphids regularly coexist, e.g. *Tuberculoides annulatus*, *Thelaxes dryophila*, *Lachnus roboris* and *Stomaphis quercus* on oak; and *Betulaphis quadrituberculata*, *Euceraphis punctipennis*, *Monaphis antennata* and *Symydobius oblongus* on birch.

The increase in structural complexity that results from increase in the size of an individual plant can sometimes result in the loss of species. *Cinara todocola* tends to infest young todo-fir and is rarely found on plants more than 14 years old. This is thought to be because the aphids feed at the tips of the branches, and so on the taller trees they are more exposed to

Figure 10.12 The percentage of todo-fir plants infested with *Cinara todocola* in relation to their age. (After Yamaguchi, 1976.)

extremes of temperature and less well protected from their natural enemies by ants (Figure 10.12) (Yamaguchi, 1976). Thus in addition to having more species of aphids overall, trees are likely to both gain and lose species as they increase in size and structural complexity.

Implicit in the above discussion is that all the potential niches are occupied by aphids. However, the fact that birch in the UK hosts 13 species of aphid yet equally abundant trees like oak only host six species does tend to argue against this notion. Similarly, the underside of the leaves of pecan host three species of aphid with the largest species feeding on the largest veins, etc., whereas the leaves of most trees only host one species. Therefore, many potential niches would appear to be unoccupied.

10.10 Size diversity

Empiricists have tended to view size as optimum for the niche a species occupies (Stanley, 1973). In terms of functional biology the size of phytophagous insects is thought to reflect the size of their host plants (e.g. Elton, 1927; Kirk, 1991; Dixon *et al.*, 1995). As such explanations often lack a plausible mechanism, theorists have turned to the fractal structure of habitats or energetics for explanations of the evolution of body size distributions. The explanation based on fractal structure (Morse *et al.*, 1985) only offers a partial explanation as in none of the body size distributions does the smallest class contain the most species. This led Blackburn and Gaston (1994) to champion the energetic model proposed by Brown *et al.* (1993), to account for the right-skewed size distributions observed in

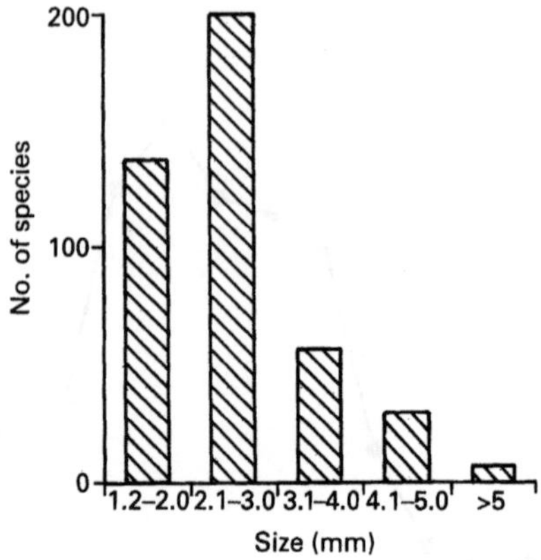

Figure 10.13 The numbers of species of British aphids classified according to their length. The slope of the right-hand side of the relationship on a log log scale is –3.2.

mammals, as offering the best basis for a realistic description of the forces structuring body size distributions. However, if it is to be the basis of a general explanation it has to be shown to apply more generally to mammals, to birds and especially to poikilothermic organisms. A constraint is the availability of measures of acquisition of energy and conversion of energy to reproduction comparable to those available for mammal and used by Brown *et al.* (1993) to develop their model. That an attempt to use this model to predict the optimum body size for aphids, for which the appropriate energetic values exist, proved unsuccessful (Kindlmann and Dixon, unpublished results), tends to throw doubt on the validity of this model as a general explanation. In the absence of a general model there is a tendency to seek the directional processes that have shaped the size diversity curves of particular groups of organisms.

The size diversity curve of aphids (Figure 10.13), in common with other groups of organisms, is best described by a distribution that is right-skewed on a logarithmic axis. The form of the relationship for aphids is similar to that for plants (Figure 10.14) the more so when the latter is expressed in terms of the estimated depth of phloem elements and corrected for the relationship between depth of phloem elements within plant tissues and the body size (*cf.* Figure 4.9) of the aphid feeding on those particular phloem elements (Figure 10.15). That is, subject to the assumed relationship between the maximum depth of phloem elements and plant size being established by direct measurement, it is possible to offer a functional interpretation of the size diversity curve of aphids. The structure of plants, in particular the depth of the phloem in specific plant structures, along with the relative surface area of these structures, determines the form of the curve (Dixon *et al.*, 1995).

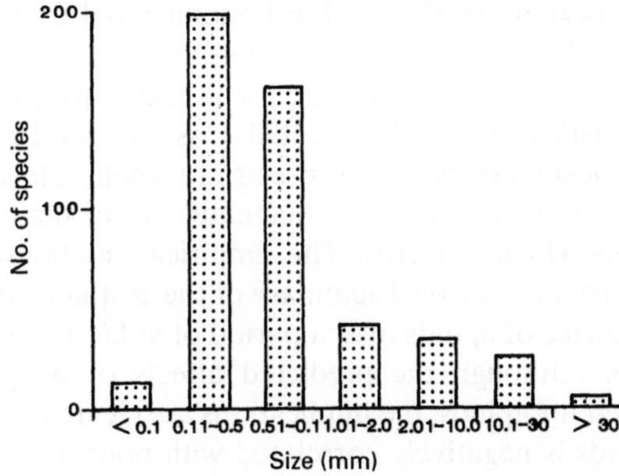

Figure 10.14 The numbers of species of British plants classified according to size (height or length). The slope of the right-hand side of the relationship on a log log scale is –0.7.

10.11 Distribution, abundance and variability

The temporal variability in population abundance is likely to depend on species interactions within a community. MacArthur (1955) argued that population densities of species should be less variable in communities with complex food webs. For example, the abundance of a monophagous species, which is dependent on the abundance of one resource, should be more variable than a polyphagous species. In contrast Watt (1964) argued that polyphagous species may be more variable because they are more likely to

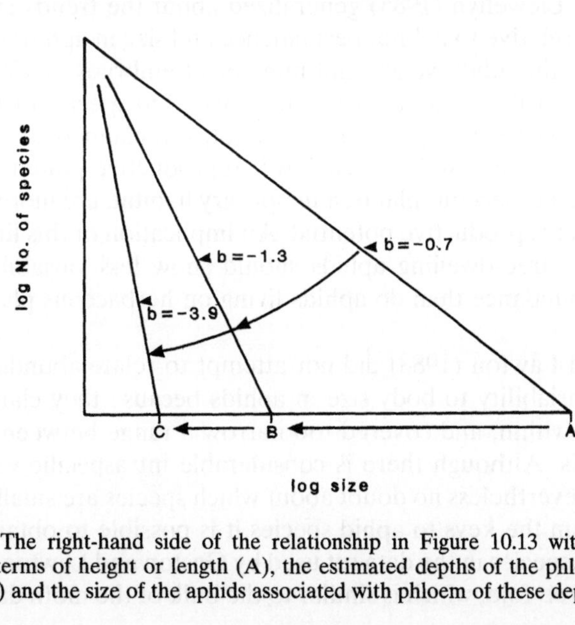

Figure 10.15 The right-hand side of the relationship in Figure 10.13 with size of plants expressed in terms of height or length (A), the estimated depths of the phloem in plants of these sizes (B) and the size of the aphids associated with phloem of these depths (C).

reach high numbers when their natural enemies fail to control them, i.e. Watt favours top-down and MacArthur bottom-up control.

Theoretical considerations and empirical relationships between population characteristics and body size led Gaston and Lawton (1988) to propose that widespread species of insects are smaller, locally more abundant and have populations that fluctuate more than geographically restricted species (Figure 10.16). The empirical relationships for aphids were obtained from a statistical summary of the Rothamsted Insect Survey catches of 97 species of aphids over a period of at least 6 years (Taylor and Woiwod, 1980). Although the predicted effects of body size were not specifically tested for aphids, the implications are clear. The intrinsic rate of increase in aphids is negatively correlated with body size, and as a consequence the local abundance of small species will tend to be more variable in time and their fast population growth makes them less subject to extinction and therefore more likely to be widely distributed. Small species are more abundant because they have a lower per capita resource requirement than large species.

Redfearn and Pimm (1988) and Pimm (1991), using essentially the same data set, addressed one aspect of the problem (Figure 10.16): the local variability in abundance from year to year shown by different species of aphids. They argue following MacArthur (1955) that this is driven by the degree of polyphagy exhibited by each species, with polyphagous species less susceptible to fluctuations in the level of a particular resource. The implication of this in the context of the Gaston and Lawton hypothesis is that there is a positive correlation between aphid size and polyphagy.

Brown and Llewellyn (1985) generalized about the trends in reproductive potential relative to habitat permanence and size in aphids on the basis of a survey of the adult weight and number of embryos of 100 species of aphids, collected throughout a year from over 120 species of plants. The species living on trees, a permanent habitat, are claimed to be more *K*-selected, which is reflected in their lower reproductive potential, whereas those living on herbaceous plants, a temporary habitat, are more *r*-selected, having a higher reproductive potential. An implication of this line of thinking is that the tree-dwelling aphids should show less variability in local population abundance than do aphids living on herbaceous plants (Figure 10.16).

Gaston and Lawton (1988) did not attempt to relate abundance, distribution and variability to body size in aphids because they claimed size is very variable within, and covered too narrow a range between, species to justify analysis. Although there is considerable intraspecific variability in size there is nevertheless no doubt about which species are small and which are large. From the keys to aphid species it is possible to obtain the body lengths of the species in the data set used by Gaston and Lawton. The range in length is three-fold, which is similar to the bulk of the moth data that was

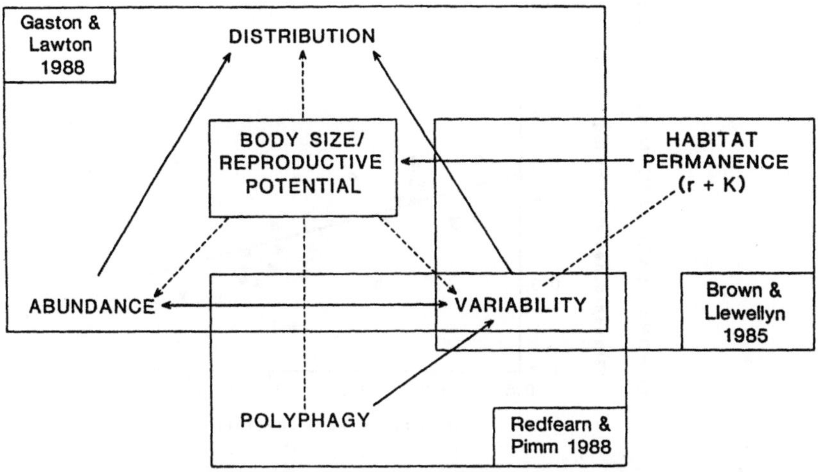

Figure 10.16 Relationships revealed (solid lines) and proposed (dashed lines) between population characteristics and body size in aphids by Gaston and Lawton (1988), population variability and polyphagy by Redfearn and Pimm (1988) and reproductive potential and habitat permanence by Brown and Llewellyn (1985), where abundance and variability refer to local populations and distribution is a measure of the number of local sites occupied by a species throughout a geographical region.

analysed. Accepting this measure of size there is no correlation between body size in aphids, abundance, distribution and variability in abundance ($r = 0.13, 0.08$ and 0.10, $n = 97$, respectively). This is not unexpected because there are no theoretical grounds for expecting (p. 79), or empirical evidence in support of, a relationship between body weight and the intrinsic rate of increase in aphids (*cf.* Figure 4.26), which is central to Gaston and Lawton's hypothesis. Similarly, there is no correlation between body size and polyphagy ($r = -0.23$, $n = 26$).

The claim that highly polyphagous species show less variability in population abundance than host specialists (Figure 10.17; Redfearn and Pimm, 1988) is based on a significant negative correlation at three out of six suction-trap sites for non-host-alternating aphids but not for host-alternating aphids However, the high levels of polyphagy attributed to species that are known to survive and reproduce on only one host plant is worrying, e.g. *Drepanosiphum platanoidis* is assigned 21 host plants and *Phyllaphis fagi* is assigned 10. Correcting for this using data on host range from aphid keys reveals weak negative relationships between polyphagy and variability in population abundance in three out of the six sites, with one site different from those recorded by Redfearn and Pimm. More surprising, host-alternating species at one of the sites show a significant positive relationship. Thus the relationship is dependent upon the measure of polyphagy used and

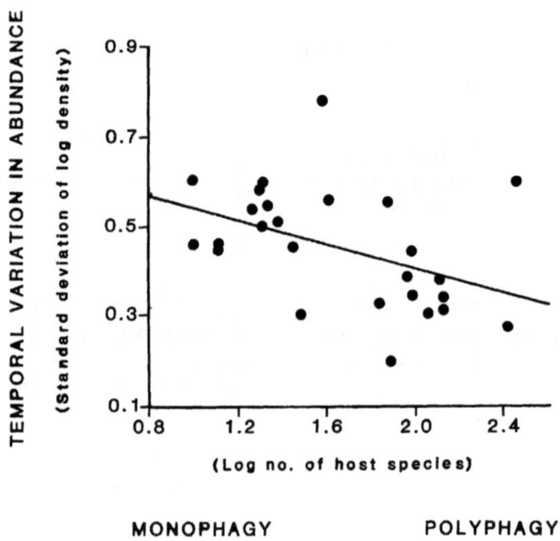

Figure 10.17 Temporal variation in population abundance decreasing with increasing polyphagy for 25 species of aphid.

the site. In addition, several of the relationships are dependent on one species (*Macrosiphum euphorbiae* or *Aphis fabae*), which when omitted leaves only one relationship in 12 significant. Therefore, even the claim of a weak negative relationship between polyphagy and local population variability is not well founded.

It is also relevant to ask whether the Rothamsted Insect Survey data used by Gaston and Lawton, and Redfearn and Pimm is suitable for testing their ideas. The suction traps used to catch aphids are sited so that they 'are free from obstructions to wind-run over gently rising, agricultural land' (Taylor and Woiwod, 1980) as the objective is to monitor the aerial abundance of pest species of aphids. Not surprisingly therefore, 92% of the 28 most commonly caught aphids are pest species and are taken at all the sites, which could account for the correlation between abundance and distribution. The high variability in the local abundance of the pest species may be a consequence of their aerial abundance being largely dependent on the extent to which they can recolonize an annual crop each year. It also remains to be shown that there is a good correlation between the aerial and terrestrial populations of aphids, especially those species that live on long-lived plants like trees. Thus even the relationships between abundance, distribution and variability revealed by Gaston and Lawton (1988) (Figure 10.16) may be spurious.

Brown and Llewellyn (1985) and Llewellyn and Brown (1985) use two measures of reproductive potential: number of embryos and the number of embryos per unit weight, which indicate different trends relative to habitat

Table 10.1 The reproductive potential expressed as number of embryos and embryos per unit weight of aphids living on herbaceous and woody plants

	Plants			
	Herbaceous		*Woody*	
	Annuals	*Perennials*	*Shrubs*	*Trees*
Number of embryos				
Brown (1984)	70.0 ± 4.4		50.2 ± 2	
Brown (1985)	58	50.9	36.5	
Brown and Llewellyn (1985)	52.3[a]	49.3[a]	50.1	35.0
Number of embryos/unit weight				
Data from Brown and Llewellyn (1985)	37.5		38.0	

[a]Excludes aphids on ferns (29.9) and grasses (29.2).

permanence (Table 10.1). The number of embryos gives an indication of an aphid's potential fecundity. However, it is a poor indicator of an aphid's intrinsic rate of increase (r_m), which is determined mainly by its developmental time and rate of production of offspring in early adult life (Dixon, 1987a). For example, for nine species of the genus *Aphis* the relationship between the number of well-developed embryos and r^m accounts for only 32% of the variation and there is no relationship between r_m and the number of embryos, embryos per unit weight or adult weight (data from Dalimin, 1982). The data available for nine species of aphids from different genera indicates no relationship between either total embryos or embryos per unit weight and r_m (Figures 10.18 and 10.19). This is also well illustrated

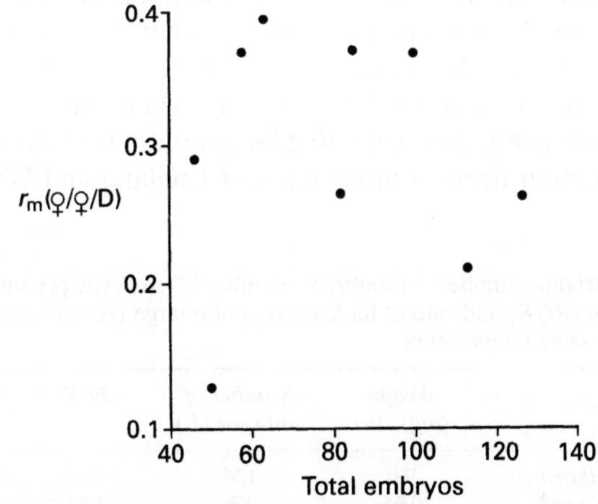

Figure 10.18 Intrinsic rate of natural increase (r_m) at 20°C in relation to total embryos for eight species of aphid.

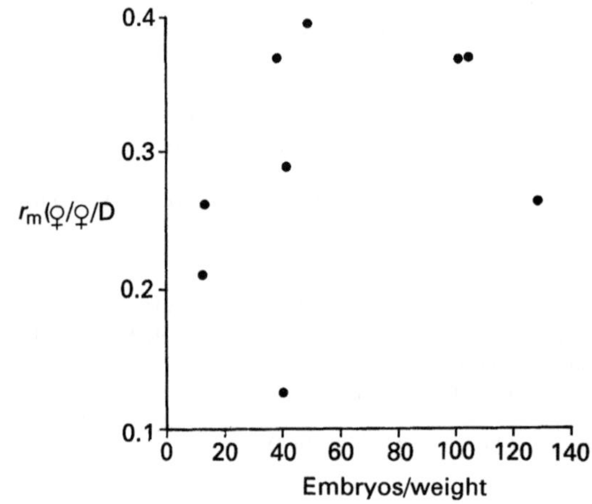

Figure 10.19 Intrinsic rate of natural increase (r_m) at 20°C in relation to number of embryos per unit weight for nine species of aphid.

by reference to the aphids *Macrosiphum albifrons* and *Rhopalosiphum padi*, which have markedly different numbers of embryos and are very different in size but can have the same intrinsic rate of increase (Table 10.2). In addition, the data set presented by Taylor and Woiwood (1980), which gives a measure of the yearly variability in abundance of 97 species of aphids, does not support a corollary of Brown and Llewellyn's (1985) hypothesis: populations of tree-dwelling aphids should show greater temporal stability than populations of aphids living on herbaceous plants (Table 10.3).

As there appears to be very little if any support for the relationships depicted in Figure 10.16 applying to aphids, what determines their distribution and abundance? Most aphids are highly host specific; therefore, particular species are likely to be found mainly where their host plants are common. A good example of this is that large numbers of *Anoecia corni* are only taken in suction traps in those parts of England and Wales where its

Table 10.2 The weights, numbers of embryos, number of embryos per unit weight (*E/W*), relative growth rate (*RGR*) and rate of increase (r_m) of a large (A) and small (B) species of aphid reared at the same temperature

Species	*Weight (mg) (W)*	*Number of embryos (E)*	*E/W*	*RGR*	r_m
A. *Macrosiphum albifrons*[a]	9.0	126	14	0.288	0.266
B. *Rhopalosiphum padi*[b]	0.65	82	127	0.275	0.268

[a]After Frazer and Gill (1981) and unpublished results of C. Brough.
[b]After Leather and Dixon (1981, 1984) and unpublished results.

Table 10.3 The year to year variability in abundnace, expressed as the standard deviation of the mean annual catch ($m_{\log}$) of monoecious aphids living on trees and herbaceous plants caught by the Rothamsted Insect Survey (data from Taylor and Woiwod, 1980)

Aphids living on	*No.*	*Range in SD*	$\overline{SD}$
Trees	22	0.21–0.59	0.39
Herbaceous plants	30	0.10–0.54	0.28

$t = 3.17$; d.f. 50; $P < 0.01$.

host dogwood (*Cornus sanguinea*) occurs and south-west Scotland where dogwood is grown as a garden shrub (p. 215). Another example is the hop aphid, *Phorodon humuli* (Figure 8.14) (Taylor *et al.*, 1979). For anholocyclic species winter temperatures, through their effect on survival, can also determine distribution (Taylor, 1977). Abundance is likely to be determined by the abundance/proportional cover of their host plants (p. 207), and variability in abundance to changes in weather (Dewar and Carter, 1984; Wellings and Dixon, 1987a), the effect of aphid spatial distribution on predator–prey interactions (Cappuccino, 1987) and the extent to which an aphid's host plant varies in abundance in time. However, the actual factors important in determining the patterns in abundance shown by aphids will only be revealed by more field and experimental studies of the type carried out by Cappuccino (1987). She revealed that of the two species of *Uroleucon* living on *Solidago altissima* the aphid with the highest r_m shows least variability in abundance. The difference in population stability shown by these two species of aphid may be largely a consequence of their different spatial distributions with the species showing the greatest variability in abundance being the more spatially aggregated.

10.12 Rarity

Most species are uncommon to rare. Plots of the number of species against individuals per species give a hollow curve (Williams, 1964). This highlights why rarity defies precise definition. A pragmatic solution is to define it as the first quartile of the frequency distribution of species abundances (Gaston, 1994). When asking why particular species are rare it has proved useful to consider rarity at three levels: geographical range, habitat (feeding) specificity and local population size. This gives eight classes (Table 10.4). The only truly common species of aphids are the few generalists aphids, like *Aphis fabae*, which have a large geographical range and are locally abundant in several habitats. The specialist species can be assigned to one of four remaining seven forms of rarity in Table 10.4.

Table 10.4 A classification of species abundance based on three characteristics, i.e. geographical range, feeding specificity and local population size with examples of the five classes of rarity applicable to aphids (after Rabinowitz, 1981)

Feeding specificity *Geographical range*	*Generalist* *Large*	*Specialist* *Large*	*Generalist* *Small*	*Specialist* *Small*
Population size				
Locally abundant	*Aphis fabae*	*Euceraphis betulae*	–[a]	*Aneocia corni*
Constantly sparse	–[a]	*Monaphis antennata*	–[a]	*Nasonovia saxifragae*

[a]Not applicable to aphids.

Several of the classes of rarity involving specialist aphids can be accounted for in terms of the abundance of their host plants (Dixon and Kindlmann, 1990; p. 207). Host-specific aphids that live on uncommon plants are likely to incur great losses in finding their host plants and as a consequence are rarer than aphids living on common plants. However, some species like *Monaphis antennata* have a very wide geographical range, most of the palaearctic live on a very abundant host, birch, but always occur at a very low density. As there are no specific parasitoids recorded for this aphid its rarity is unlikely to be attributable to the searching efficiency of its natural enemies. Accounting for the rarity of this and several other similarly rare species is a major challenge.

10.13 Why are there so few species of aphids, especially in the tropics?

Most aphids are monophagous, have similar sensory and dispersal powers and it is likely they do not locate their host plants from a distance (p. 28), i.e. host finding tends to be a random process. Thus the concept of plant apparency is useful when considering why relatively few plants have aphids, and also the somewhat surprising paucity of aphids in the tropics. A plant that is plentiful for long periods of time, and over a large area, is apparent, and is more likely to have aphids than less apparent species of plants.

About 4000 species of aphids have so far been described and most of these come from the temperate regions of the world. In the tropics and subtropics Aleyrodidae and Coccoidea seem to replace the aphids. The respective distributions of all three groups are assumed to reflect their ability to survive in the physical conditions that prevail in the temperate and tropical regions (Bodenheimer and Swirski, 1975). Alternatively, the distribution of aphids is seen as constrained by cyclical parthenogenesis, a very successful way of exploiting the short-lived growth flushes of temperate plants, which cannot be adapted to tropical conditions (Blackman and Eastop, 1994). However, aphids associated with crops have often retained their pest status when introduced from the temperate regions into tropical and subtropical regions of the world. There are also holocyclic endemic species in the tropics and subtropics, well adapted to the climatic conditions that prevail there. For example, in Australia species of *Schoutedenia* (Greenideinae), *Sensoriaphis* and *Neophyllaphis* (Drepanosiphidae) avoid the rigours of the summer by producing eggs in spring (Hales, 1976). In contrast, the anholocyclic *Cervaphis shouteniae* (Greenideinae) in the vicinity of Calcutta, India, achieves its highest rates of increase during the hottest (30°C and above) and driest period of the year (Agarwala and Dixon, 1986). Indeed species from all the aphid subfamilies are found in the tropics and subtropics. Although continuous parthenogenesis is commoner in species living in the tropics and subtropics than in the temperate regions, nevertheless sexual forms have been recorded for 20% of the Indian aphids (Agarwala, personal communication). Thus as some species of aphids are able to survive and thrive in the 'harsh' environment of the tropics and subtropics, and speciation is unlikely to have been limited by a lack of genetic recombination, it is difficult to understand why aphids have not flourished as a group on the rich tropical and subtropical floras.

Aphids differ from most other groups of insects by showing an inverse relationship between the number of aphid species and the number of plant species in different parts of the world. In the temperate regions there are more aphid species per 1000 species of plants than in the tropics and subtropics (Table 12 in Eastop, 1973). The aphid faunas and floras of several countries are sufficiently well known to define the form of the relationship between the number of aphid species per plant species, relative to the number of plant species per unit area of land (Figure 10.20).

A similar relationship can be derived mathematically from the observations that aphids are mostly monophagous and find their host plants by random search (Dixon *et al.*, 1987):

$$S_a/S_p \cong K \exp(-C_{crit})S_p \tag{10.1}$$

where S_a is the number of aphid species, S_p is the number of plant species, K is the average number of aphid species per plant species whose cover is equal to or greater than C_{crit}, which is the proportional ground cover a plant species must achieve if it is to support an aphid species. Thus in the tropics and subtropics, where the floral diversity is very high, few species of plants are apparent enough to sustain an aphid species. In this context it is interesting to note also that proportionately more aphids are polyphagous in the tropics than in the temperate regions (Holman, 1971). By living on a number of uncommon plant species, whose collective apparency is above the threshold necessary to sustain an aphid species, polyphagous aphids possibly compensate for the 'rarity' of their individual host plant species. In the temperate regions there are fewer plant species than in the tropics, and the commoner ones are apparent enough to sustain one or more species of

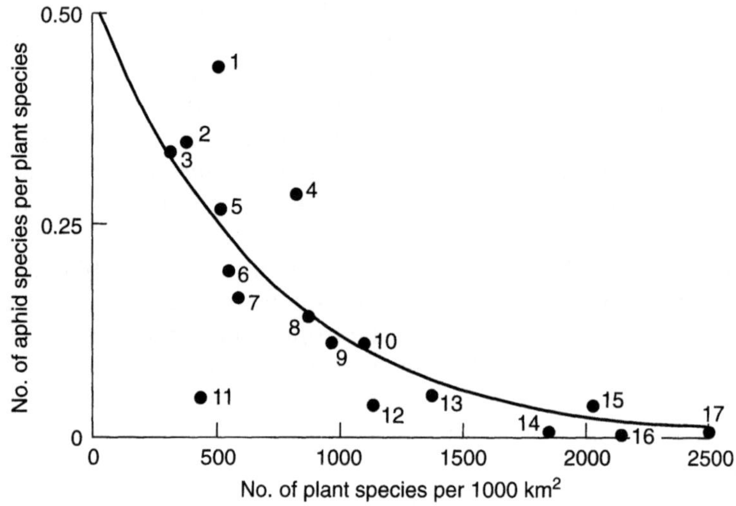

Figure 10.20 The number of aphid species per plant species in relation to the number of plant species per unit area of land surface for 17 countries. (1) The Netherlands, (2) UK, (3) Nordic countries, (4) Czechoslovakia, (5) Poland, (6) Azores, (7) Russia (European), (8) Romania, (9) Portugal, (10) Madeira, (11) New Zealand, (12) Turkey, (13) Spain, (14) Cuba, (15) India, (16) Australia, (17) South Africa.

aphids. This possibly accounts for the greater number of species of aphids that have evolved in temperate regions.

It has been suggested that aphids originated in the northern hemisphere and have not achieved an equivalent diversity in the southern hemisphere due to the barrier posed by the tropics. This can been tested by determining each country's deviation in number of species of aphids from that predicted by the above model. The relationship between each country's deficit or excess of aphid species and each country's mean latitude is shown in Figure 10.21. Although there are aphid faunas for very few countries in the southern hemisphere, nevertheless there is a significant positive correlation indicating a deficit of aphid species in southern latitudes (Mackenzie *et al.*, 1994).

The model is a good predictor of aphid species diversity when averaged across countries and strongly suggests that aphid speciation has been determined by the limited host location abilities of the taxon. There are, however, local factors that give rise to variation from this general picture, one of which appears to be latitude, due to constraints imposed by the area of evolutionary origin.

Consideration of the world-wide distribution of aphids not only provides an answer for the paucity of aphids in the tropics and subtropics, in terms of aphid biology, but also for why so few species of plants are able to sustain aphid species. This is confirmed by an analysis of the occurrence of

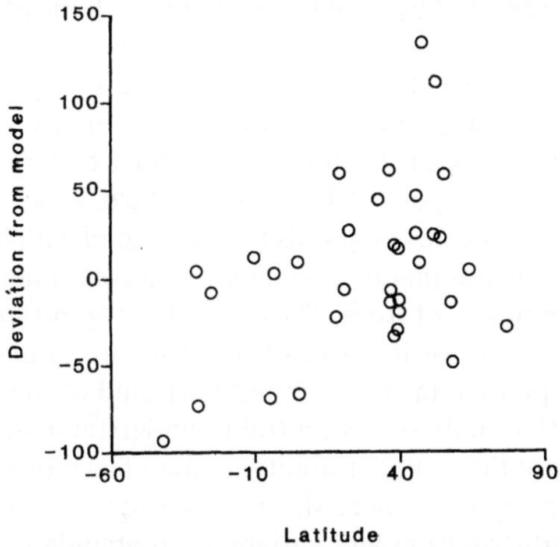

Figure 10.21 The effect of latitude of the country on the deviation from the relationship between average area-adjusted species diversity in aphids (S_a) and plants (S_p) predicted by equation (10.1).

monophagous species of aphids on species of plants belonging to the Compositae, Ranunculaceae and Umbelliferae in the Czech Republic and Slovakia, where the aphid fauna and flora are well known (Table 10.5). Proportionately more monophagous species of aphids are associated with common than with scattered or rare species of plants. The presence of aphids on the rare species of plants does not invalidate the idea because many of these plants are common in adjacent countries (Dixon *et al.*, 1987).

Summarizing, there is no evidence that plants infested with aphids are usually fitter than uninfested plants, in fact quite the reverse. However, some plants are thought to use aphids to attract ants and so gain protection from insect defoliators. The mutualism between ants and some aphids, however, is well established, with the large species of aphids and those feeding on plant roots most likely to be ant-attended. Most of the large

Table 10.5 Percentage of common, scattered and rare plants of three plant families that are host to monophagous species of aphids in the Czech Republic and Slovakia

Plant family	*No. of species*	*Abundance*		
		Common	*Scattered*	*Rare*
Compositae	226	71	31	14
Ranunculaceae	72	33	30	8
Umbelliferae	78	65	43	13
Average percentage		65	33	12

amount of sugar ingested by aphids is excreted as honeydew, which mainly passes directly to the decomposers. Although relatively little of the energy ingested by aphids is used for growth, nevertheless they can become abundant and so support large numbers of natural enemies. Interspecific competition between aphids has been recorded under experimental conditions and is likely to be an important factor in the field, especially when aphids are abundant. More species of aphids are associated with individual species of trees than herbaceous plants, which may be a consequence of the greater architectural complexity of trees. The diversity of aphid species of different sizes can be accounted for in terms of the structure of plants, in particular the depth of the phloem in specific plant parts and of their relative surface area. The claim that body size is central to understanding what determines abundance, distribution and variability in abundance of a species does not appear to hold for aphids, where size is a consequence of where they feed, and abundance, distribution and variability in abundance depend on host plant abundance and distribution, winter temperatures and variation in weather. The reason why a few aphids that live on very common plants are rare, however, is very intriguing and remains to be resolved. The greater number of species of aphids in the temperate regions can possibly be accounted for mainly in terms of floral diversity and aphid biology. In the tropics the high floral diversity means that relatively few species of plants are common enough to sustain an aphid species. In addition, assuming that aphids originated in the northern hemisphere, there is evidence to support the idea that aphids have not achieved an equivalent density in the southern hemisphere due to its distance from the area of evolutionary origin of aphids.

11 Epilogue

Cyclical parthenogenesis, which developed early in the evolution of aphids, has been a major factor shaping their ecology. It led to the evolution of the telescoping of generations and prodigious rates of population increase. The latter they fuel by efficiently extracting all the amino nitrogen they need from large quantities of low quality phloem sap and utilizing their symbionts to upgrade and recycle nitrogen. Their short generation times enable aphids to track very closely the seasonal trend in resources. By anticipating this trend and by means of phenotypic plasticity they develop forms that are close to optimal for the conditions they each experience. In this context the function of sex is to reset the biological clocks that enable them to anticipate the trends. Accepting that there is a species specific minimum size at birth determined by the depth of phloem within the plant parts fed on by aphids, then the optimum adult size for maximizing r_m is 15 times the birth size, which for most aphids is relatively small.

The simultaneous commitment to growth and reproduction, which accompanies the telescoping of generations, puts a severe constraint on the length of time for which aphids can survive without feeding, and thus the time they can spend searching for host plants. Because of their prodigious rates of population increase it is adaptive to seek out the species of plant on which they perform best, which favours host specialization. However, the extent to which trade-offs in performance on different species of plants, or the need to rendezvous on a particular host for sex, additionally account for the very high degree of host specificity shown by aphids is debatable. Small size and host specificity also impose constraints on host location. The short period for which aphids can survive without food, their high degree of host specificity and random location of host plants appear to have limited the number of species of aphids and restricted speciation in areas where plant species diversity is high, i.e. the tropical regions.

Their high degree of host specificity and close association with plants leads one to expect co-evolution is a major force generating patterns in aphid ecology. Although there is abundant evidence of how aphids adapt to plants there is little unequivocal evidence of how plants have adapted to aphids, possibly because plants are subject to attack by a very wide range of organisms and selection has favoured the development of plant defences effective against generalist phytophages rather than specialists. One possible example is the use of aphids by plants to attract ants, which afford the plant some protection from specific insect defoliators.

A good understanding of the functional biology of aphids has facilitated the study of their ecology. However, morphological, physiological, etc., studies can only identify the constraints that need to be considered when studying ecological patterns. The relative role of optimization, the approach favoured here and phylogenetic constraints in determining ecological patterns need further study. The resolution of this debate is in part dependent on accurate phylogenies, which should shortly become available. Another approach, still largely unexplored, is to compare aphid ecology with that of the closely related adelgids, phylloxerids and, above all, coccids.

References

Addicott, J. F. (1978a) Niche relationships among species of aphids feeding on fireweed. *Canadian Journal of Zoology*, **56**, 1837–1841.

Addicott, J. F. (1978b) Competition for mutualists: aphids and ants. *Canadian Journal of Zoology*, **56**, 2093–2096.

Addicott, J. F. (1979) A multispecies aphid–ant association: density dependence and species-specific effects. *Canadian Journal of Zoology*, **57**, 558–569.

Agarwala, B. K. and Dixon, A. F. G. (1986) Population trends of *Cervaphis shouteniae* van der Goot on *Microcus paniculata* and its relevance to the paucity of aphid species in India. *Indian Biologist*, **18**, 37–39.

Akimoto, S. (1981) Gall formation by *Eriosoma* fundatrices and gall parasitism in *Eriosoma yangi* (Homoptera, Pemphigidae). *Kontyu*, **49**, 426–436.

Akimoto, S. (1983) A revision of the genus *Eriosoma* and its allied genera in Japan (Homoptera: Aphidoidea). *Insecta Matsumurana NS*, **27**, 37–106.

Akimoto, S. (1988) Competition and niche relationships among *Eriosoma* aphids occurring on the Japanese elm. *Oecologia (Berlin)*, **75**, 44–53.

Akimoto, S. (1992) Shift in life-history strategy from reproduction to defense with colony age in the galling aphid *Hemipodaphis persimilis* producing defensive first-instar larvae. *Research in Population Ecology*, **34**, 359–372.

Akimoto, S. (1995) Ecological factors promoting the evolution of colony defense in aphids: computer simulations. *Insectes Sociaux*, **41**, 1–11.

Akimoto, S. and Yamaguchi, Y. (1994) Phenotypic selection on the process of gall formation of *Tetraneara* aphid (Pemphigidae). *Journal of Animal Ecology*, **63**, 727–738.

Alosi, M. C., Melroy, D. L. and Park, R. B. (1988) The regulation of gelation of phloem exudate from *Cucurbita* fruit by dilution, glutathione, and glutathione reductase. *Plant Physiology*, **86**, 1089–1094.

Amiressami, M. (1980) Investigation of the light microscopical and ultrastructure of the demeton-s-methyl resistance aphids under consideration of the mycetome symbionts of the *Phorodon humuli* Schrank. In W. Schwemmler and H. E. A. Schenk (eds), *Endocytobiology Endosymbiosis and Cell Biology*, Vol. 1, pp. 425–443. Berlin: Walter de Gruyter.

Amiressami, M. and Petzold, H. (1976) Licht- und elektronenmikroskopische Untersuchungen uber das Verhalten der Mycetomsymbionten bei insektizidresistenten und normal – sensiblen Pfirsichblattlausen *Myzus persicae* Sulz. *Zeitschrift für Angewandte Zoologie*, **63**, 273–289.

Amiressami, M. and Petzold, H. (1977) Symbioseforschung und Isektizidresistenz ein licht- und elektronenmikroskopischer Beitrag zur Klärung der Insektizidresistenz von Aphiden unter Berücksichtigung der Mycetom – Symbionten bei *Myzus persicae* Sulz. *Zeitschrift für Angewandte Entomologie*, **82**, 252–259.

Anderson, R. M., Gordon, D. M., Crawley, M. J. and Hassell, M. P. (1982) Variability in the abundance of animal and plant species. *Nature*, **296**, 245–248.

Antolin, M. F. and Addicott, J. F. (1988) Habitat selection and colony survival of *Macrosiphum valeriani* Clarke (Homoptera: Aphididae). *Annals of the Entomological Society of America*, **81**, 245–251.

Aoki, S. (1975) Descriptions of the Japanese species of *Pemphigus* and its allied genera (Homoptera: Aphidoidea). *Insecta Matsumurana NS*, **5**, 1–63.

Aoki, S. (1977) *Colophina clematis* (Homoptera, Pemphigidae), an aphid species with 'soldiers'. *Kontyu*, **45**, 276–282.

Aoki, S. (1979) Further observations on *Astegopteryx styracicola* (Homoptera: Pemphigidae), an aphid species with soldiers biting man. *Kontyu*, **47**, 99–104.

Aoki, S. and Makino, S. (1982) Gall usurpation and lethal fighting among fundatrices of the aphid *Epipemphigus niisimae* (Homoptera, Pemphigidae). *Kontyu*, **50**, 365–376.

Aoki, S. and Kurosu, U. (1991) Galls of the soldier-producing aphid *Ceratoglyphina bambusae* broken by vertebrates (Homoptera, Aphidoidea). *Japanese Journal of Entomology*, **59**, 743–746.

Aoki, S., Kurosu, U. and Stern, D. L. (1991) Aphid soldiers discriminate between soldiers and non-soldiers rather than between kin and non-kin, in *Ceratoglyphina bambusae. Animal Behaviour*, **42**, 865–866.

Aoki, S., Yamane, S. and Kiuchi, M. (1977) On the biters of *Astegopteryx styracicola* (Homoptera, Aphidoidea). *Kontyu*, **45**, 563–570.

Arakaki, N. (1989) Alarm pheromone eliciting attack and escape responses in the sugar cane woolly aphid *Ceratovacuna lanigera* (Homoptera: Pemphigidae). *Journal of Ethology*, **7**, 83–90.

Argandona, V. H., Corcuera, L. J., Niemeyer, H. M. and Campbell, B. C. (1983) Toxicity and feeding deterrency of hydroxamic acids from gramineae in synthetic diets against the green bug, *Schizaphis graminum. Entomologia Experimentalis et Applicata*, **34**, 134–138.

Argandona, V. H., Luza, J. G., Niemeyer, H. M. and Corcuera, L. J. (1980) Role of hydroxamic acids in the resistance of cereals to aphids. *Phytochemistry*, **19**, 1665–1668.

Avé, D. A., Gregory, P. and Tingey, W. M. (1987) Aphid repellent sesquiterpenes in glandular trichomes of *Solanum berthaultii* and *S. tuberosum. Entomologia Experimentalis et Applicata*, **44**, 131–138.

Ayal, Y. and Green, R. F. (1993) Optimal egg distribution among host patches for parasitoids subject to attack by hyperparasitoids. *American Naturalist*, **141**, 120–138.

Bacon, J. S. D. and Dickinson, B. (1957) The origin of melezitose: a biochemical relationship between the lime tree (*Tilia* spp.) and an aphid (*Eucallipterus tiliae* L.). *Biochemical Journal*, **66**, 289–299.

Bale, J. S., Harrington, R. and Clough, M. S. (1988) Low temperature mortality of the peach potato aphid *Myzus persicae. Ecological Entomology*, **13**, 121–129.

Ball, B. V. and Bailey, L. (1978) The symbiotes of *Myzus persicae* (Sulz.) in strains resistant and susceptible to demeton-s-methyl. *Pesticide Science*, **9**, 522–524.

Banks, C. J. (1958) Effects of the ant, *Lasius niger* (L.), on the behaviour and reproduction of the black bean aphid, *Aphis fabae* Scop. *Bulletin of Entomological Research*, **49**, 701–714.

Banks, C. J. (1962) Effects of the ant, *Lasius niger* (L.), on insects preying on small populations of *Aphis fabae* Scop. on bean plants. *Annals of Applied Biology*, **50**, 669–679.

Banks, C. J. (1965) Aphid nutrition and reproduction. *Reports of the Rothamsted Experimental Station*, **1964**, 299–309.

Banks, C. J. and Macaulay, E. D. M. (1967) Effects of *Aphis fabae* Scop. and its attendant ants and insect predators on yields of field beans (*Vicia faba* L.). *Annals of Applied Biology*, **60**, 445–453.

Banks, C. J. and Nixon, H. L. (1958) Effects of the ant, *Lasius niger* (L.), on the feeding and excretion of the bean aphid, *Aphis fabae* Scop. *Journal of Experimental Biology*, **35**, 703–711.

Banks, C. J., Macaulay, E. D. M. and Holman, J. (1968) Cannibalism and predation by aphids. *Nature*, **218**, 491.

Barbagallo, S. and Patti, I. (1985) Remarks on and a check-list of the Italian fauna. In H. Szelegiewiez (ed.), *Evolution and Biosystematics of Aphids*, pp. 393–424. Wasawa: Polska Akademia Nauk.

Barlow, C. A. (1962) The influence of temperature on the growth of experimental populations of *Myzus persicae* (Sulzer) and *Macrosiphum euphorbiae* (Thomas) (Aphididae). *Canadian Journal of Zoology*, **40**, 145–156.

Barlow, N. D. and Dixon, A. F. G. (1980) *Simulation of Lime Aphid Population Dynamics*. Wageningen: Pudoc.

Baumann, L. and Baumann, P. (1994) Growth kinetics of the endosymbiont *Buchnera aphidicola*, in the aphid, *Schizaphis graminum. Applied Environmental Microbiology*, **60**, 3440–3443.

Behrendt, K. (1966) Population dynamics of *Aphis fabae* Scop. and the influences of Coccinellidae. In I. Hodek (ed.), *Ecology of Aphidphagous Insects*, pp. 259–262. Prague: Academia.

Behrendt, K. (1969) Über langjährige Massenwechselbeobachtung an der Schwarzen Bohnenblattlaus, *Aphis fabae* Scopoli (Homoptera Aphididae). *Bericht über die 10. Wanderversammlung Deutscher. Entomologen*, **1965**, 335–354.

Behrendt, K. (1971) Zur Bedeutung der Dispersion für die Abundanz-dynamik von *Aphis fabae* Scop. (Homoptera: Aphididae) auf Zuckerrübenbestanden. *Zoologische Jahrbücher Abteilung Systematik Oekologie Geographie Tiere*, **98**, 418–454.

Bejer-Petersen, B. (1962) Peak years and regulation of numbers in the aphid *Neomyzaphis abietina* Walker. *Oikos*, **13**, 155–168.

Bell, G. (1980) The costs of reproduction and their consequences. *American Naturalist*, **116**, 45–76.

Bernays, R. and Graham, M. (1988) On the evolution of host specificity in phytophagous arthropods. *Ecology*, **69**, 886–892.

Berryman, A. A. (1991) Stabilisation or regulation: what it all means! *Oecologia (Berlin)*, **86**, 140–143.

Bevan, D. and Carter, C. I. (1980) Frost proofed aphids. *Antenna*, **4**(1), 6–8.

Birch, A. N. E., Jones, A. T., Woodford, J. A. T. and Jones, D. R. (1990) Behavioural responses of the virus vector aphid *Amphorophora idaei* to resistance in red raspberry. In D. C. Peters, J. A. Webster and C. S. Chlouber (eds), *Aphid–Plant Interactions: from Populations to Molecules*, p. 265. Stillwater, OH: Oklahoma State University.

Birch, A. N. E., Fenton, B., Malloch, G., Jones, A. T., Phillips, M. S., Harrower, B. E., Woodford, J. A. T. and Catley, M. A. (1994) Ribosomal spacer length variability in the large raspberry aphid, *Amphorophora idaei* (Aphidinae: Macrosiphini). *Insect Molecular Biology*, **3**, 239–245.

Blackburn, T. M. and Gaston, K. J. (1994) Animal body size distributions: patterns, mechanisms and implications. *Trends in Ecology and Evolution*, **9**, 471–474.

Blackman, R. L. (1976) Biological approaches to the control of aphids. *Philosophical Transactions of the Royal Society of London B*, **274**, 473–488.

Blackman, R. L. (1979) Stability and variation in aphid clonal lineages. *Biological Journal of the Linnean Society*, **11**, 259–277.

Blackman, R. L. (1980) Chromosomes and parthenogenesis in aphids. In R. L. Blackman, G. M. Hewitt and M. Ashburner (eds), *Insect Cytogenetics*, pp. 133–148. 10th Symposium of the Royal Entomological Society of London, Oxford: Blackwell.

Blackman, R. L. (1981) Species, sex and parthenogenesis in aphids. In P. L. Forey (ed.), *The Evolving Biosphere*, pp. 75–85. Cambridge: Cambridge University Press.

Blackman, R. L. (1995) Sex determination in insects. In S. R. Leather and J. Hardie (eds), *Insect Reproduction*, pp. 57–94. Boca Raton: CRC Press.

Blackman, R. L. and Eastop, V. F. (1984) *Aphids on the World's Crops: An Identification and Information Guide*. Chichester: John Wiley.

Blackman, R. L. and Eastop, V. F. (1994) *Aphids on the World's Trees*. Wallingford: CAB International.

Blackman, R. L. and Hales, D. F. (1986) Behaviour of the X chromosomes during growth and maturation of parthenogenetic eggs of *Amphorophora tuberculata* (Homoptera, Aphididae), in relation to sex determination. *Chromosoma*, **94**, 58–64.

Blackman, R. L. and Takada, H. (1975) A naturally occurring chromosomal translocation in *Myzus persicae* (Sulzer). *Journal of Entomology (A)*, **50**, 147–156.

Blair, W. F. (1955) Mating call and stage of speciation in the *Microhyla olivacea–M. carolinensis* complex. *Evolution*, **9**, 469–480.

Blakley, N. (1982) Biotic unpredictability and sexual reproduction: Do aphid genotypic-host genotype interactions favour aphid sexuality? *Oecologia (Berlin)*, **52**, 396–399.

Bodenheimer, F. S. and Swirski, E. (1957) *The Aphidoidea of the Middle East*. Jerusalem: Weizmann Science Press.

Boeckh, J., Kaissling, K.-E. and Schneider, D. (1965) Insect olfactory receptors. *Cold Spring Harbour Symposium in Quantitative Biology*, **30**, 263–280.

Bonnemaison, L. (1951) Contribution à l'étude des facteurs provoquant l'apparition des formes ailées et sexuées chez les Aphidinae. *Annales des Épiphyties*, **2**, 1–380.

Bonnemaison, L. (1956) Déterminisme de l'apparition des larves estivales de *Periphyllus* (Aphidinae). *Compte rendu hebdomadaire des séances de l'Académie des sciences Paris*, **243**, 1166–1168.

Bonner, J. T. (1988) *The Evolution of Complexity by Means of Natural Selection*. Princeton: Princeton University Press.

Bonnet, C. (1745) *Traité d'Insectologie au observations sur les Pucerons*. Paris: Chez Durand.

Börner, C. (1939) Anfälligkeit, Resistenz und Immunität der Reben gegen Reblaus. Allgemeine Gesichtspunkte zur Frage der Spezialisierung von Parasiten; die harmonische Beschränkung des Lebensraums. *Zeitschrift fur Hygiene Zoologie Schädlingsbekampfung* **31**, 274–285, 301–308, 325–334.

Börner, C. (1952) Europae centralis Aphides. *Mitteilungen der Thüringischen Botanischen Gesellschaft*, **3**, Weimar.

Botha, C. E. J., Evert, R. F. and Walmsley, R. D. (1975a) Studies on *Gomphocarpus physocarpus*: further evidence of preferential feeding by the aphid, *Aphid nerii* on the internal phloem. *Protoplasma*, **84**, 345–356.

Botha, C. E. J., Evert, R. F. and Walmsley, R. D. (1975b) Observations of the penetration of the phloem in leaves of *Nerium oleander* (Linn.) by stylets of the aphid *Aphis nerii* (B. de F.). *Protoplasma*, **86**, 309–319.

Bradley, G. A. and Hinks, J. D. (1968) Ants, aphids, and jack pine in Manitoba. *Canadian Entomologist*, **100**, 40–50.

Breen, J. P. (1993) Enhanced resistance to three species of aphids (Homoptera: Aphididae) in *Acremonium* endophyte-infected turfgrasses. *Journal of Economic Entomology*, **86**, 1279–1286.

Breton, L. M. and Addicott, J. F. (1992a) Density-dependent mutualism in an aphid–ant interaction. *Ecology*, **73**, 2175–2180.

Breton, L. M. and Addicott, J. F. (1992b) Does host-plant quality mediate aphid–ant mutualism? *Oikos*, **63**, 253–259.

Briggs, J. B. (1965) The distribution, abundance, and genetic relationships of four strains of the rubus aphid (*Amphorophora rubi* (Kalt.)) in relation to raspberry breeding. *Journal of Horticultural Science*, **40**, 109–117.

Bristow, C. M. (1991) Why are so few aphids ant-tended? In C. R. Huxley and D. F. Cutler (eds), *Ant–Plant Interactions*, pp. 104–119. Oxford: Oxford University Press.

Bromley, A. K. and Anderson, M. (1982) An electrophysiological study of olfaction in the aphid, *Nasonovia ribis-nigri*. *Entomologia Experimentalis et Applicata*, **32**, 101–110.

Bromley, A. K., Dunn, J. A. and Anderson, M. (1979) Ultrastructure of the antennal sensilla of aphids. I. Coeloconic and placoid sensilla. *Cell and Tissue Research*, **203**, 427–442.

Bromley, A. K., Dunn, J. A. and Anderson, M. (1980) Ultrastructure of the antennal sensilla of aphids. II. Trichoid chordotonal and campaniform sensilla. *Cell and Tissue Research*, **205**, 493–511.

Brooks, D. R. and McLennan, D. A. (1991) *Phylogeny, Ecology and Behaviour*. Chicago: University of Chicago Press.

Brough, C. N. and Dixon, A. F. G. (1989) Intraclonal trade-off between reproductive investment and size of fat body in the vetch aphid, *Megoura viciae* Buckton. *Functional Ecology*, **3**, 747–751.

Brough, C. N., Dixon, A. F. G. and Kindlmann, P. (1990) Pattern of growth and fat content of somatic and gonadal tissues of virginoparae of the vetch aphid, *Megoura viciae*. *Entomologia Experimentalis and Applicata*, **56**, 269–275.

Brown, J. H., Marquet, P. A. and Taper, M. L. (1993) Evolution of body size: consequences of an energetic definition of fitness. *American Naturalist*, **142**, 573–584.

Brown, M. I. (1975) *Intra-specific mechanisms regulating the numbers of the lime aphid*. PhD Thesis, University of Glasgow.

Brown, M. W., Hogmire, H. W. and Schmitt, J. J. (1995) Competitive displacement of apple aphid by spirea aphid (Homoptera: Aphididae) on apple as mediated by human activities. *Environmental Entomology*, **24**, 1581–1591.

Brown, V. K. (1984) Secondary succession: insect–plant relationships. *Bioscience*, **34**, 710–716.

Brown, V. K. (1985) Insect herbivores and plant succession. *Oikos*, **44**, 17–22.

Brown, V. K. and Llewellyn, M. (1985) Variation in aphid weight and reproductive potential in relation to plant growth form. *Journal of Animal Ecology*, **54**, 651–661.

Buchner, P. (1965) *Endosymbiosis of Animals with Plant Microorganisms*. New York: Interscience Publishers.
Buckton, G. B. (1879) *Monograph of British Aphids*. London: Royal Society.
Bumroongsook, S. and Harris, M. K. (1991) Nature of the conditioning effect on pecan by blackmargined aphid. *Southwest Entomology*, **16**, 267–275.
Burns, M. (1971) *Flight in the vetch aphid, Megoura viciae Buckton*. PhD Thesis, University of Glasgow.
Burstein, M. and Wool, D. (1993) Gall aphids do not select optimal galling sites (*Smynthurodes betae*; Pemphigidae). *Ecological Entomology*, **18**, 155–164.
Bush, B. L. (1969) Sympatric host race formation and speciation in frugivorous flies of the genus *Rhagoletis* (Diptera: Tephritidae). *Evolution*, **23**, 237–251.
Bush, G. L. (1975) Sympatric speciation in phytophagous parasite insects. In P. W. Price (ed.), *Evolutionary Strategies of Parasitic Insects and Mites*, pp. 187–206. New York: Plenum.
Butlin, R. K. (1990) Divergence in emergence time of host races due to differential gene flow. *Heredity*, **65**, 47–50.
Butlin, R. K. (1995) Reinforcement: an idea evolving. *Trends in Ecology and Evolution*, **10**(11), 432–434.
Calder, W. A. (1984) *Size, Function and Life History*. Cambridge, MA: Harvard University Press.
Calow, P. (1978) *Life Cycles*. London: Chapman & Hall.
Campbell, A. and Mackauer, M. (1977) Reproduction and population growth of the pea aphid (Homoptera: Aphididae) under laboratory and field conditions. *Canadian Entomologist*, **109**, 277–284.
Campbell, B. C. and Nes, W. D. (1983) A reappraisal of sterol biosynthesis and metabolism in aphids. *Journal of Insect Physiology*, **29**, 149–156.
Campbell, A., Frazer, B. D., Gilbert, N., Gutierrez, A. P. and Mackauer, M. (1974) Temperature requirements of some aphids and their parasites. *Journal of Applied Ecology*, **11**, 431–438.
Campbell, C. A. M., Dawson, G. W., Griffiths, D. C., Pettersson, J., Pickett, J., Wadhams, L. J. and Woodcock, C. M. (1990) The sex attractant pheromone of the damson-hop aphid *Phorodon humuli* (Homoptera, Aphididae). *Journal of Chemical Ecology*, **16**, 3455–3465.
Cappuccino, N. (1987) Comparative population dynamics of two goldenrod aphids: spatial patterns and temporal constancy. *Ecology*, **68**, 1634–1646.
Carter, C. I. (1972) Winter temperatures and survival of the green spruce aphid. *Forest. Commission Forest Record*, **84**, 1–10.
Carter, C. I. (1982) Susceptibility of *Tilia* species to the aphid *Eucallupterus tiliae*. In *Proceedings of the 5th International Symposium on Insect–Plant Relationships*, pp. 421–423. Wageningen.
Carter, M. C. and Dixon, A. F. G. (1982) Habitat quality and the foraging behaviour of coccinellid larvae. *Journal of Animal Ecology*, **51**, 865–878.
Carter, N., Dixon, A. F. G. and Rabbinge, R. (1982) *Cereal Aphid Populations: Biology, Simulation and Prediction*. Wageningen: Pudoc.
Carvalho, G. R., Maclean, N., Wratten, S. D., Carter, R. E. and Thurston, J. P. O. (1991) Differentiation of aphid clones using DNA fingerprints from individual aphids. *Proceedings of the Royal Society of London B*, **243**, 109–114.
Chambers, R. J. (1979) *Simulation modelling of a sycamore aphid population*. PhD Thesis, University of East Anglia.
Chambers, R. J. and Sunderland, K. D. (1983) The abundance and effectiveness of natural enemies of cereal aphids on two farms in Southern England. In R. Calvalloro (ed.), *Aphid Antagonists*, pp. 83–87. Rotterdam: Balkema.
Chambers, R. J., Wellings, P. W. and Dixon, A. F. G. (1985) Sycamore aphid numbers and population density. II. Some Processes. *Journal of Animal Ecology*, **54**, 425–442.
Chapman, R. F., Bernays, E. A. and Simpson, S. J. (1981) Attraction and repulsion of the aphid, *Cavariella aegopodi*, by plant odours. *Journal of Chemical Ecology*, **7**, 881–888.
Charnov, E. L. (1982) *The Theory of Sex Allocation*. Princeton: Princeton University Press.
Charnov, E. L. (1993) *Life History Invariants*. Oxford: Oxford University Press.
Chongrattanameteekul, W., Foster, J. E. and Araya, J. E. (1991) Biological interactions between the cereal aphids *Rhopalosiphum padi* (L.) and *Sitobion avenae* (F.) (Hom., Aphididae) on wheat. *Journal Applied Entomology*, **111**, 249–253.

Cognetti, G. (1961) Citogenetica della partenogenesi negli Afidi. *Archivo zoologico italiano*, **46**, 89–122.
Cole, R. A. (1994a) Locating a resistance mechanism to the cabbage aphid in two wild brassicas. *Entomologia Experimentalis et Applicata*, **71**, 23–31.
Cole, R. A. (1994b) Isolation of a chitin-binding lectin, with insecticidal activity in chemically-defined diets from two wild brassica species with resistance to cabbage aphid *Brevicoryne brassicae*. *Entomologia Experimentalis et Applicata*, **72**, 181–187.
Collins, M. D. and Dixon, A. F. G. (1986) The effect of egg depletion on the foraging behaviour of an aphid parasitoid. *Zeitschrift für Angewandte Entomologie*, **102**, 342–352.
Connell, J. H. (1980) Diversity and the coevolution of competitors, or the ghost of competition past. *Oikos*, **35**, 131–138.
Cottier, W. (1953) Aphids of New Zealand. *Bulletin of the New Zealand Department of Scientific and Industrial Research*, **106**.
Crane, P. R., Friis, E. M. and Pedersen, K. R. (1995) The origin and early diversification of angiosperms. *Nature*, **374**, 27–33.
Cuellar, O. (1977) Animal parthenogenesis. *Science*, **197**, 837–843.
Cushman, J. H. and Addicott, J. F. (1989) Intra- and interspecific competition for mutualists: ants as a limited and limiting resource for aphids. *Oecologia (Berlin)*, **79**, 315–321.
Cushman, J. H. and Addicott, J. F. (1991) Conditional interactions in ant-plant herbivore mutualisms. In C. R. Huxley and D. F. Cutler (eds), *Ant–Plant Interactions*, pp. 92–103. Oxford: Oxford University Press.
Dahl, M. L. (1968) Biologische und morphologische Untersuchungen bei den Formenkreis der schwarzen Kirchenlaus *Myzus cerase* F. (Homoptera: Aphididae). *Deutsche Entomologische Zeitschrift, NF*, **15**, 281–312.
Dalimin, M. N. D. (1982) *The biology and morphometrics of some species of the genus Aphis L. (Homoptera: Aphididae)*. PhD Thesis, University of London.
Davatchi, A. (1958) *Étude biologique de la faune entomologique des Pistacia sauvages et cultivés*. Université de Paris.
David, C. T. and Hardie, J. (1988) The visual responses of free-flying summer and autumn forms of the black bean aphid, *Aphis fabae*, in an automated flight chamber. *Physiological Entomology*, **13**, 277–284.
Davidson, J. (1927) The biological and ecological aspects of migration in aphids. *Scientific Progress in the Twentieth Century*, **21**, **22**, 57–69, 641–658.
Dawkins, R. (1982) *The Extended Phenotype*. Oxford: Oxford University Press.
de Geer, C. (1773) *Mémoires pour servir a l'Histoire des Insectes: Trois Mem. Suite des Pucerons*. Stockholm: P. Hasselberg.
DeLoach, C. J. (1974) Rate of increase of populations of cabbage, green peach, and turnip aphids at constant temperatures. *Annals of the Entomological Society of America*, **67**, 332–340.
Dempster, J. P. (1983) The natural control of populations of butterflies and moths. *Biological Reviews*, **58**, 461–481.
Dewar, A. M. and Carter, N. (1984) Decision trees to assess the risk of cereal aphid outbreaks in summer in England. *Bulletin of Entomological Research*, **74**, 387–398.
Dickson, R. C. (1962) Development of the spotted alfalfa aphid population in North America. *International Kongress für Entomologie (Wien)*, **1960**, **2**, 26–28.
Dighton, J. (1978a) Effects of synthetic lime aphid honeydew on populations of soil organisms. *Soil Biology and Biochemistry*, **10**, 369–76.
Dighton, J. (1978b) *In vitro* experiments simulating the possible fates of aphid honeydew sugars in soil. *Soil Biology and Biochemistry*, **10**, 53–57.
Dixon, A. F. G. (1958) Escape responses shown by certain aphids to the presence of the coccinellid, *Adalia decempunctata* (L.). *Transactions of the Royal Entomological Society of London*, **10**, 319–334.
Dixon, A. F. G. (1959) An experimental study of the searching behaviour of the predatory coccinellid beetle *Adalia decempunctata* (L.). *Journal of Animal Ecology*, **28**, 259–281.
Dixon, A. F. G. (1966) The effect of population density and nutritive status of the host on the summer reproductive activity of the sycamore aphid, *Drepanosiphum platanoides* (Schr.). *Journal of Animal Ecology*, **35**, 105–112.
Dixon, A. F. G. (1969) Population dynamics of the sycamore aphid *Drepanosiphum platanoidis* (Schr.) (Hemiptera: Aphididae): migratory and trivial flight activity. *Journal of Animal Ecology*, **38**, 585–606.

Dixon, A. F. G. (1970a) Quality and availability of food for a sycamore aphid population. In A. Waton (ed.), *Animal Populations in Relation to their Food Resources*, pp. 271–287. Oxford: Blackwell.

Dixon, A. F. G. (1970b) Stabilization of aphid populations by an aphid induced plant factor. *Nature*, **227**, 1368–1369.

Dixon, A. F. G. (1971a) The 'interval timer' and photoperiod in the determination of parthenogenetic and sexual morphs in the aphid, *Drepanosiphum platanoidis*. *Journal of Insect Physiology*, **17**, 251–260.

Dixon, A. F. G. (1971b) Migration in aphids. *Science Progress*, **59**, 41–53.

Dixon, A. F. G. (1971c) The life-cycle and host preferences of the bird cherry-oat aphid, *Rhopalosiphum padi* L. and their bearing on theories of host alternation in aphids. *Annals of Applied Biology*, **68**, 135–147.

Dixon, A. F. G. (1971d) The role of intra-specific mechanisms and predation in regulating the numbers of the lime aphid *Eucallipterus tiliae* L. *Oecologia (Berlin)*, **8**, 179–193.

Dixon, A. F. G. (1971e) The role of aphids in wood formation. I The effect of the sycamore aphid, *Drepanosiphum platanoides* (Schr.) (Aphididae) on the growth of sycamore, *Acer pseudoplatanus* (L.). *Journal of Applied Ecology*, **8**, 165–179.

Dixon, A. F. G. (1971f) The role of aphids in wood formation. II The effect of the lime aphid, *Eucallipterus tiliae* L. (Aphididae), on the growth of lime *Tilia × vulgaris* Hayne. *Journal of Applied Ecology*, **8**, 393–399.

Dixon, A. F. G. (1972a) Crowding and nutrition in the induction of macropterous alatae in *Drepanosiphum dixoni*. *Journal of Insect Physiology*, **18**, 459–464.

Dixon, A. F. G. (1972b) Fecundity of brachypterous and macropterous alatae in *Drepanosiphum dixoni* (Callaphididae, Aphididae). *Entomologia Experimentalis et Applicata*, **15**, 335–340.

Dixon, A. F. G. (1972c) The 'interval timer' photoperiod and temperature in the seasonal development of parthenogenetic and sexual morphs in the lime aphid, *Eucallipterus tiliae* L. *Oecologia (Berlin)*, **9**, 301–310.

Dixon, A. F. G. (1972d) Control and significance of the seasonal development of colour forms in the sycamore aphid *Drepanosiphum platanoides* (Schr.). *Journal of Animal Ecology*, **41**, 689–697.

Dixon, A. F. G. (1973) Metabolic acclimatization to seasonal changes in temperature in the sycamore aphid, *Drepanosiphum platanoides* (Schr.) and lime aphid, *Eucallipterus tiliae* L. *Oecologia (Berlin)*, **13**, 205–210.

Dixon, A. F. G. (1974a) Wing loading and flight activity in the sycamore aphid, *Drepanosiphum platanoides*. *Entomologia Experimentalis et Applicata*, **17**, 157–162.

Dixon, A. F. G. (1974b) Changes in the length of the appendages and the number of rhinaria in young clones of the sycamore aphid, *Drepanosiphum platanoides*. *Entomologia Experimentalis et Applicata*, **17**, 1–8.

Dixon, A. F. G. (1975a) Aphids and translocation. In M. H. Zimmerman and J. A. Milburn (eds), *Transport in Plants I. Phloem Transport*, pp. 154–170. Berlin: Springer.

Dixon, A. F. G. (1975b) Seasonal changes in fat content, form, state of gonads and length of adult life in the sycamore aphid, *Drepanosiphum platanoides* (Schr.). *Transactions of the Royal Entomological Society of London*, **127**, 87–99.

Dixon, A. F. G. (1976) Timing of egg hatch and viability of the sycamore aphid, *Drepanosiphum platanoides* (Schr.), at bud burst of sycamore, *Acer pseudoplatanus* L. *Journal of Animal Ecology*, **45**, 593–603.

Dixon, A. F. G. (1977) Aphid ecology: life cycles, polymorphism and population regulation. *Annual Review of Ecological Systems*, **8**, 329–353.

Dixon, A. F. G. (1979) Sycamore aphid numbers: the role of weather, host and aphid. In R. M. Anderson, B. D. Turner and L. R. Taylor (eds), *Population Dynamics*, pp. 105–121. Oxford: Blackwell.

Dixon, A. F. G. (1983) Insect-induced phytotoxemias: damage, tumors, and galls. *Current Topics in Vector Research*, **1**, 297–314.

Dixon, A. F. G. (1985) Structure of aphid populations. *Annual Review of Entomology*, **30**, 155–174.

Dixon, A. F. G. (1987a) Parthenogenetic reproduction and the intrinsic rate of increase of aphids. In P. Harrewijn and A. Minks (eds), *Aphids, their Biology, Natural Enemies and Control*, pp. 269–287. Amsterdam: Elsevier.

Dixon, A. F. G. (1987b) Adaptive significance of cyclical parthnogenesis in aphids. In P. Harrewijn and A. Minks (eds), *Aphids, their Biology, Natural Enemies and Control*, pp. 289–297. Elsevier: Amsterdam.

Dixon, A. F. G. (1987c) The way of life aphids: host specificity, speciation and distribution. In A. K. Minks and P. Harrewijn (eds), *Aphids, their Biology, Natural Enemies and Control*, Vol. 2, pp. 197–207. Amsterdam: Elsevier.

Dixon, A. F. G. (1987d) Aphid reproductive tactics. In J. Holman, J. Pelikan, A. F. G. Dixon and L. Weismann (eds), *Population Structure, Genetics and Taxonomy of Aphids and Thysanoptera*, pp. 3–18. The Hague: SPB Academic Publishing.

Dixon, A. F. G. (1990a) Evolutionary aspects of parthenogenetic reproduction in aphids. *Acta Phytopathology and Entomology Hungaria*, **25**, 41–56.

Dixon, A. F. G. (1990b) Ecological interactions of aphids and their host plants. In R. K. Campbell and R. D. Eikenbary (eds), *Aphid–Plant Genotype Interactions*, pp. 7–19. Amsterdam: Elsevier.

Dixon, A. F. G. (1990c) Population dynamics and abundance of deciduous tree-dwelling aphids. In M. Hunter and N. Kidd (eds), *Population Dynamics of Forest Insects*, pp. 11–23. Andover: Intercept.

Dixon, A. F. G. (1994) Individuals, populations and patterns. In S. R. Leather, A. D. Watt, W. J. Mills and K. F. A. Walters (eds), *Individuals, Populations and Patterns in Ecology*, pp. 449–476. Andover: Intercept.

Dixon, A. F. G. and Dharma, T. R. (1980a) 'Spreading of the risk' in developmental mortality: size, fecundity and reproductive rate in the black bean aphid. *Entomologia Experimentalis et Applicata*, **28**, 301–312.

Dixon, A. F. G. and Dharma, T. D. (1980b) Number of ovarioles and fecundity in the black bean aphid, *Aphis fabae*. *Entomologia Experimentalis et Applicata*, **28**, 1–14.

Dixon, A. F. G. and Glen, D. M. (1971) Morph determination in the bird cherry-oat aphid, *Rhopalosiphum padi* L. *Annals of Applied Biology*, **68**, 11–21.

Dixon, A. F. G. and Guo, Y. (1993) Egg and cluster size in ladybird beetles (Coleoptera: Coccinellidae): the direct and indirect effects of aphid abundance. *European Journal of Entomology*, **90**, 457–463.

Dixon, A. F. G. and Howard, M. T. (1986) Dispersal in aphids, a problem in resource allocation. In W. Danthanaryana (ed.), *Insect Flight: Dispersal and Migration*, pp. 145–151. Berlin: Springer.

Dixon, A. F. G. and Kindlmann, P. (1990) Role of plant abundance in determining the abundance of herbivorous insects. *Oecologia (Berlin)*, **83**, 281–283.

Dixon, A. F. G. and Kindlmann, P. (1994) Optimum body size in aphids. *Ecological Entomology*, **19**, 121–126.

Dixon, A. F. G. and Kindlmann, P. (1998) Cost of flight apparatus and optimum body size of aphid migrants. *Ecology* (in press).

Dixon, A. F. G. and Kundu, R. (1994) Ecology of host alternation in aphids. *European Journal of Entomology*, **91**, 63–70.

Dixon, A. F. G. and Logan, M. (1973) Leaf size and availability of space to the sycamore aphid *Drepanosiphum platanoides*. *Oikos*, **24**, 58–63.

Dixon, A. F. G. and McKay, S. (1970) Aggregation in the sycamore aphid *Drepanosiphum platanoides* (Schr.) (Hemiptera: Aphididae) and its relevance to the regulation of population growth. *Journal of Animal Ecology*, **39**, 439–454.

Dixon, A. F. G. and Mercer, D. R. (1983) Flight behaviour in the sycamore aphid: Factors affecting take-off. *Entomologia Experimentalis et Applicata*, **33**, 43–49.

Dixon, A. F. G. and Stewart, W. A. (1975) Function of the siphunculi in aphids with particular reference to the sycamore aphid, *Drepanosiphum platanoides*. *Journal of Zoology (London)*, **175**, 279–289.

Dixon, A. F. G. and Wellings, P. W. (1982) Seasonality and reproduction in aphids. *International Journal of Invertebrate Production*, **5**, 83–89.

Dixon, A. F. G. and Wratten, S. D. (1971) Laboratory studies on aggregation, size and fecundity in the black bean aphid, *Aphis fabae* Scop. *Bulletin of Entomological Research*, **61**, 97–111.

Dixon, A. F. G., Burns, M. D. and Wangboonkong, S. (1968) Migration in aphids: response to current adversity. *Nature*, **220**, 1337–1338.

Dixon, A. F. G., Chambers, R. J. and Dharma, T. R. (1982) Factors affecting size in aphids with particular reference to the black bean aphid, *Aphis fabae*. *Entomologia Experimentalis et Applicata*, **32**, 123–128.

Dixon, A. F. G., Kindlmann, P., Leps, J. and Holman, J. (1987) Why are there so few species of aphids, especially in the tropics? *American Naturalist*, **29**, 580–592.

Dixon, A. F. G., Croghan, P. C. and Gowing, R. P. (1990) The mechanism by which aphids adhere to smooth surfaces. *Journal of Experimental Biology*, **152**, 243–253.

Dixon, A. F. G., Horth, S. and Kindlmann, P. (1993a) Migration in insects: cost and strategies. *Journal of Animal Ecology*, **62**, 182–190.

Dixon, A. F. G., Wellings, P. W., Carter, C. and Nichols, J. F. A. (1993b) The role of food quality and competition in shaping the seasonal cycle in the reproductive activity of the sycamore aphid. *Oecologia (Berlin)*, **95**, 89–92.

Dixon, A. F. G., Kindlmann, P. and Jarosik, V. (1995) Body size distribution in aphids: relative surface area of specific plant structures. *Ecological Entomology*, **20**, 111–117.

Dixon, A. F. G., Kindlmann, P. and Sequeira, R. (1996) Population regulation in aphids. In R. B. Floyd, A. W. Sheppard and P. J. D. Barro (eds), *Frontiers of Population Biology*, pp. 77–88. Melbourne: CSIRO Publishing.

Dobzhansky, T. (1940) Speciation as a stage in evolutionary divergence. *American Naturalist*, **74**, 312–321.

Dobzhansky, T. (1951) *Genetics and the Origin of Species*, 3rd edn. New York: Columbia University Press.

Douglas, A. E. (1988a) On the source of sterols in the green peach aphid, *Myzus persicae*, reared on holidic diets. *Journal of Insect Physiology*, **34**, 403–408.

Douglas, A. E. (1988b) Sulphate utilisation in an aphid symbiosis. *Insect Biochemistry*, **18**, 599–605.

Douglas, A. E. (1989) Mycetocyte symbiosis in insects. *Biological Reviews*, **64**, 409–434.

Douglas, A. E. (1992) Requirements of pea aphids (*Acyrthosiphon pisum*) for their symbiotic bacteria. *Entomologia Experimentalis and Applicata*, **65**, 195–198.

Douglas, A. E. (1993) The nutritional quality of phloem sap utilised by natural aphid populations. *Ecological Entomology*, **18**, 31–38.

Douglas, A. E. and Dixon, A. F. G. (1987) The mycetocyte symbiosis of aphids: variation with age and morph in virginoparae of *Megoura viciae* and *Acyrthosiphon pisum*. *Journal of Insect Physiology*, **33**, 109–113.

Douglas, A. E. and Prosser, W. A. (1992) Synthesis of the essential amino acid tryptophan in the pea aphid (*Acyrthosiphon pisum*) symbiosis. *Journal of Insect Physiology*, **38**, 565–568.

Doumbia, M., Hemptinne, J.-L. and Dixon, A. F. G. (1997) Assessment of patch quality by ladybirds: role of larval tracks. *Oecologia* (in press).

Downing, N. (1978) Measurements of the osmotic concentrations of stylet sap, haemolymph and honeydew from an aphid under osmotic stress. *Journal of Experimental Biology*, **77**, 247–250.

Dreyer, D. L., Jones, K. C. and Molyneux, R. J. (1985) Feeding deterrency of some pyrrolizidine, indolizidine, and quinolizidine alkaloids towards the pea aphid (*Acyrthosiphon pisum*) and evidence for phloem transport of indolizidine alkaloid swainsonine. *Journal of Chemical Ecology*, **11**, 1045–1051.

Dritschilo, W., Krummel, J., Nafus, D. and Pimentel, D. (1979) Herbivorous insects colonising cyanogenic and acyanogenic *Trifolium repens*. *Heredity*, **42**, 49–56.

Dunn, J. A. (1959) The survival in soil of apterae of lettuce root aphid. *Annals of Applied Biology*, **47**, 766–771.

Dunn, J. A. (1960) The formation of galls by some species of *Pemphigus* (Homoptera: Aphididae). *Marcellia*, **30**, 155–167.

Dussourd, D. E. (1995) Entrapment of aphids and whiteflies in lettuce latex. *Annals of the Entomological Society of America*, **88**, 163–172.

Eastop, V. F. (1958) *A study of the Aphididae (Homoptera) of East Africa*. London: HMSO.

Eastop, V. F. (1961) *A study of the Aphididae (Homoptera) of West Africa*. London: HMSO.

Eastop, V. F. (1966) A taxonomic study of Australian Aphidoidea (Homoptera). *Australian Journal of Zoology*, **14**, 399–592.

Eastop, V. F. (1972) Deductions from the present day host plants of aphids and related insects. In *Insect/Plant Relationships*, Vol. 6, pp. 157–178. London: Symposium of the Royal Entomological Society.

Eastop, V. F. (1977) Worldwide importance of aphids as virus vectors. In K. F. Harris and K. Maramorosch (eds), *Aphids as Virus Vectors*, pp. 3–62. London: Academic Press.

Eastop, V. F. (1978) Diversity of the Sternorrhyncha within major climate zones. In *Diversity of Insect Faunas*, pp. 71–88. London: Symposium of the Royal Entomological Society.

Eastop, V. F. (1986) Aphid plant associations. In A. R. Stone and D. L. Hawksworth (eds), *Coevolution and Systematics*, pp. 35–54. Oxford: Clarendon Press.

Eastop, V. F. (1991) Host plant range and virus transmission by aphids. *Fitopatologia Brasiliansis*, **16**, 241–245.

Edinger, B. B. (1985) Conditional mutualism in three aphid-tending ants. *Bulletin of the Ecological Society of America*, **66**, 168.

Edmunds, G. F. and Alstad, D. N. (1978) Coevolution in insect herbivores and conifers. *Science*, **199**, 941–945.

Edson, J. L. (1985) The influences of predation and resource subdivision on the coexistence of goldenrod aphids. *Ecology*, **66**, 1736–1743.

Ehrlich, P. R. and Raven, P. H. (1964) Butterflies and plants: a study in coevolution. *Evolution*, **18**, 586–608.

El-Ziady, S. (1960) Further effects of *Lasius niger* L. on *Aphis fabae* Scopoli. *Proceedings of the Royal Entomological Society of London (A)*, **35**, 30–38.

Elton, C. S. (1925) The dispersal of insects to Spitsbergen. *Transactions of the Entomological Society of London*, **1925**, 289–299.

Elton, C. S. (1927) *Animal Ecology*. London: Sidgewick and Jackson.

Entwistle, J. C. and Dixon, A. F. G. (1986) Short-term forecasting of the peak population density of the grain aphid (*Sitobion avenae*) on wheat. *Annals of Applied Biology*, **109**, 215–222.

Entwistle, J. C. and Dixon, A. F. G. (1987) Short-term forecasting of wheat yield loss caused by the grain aphid (*Sitobion avenae*) in summer. *Annals of Applied Biology*, **111**, 489–508.

Ernst, W. H. O. (1987) Impact of the aphid *Aulocorthum solani* KLTB on growth and reproduction of winter and summer annual life forms of *Senecio sylvaticus* L. *Acta Oecologia/Oecologia Generalis*, **8**, 537–547.

Esau, K. (1965) *Plant Anatomy*. New York: Wiley.

Eschrich, W. (1970) Biochemistry and fine structure of phloem in relation to transport. *Annual Review of Plant Physiology*, **21**, 30–38.

Evans, E. W. and Dixon, A. F. G. (1986) Cues for oviposition by ladybird beetles (Coccinellidae): response to aphids. *Journal of Animal Ecology*, **55**, 1027–1034.

Ewing, H. E. (1916) Eighty-seven generations in a parthenogenetic pure line of *Aphis avenae* Fab. *Biological Bulletin*, **31**, 53–112.

Fahn, A. (1979) *Secretary Tissues in Plants*. New York: Academic.

Faith, D. P. (1979) Strategies of gall formation in *Pemphigus* aphids. *Journal of the New York Entomological Society*, **87**, 21–37.

Fenchel, T. (1974) Intrinsic rate of natural increase: the relationship with body size. *Oecologia (Berlin)*, **14**, 317–326.

Fisher, D. B., Wright, J. P. and Mittler, T. E. (1984) Osmoregulation by the aphid *Myzus persicae:* a physiological role for honeydew oligosaccharides. *Journal of Insect Physiology*, **30**, 387–.

Fisher, M. (1982) *Morph determination in Elatobium abietinum (Walk.) the green spruce aphid.* PhD Thesis, University of East Anglia.

Fisher, R. A. (1930) *The Genetical Theory of Natural Selection*. Oxford: Oxford University Press.

Foldi-Hope, C. (1990) Ultrastructure des cornicules de *Rhopalosiphum padi* L. (Homoptera: Aphididae). *International Journal of Insect Morphology and Embryology*, **19**, 25–33.

Forbes, A. R. (1966) Electron microscope evidence for nerves in the mandibular stylets of the green peach aphid. *Nature*, **212**, 726.

Forrest, J. M. S. (1970) The effects of maternal and larval experience on morph determination in *Dysaphis devecta*. *Journal of Insect Physiology*, **16**, 2281–2282.

Forrest, J. M. S. (1971) The growth of *Aphis fabae* as an indicator of the nutritional advantages of galling to the apple aphid *Dysaphis devecta*. *Entomologia Experimentalis et Applicata*, **14**, 477–483.

Forrest, J. M. S. and Dixon, A. F. G. (1975) The induction of leaf-roll galls by the apple aphids *Dysaphis devecta* and *D. plantaginea*. *Annals of Applied Biology*, **81**, 281–288.

Foster, W. A. (1978) Dispersal behaviour of an intertidal aphid. *Journal of Animal Ecology*, **47**, 653–659.

Foster, W. A. (1984) The distribution of the sea-lavender aphid *Staticobium staticis* on a marine saltmarsh and its effect on host plant fitness. *Oikos*, **42**, 97–104.

Foster, W. A. (1990) Experimental evidence for effective and altruistic colony defence against natural predation by soldiers of the gall-forming aphid *Pemphigus spyrothecae* (Hemiptera: Pemphigidae). *Behavioural Ecology and Sociobiology*, **27**, 421–430.

Foster, W. A. and Benton, T. G. (1992) Sex ratio, local mate competition and mating behaviour in the aphid *Pemphigus spyrothecae*. *Behaviour, Ecology and Sociobiology*, **30**, 297–307.

Foster, W. A. and Treherne, J. E. (1978) Dispersal mechanisms in an intertidal aphid. *Journal of Animal Ecology*, **47**, 205–217.

Frank, A. B. (1896) *Die Krankheiten der Pflanzen Bd III: Die tierparasitären Krankheiten der Pflanzen*, 2nd edn. Breslau: Trewendt.

Frazer, B. D. and Gill, B. (1981) Age, fecundity, weight, and the intrinsic rate of increase of the lupine aphid *Macrosiphum albifrons* (Homoptera: Aphididae). *Canadian Entomologist*, **113**, 739–745.

Fretwell, S. D. and Lucas, H. L. (1969) On territorial behaviour and other factors influencing habitat distribution in birds I. Theoretical development. *Acta Biotheoretica*, **19**, 16–36.

Fukatsu, T. (1994) Endosymbiosis of aphids with microorganisms: a model case of dynamic endosymbiotic evolution. *Plant Species Biology*, **9**, 145–154.

Fukatsu, T. and Ishikawa, H. (1992a) Soldier and male of an eusocial aphid *Colophina arma* lack endosymbiont: implications for physiological and evolutionary interaction between host and symbiont. *Journal of Insect Physiology*, **38**, 1033–1042.

Fukatsu, T. and Ishikawa, H. (1992b) Synthesis and localisation of symbionin, an aphid endosymbiont protein. *Insect Biochemistry and Molecular Biology*, **22**, 167–174.

Fukatsu, T. and Ishikawa, H. (1993) Occurence of Chaperonin 60 and Chaperonin 10 in primary and secondary bacterial symbionts of aphids: implications for the evolution of an endosymbiotic system in aphids. *Journal of Molecular Evolution*, **36**, 568–577.

Fukatsu, T. and Ishikawa, H. (1994) Differentiation of aphid clones by arbitrarily primed polymerase chain reaction (AP-PCR) DNA fingerprinting. *Molecular Ecology*, **3**, 187–192.

Fukatsu, T., Aoki, S., Kurosu, U. and Ishikawa, H. (1994) Phylogeny of Cerataphidini aphids revealed by their symbiotic microorganisms and basic structure of their galls: implications for host–symbiont coevolution and evolution of sterile soldier castes. *Zoological Science*, **11**, 613–623.

Futuyma, D. J. and Moreno, G. (1988) The evolution of ecological specialisation. *Annual Review of Ecological Systems*, **19**, 207–233.

Futuyma, D. J. and Phillippi, T. E. (1987) Genetic variation and covariation in responses to host plants by *Alsophila pometaria* (Lepidoptera: Geometridae). *Evolution*, **41**, 269–279.

Gange, A. C. (1985) *The ecology of the alder aphid (Pterocallis alni (de Geer)) and its role in integrated orchard pest management.* PhD Thesis, University of London.

Gange, A. C. (1995) Aphid performance in an alder (*Alnus*) hybrid zone. *Ecology*, **76**, 2074–2083.

Gaston, K. J. (1988) The intrinsic rates of increase of insects of different sizes. *Ecological Entomology*, **14**, 395–409.

Gaston, K. J. (1994) *Rarity*. London: Chapman & Hall.

Gaston, K. J. and Lawton, J. H. (1988) Patterns in the distribution and abundance of insect populations. *Nature*, **331**, 709–712.

Geiler, H. (1956) Eichhörnchen, *Sciurus vulgaris fuscoater* Altum, 1876, als Vertilger von Blattläusen (Aphidae). *Saugentierkundliche Mitteilungen*, **4**, 13–15.

Ghosh, A. K. (1980) *The Fauna of India and the Adjacent Countries. Homoptera, Aphidoidea Part I. General Introduction and Chaitophorinae*. Madras: Zoological Survey of India.

Ghosh, A. K. (1982) *The Fauna of India and the Adjacent Countries. Homoptera, Aphidoidea Part II. Subfamily Lachninae*. Madras: Zoological Survey of India.

Ghosh, A. K. (1984) *The Fauna of India and the Adjacent Countries. Homoptera, Aphidoidea Part III. Subfamily Pemphiginae*. Madras: Zoological Survey of India.

Gibson, R. W. and Pickett, J. A. (1983) Wild potato repels aphids by release of aphid alarm pheromone. *Nature*, **302**, 608–609.

Gilbert, L. E. and Singer, M. C. (1975) Butterfly ecology. *Annual Review of Ecological Systems*, **6**, 365–397.

Girousse, C., Febvay, G., Rhabe, Y. and Bournoville, R. (1990) Reproductive rate of the pea aphid related to phloem sap composition of alfalfa. In D. C. Peters, J. A. Webster and C. S.

Chlouber (eds), *Aphid–Plant Interactions: from Populations to Molecules*, pp. 304. Stillwater, OH: Oklahoma State University.

Givovich, A., Sandstrom, J., Niemeyer, H. M. and Pettersson, J. (1994) Presence of a hydroxamic acid glucoside in wheat phloem sap, and its consequences for performance of *Rhopalosiphum padi* (L.) (Homoptera: Aphididae). *Journal of Chemical Ecology*, **20**, 1923–1930.

Goszczynski, W. and Cichocka, E. (1978) Mszyce Zerujace na czesciach podziemnych marchwi. *Zeszyty Problemowe Postepow Nauk Rolniezych*, **208**, 46–65.

Gould, F. (1979) Rapid host range evolution in a population of the polyphagous mite *Tetranychus urticae*. *Evolution*, **33**, 791–802.

Grier, C. C. and Vogt, D. J. (1990) Effects of aphid honeydew on soil nitrogen availability and net primary production in an *Alnus rubra* plantation in Western Washington. *Oikos*, **57**, 114–118.

Grubb, P. J., Kelly, D. and Mitchley, J. (1982) The control of relative abundance in communities of herbaceous plants. In E. I. Newman (ed.), *The Plant Community as a Working Mechanism*, pp. 79–97. Oxford: Blackwell.

Gruppe, A. (1988) Elektrophoretische Untersuchungen zur Unterscheidung der Subspecies von *Myzus cerasi* F. (Hom., Aphididae). *Zeitschrift für Angewandte Entomologie*, **105**, 460–465.

Guldemond, J. A. (1987) Biosystematics of the *Cryptomyzus* complex: preliminary results on host plant acceptance, electrophoretic data and sexual isolation. In J. Holman, J. Pelikan, A. F. G. Dixon and L. Weismann (eds), *Population Structure, Genetics and Taxonomy of Aphids and Thysanoptera*, pp. 29–38. Den Haag: SPB Academic Publishing.

Guldemond, J. A. (1990a) *On aphids, their host plants and speciation.* PhD Thesis, Wageningen University, The Netherlands.

Guldemond, J. A. (1990b) Choice of host plant as a factor in reproductive isolation of the aphid genus *Cryptomyzus* (Homoptera, Aphididae). *Ecological Entomology*, **15**, 43–51.

Guldemond, J. A. (1990c) Evolutionary genetics of the aphid *Cryptomyzus* with a preliminary analysis of the inheritance of host preference, reproductive performance and host alternation. *Entomologia Experimentalis et Applicata*, **57**, 65–76.

Guldemond, J. A. (1991a) Biosystematic and morphometric study of the *Cryptomyzus galeopsidis alboapicalis* complex (Homoptera, Aphididae), with a key to and notes on the Cryptomyzus species of Europe. *Netherlands Journal of Zoology*, **41**, 1–31.

Guldemond, J. A. (1991b) Host plant relationships and life cycles in the aphid genus *Cryptomyzus*. *Entomologia Experimentalis et Applicata*, **58**, 21–30.

Guldemond, J. A. and Dixon, A. F. G. (1994) Specificity and daily cycle of release of sex pheromones in aphids: a case of reinforcement? *Biological Journal of the Linnean Society*, **52**, 287–303.

Guldemond, J. A., Dixon, A. F. G., Pickett, J. A., Wadhams, L. J. and Woodcock, C. M. (1993) Specificity of sex pheromones and the role of host plant odour in the olfactory attraction of males and mate recognition in the aphid *Cryptomyzus*. *Physiological Entomology*, **18**, 137–143.

Guldemond, J. A. and Mackenzie, A. (1994) Sympatric speciation in aphids. I. Host race formation by escape from gene flow. In S. R. Leather, A. D. Watt, N. J. Mills and K. F. A. Walters (eds), *Individuals, Populations and Patterns in Ecology*, pp. 367–378. Andover: Intercept.

Guldemond, J. A., Dixon, A. F. G. and Tigges, W. T. (1994) Mate recognition in *Cryptomyzus* aphids: copulation and insemination. *Entomologia Experimentalis et Applicata*, **73**, 67–75.

Haine, E. (1955) Aphid take-off in controlled wind speeds. *Nature*, **175**, 474.

Hairston, M. G., Smith, F. E. and Slobodkin, L. B. (1960) Community structure, population control, and competition. *American Naturalist*, **94**, 421–425.

Hajek, A. E. and Dahlsten, D. L. (1986) Coexistence of three species of leaf-feeding aphids (Homoptera) on *Betula pendula*. *Oecologia*, **68**, 380–386.

Hales, D. F. (1976) Biology of *Neophyllaphis brimblecombei* Carver (Homoptera: Aphididae) in the Sydney Region. *Australian Zoologist*, **19**, 77–83.

Hales, D. F. and Mittler, T. E. (1983) Precocene causes male determination in the aphid *Myzus persicae*. *Journal of Insect Physiology*, **29**, 819–823.

Hamilton, P. A. (1973) The biology of *Aphelinus flavus* [Hym. Aphelinidae], a parasite of the sycamore aphid *Drepanosiphum platanoidis* [Hemipt. Aphididae]. *Entomophaga*, **18**, 449–462.

Hamilton, P. A. (1974) The biology of *Monoctonus pseudoplatani*, *Trioxys cirsii* and *Dyscritulus planiceps*, with notes on their effectiveness as parasites of the sycamore aphid, *Drepanosiphum platanoides*. *Annales de la Société Entomologique de France*, **10**, 821–840.

Hamilton, W. D. (1967) Extraordinary sex ratios. *Science*, **156**, 477–488.

Hamilton, W. D. (1987) Kinship, recognition and disease: constraints of social evolution. In Y. Itô, J. L. Brown and J. Kikkawa (eds), *Animal Societies: Theories and Facts*, pp. 81–102. Tokyo: Japan Scientific Societies Press.

Hanski, I. (1980) Spatial patterns and movements in coprophagous beetles. *Oikos*, **34**, 293–310.

Hanski, I. (1982) On patterns of temporal and spatial variation in animal populations. *Annales Zoologie Fennici*, **19**, 21–37.

Harada, H. and Ishikawa, H. (1993) Gut microbe of aphid closely related to its intracellular symbiont. *Bio Systems*, **31**, 185–191.

Hardie, J. (1991) Contribution of sex pheromone to mate location and reproductive isolation in aphid species (Homoptera: Aphidinae). *Entomologia Generalis*, **16**, 249–256.

Hardie, J., Storer, J. R., Nottingham, S. F., Peace, K., Harrington, R., Merritt, L. A., Wadhams, L. J. and Wood, D. K. (1994) The interaction of sex pheromone and plant volatiles for field attraction of male bird-cherry aphid, *Rhopalosiphum padi*. *Brighton Crop Protection Conference*, pp. 1223–1230.

Hardie, J., Visser, J. H. and Piron, P. G. M. (1995) Peripheral odour perception by adult aphid forms with the same genotype but different host-plant preferences. *Journal of Insect Physiology*, **41**, 91–97.

Hare, J. D. and Kennedy, G. G. (1986) Genetic variation in plant–insect associations: survival of *Leptinotarsa decemlineata* populations on *Solanum carolinense*. *Evolution*, **40**, 1031–1043.

Harrewijn, P. (1990) Resistance mechanisms of plant genotypes to various aphid species. In R. K. Campbell and E. D. Eikenbary (eds), *Aphid–Plant Genotype Interactions*, pp. 117–130. Amsterdam: Elsevier.

Harrington, R. and Taylor, L. R. (1990) Migration for survival: fine-scale population redistribution in an aphid *Myzus persicae*. *Journal of Animal Ecology*, **59**, 1177–1193.

Harrington, R., Dewar, A. M. and George, B. (1989) Forecasting the incidence of virus yellows in sugar beet in England. *Annals of Applied Biology*, **114**, 459–469.

Harris, K. F. (1977) An ingestion–egestion hypothesis of non-circulative virus transmission. In K. F. Harris and K. Maramorosch (eds), *Aphids as Virus Vectors*, pp. 165–220. New York: Academic Press.

Hartmann, T., Ehmke, A., Eilert, U., von Borstel, K. and Theuring, C. (1989) Sites of synthesis, translocation and accumulation of pyrrolizidine alkaloid N-oxides in *Senecio vulgaris* L. *Planta*, **177**, 98–107.

Harvey, P. H. and Pagel, M. D. (1991) *The Comparative Method in Evolutionary Biology*. Oxford: Oxford University Press.

Haukioja, E. (1991) Cyclic fluctuations in density: interactions between a defoliator and its host tree. *Acta Oecologia*, **12**, 77–88.

Hayhurst, P. (1909) Observations on a gall aphid (*Aphis atriplicis* L.). *Annals of the Entomological Society of America*, **2**, 88–99.

Heathcote, G. D. (1974) Aphids caught on sticky traps in eastern England in relation to the spread of yellowing viruses of sugar beet. *Bulletin of Entomological Research*, **64**, 669–676.

Heie, O. E. (1964) Aphids collected in Iceland in August, 1961 (Homoptera, Aphididae). *Entomologiske Meddelser*, **32**, 220–235.

Heie, O. E. (1967) Studies on fossil aphids (Homoptera: Aphidoidea). *Spolia Zoologica Musei Hauniensis*, **26**, 1–274.

Heie, O. E. (1972) Bladlus på birk i Danmark (Hom., Aphidoidea). *Entomologiske Meddelser*, **40**, 81–105.

Heie, O. E. (1980) *The Aphidoidea (Hemiptera) of Fennoscandia and Denmark. I. General Part. The Families Mindaridae, Hormaphididae, Thelaxidae, Anoeciidae and Pemphigidae*, Vol. 9. Fauna Entomologica Scandinavica.

Heie, O. E. (1982) *The Aphidoidea (Hemiptera) of Fennoscandia and Denmark. II.* Vol. 11. Fauna Entomologica Scandinavica.

Heie, O. E. (1986) *The Aphidoidea (Hemiptera) of Fennoscandia and Denmark. III. Family Aphididae: subfamily Pterocomatinae and tribe Aphidinae of subfamily Aphidinae*, Vol. 17. Fauna Entomologica Scandinavica.

Heie, O. (1987) Palaeontology and phylogeny. In A. K. Minks and P. Harrewijn (eds), *Aphids, their Biology, Natural Enemies and Control*, Vol. 2A, pp. 367–391. Amsterdam: Elsevier.

Heie, O. E. (1992) *The Aphidoidea (Hemiptera) of Fennoscandia and Denmark. IV*, Vol. 25. Fauna Entomologica Scandinavica.

Heie, O. (1994a) Aphid ecology in the past and a new view on the evolution of Macrosiphini. In S. R. Leather, A. D. Watt, N. J. Mills and K. F. A. Walters (eds), *Individuals, Populations and Patterns in Ecology*, pp. 409–418. Andover: Intercept.

Heie, O. (1994b) *The Aphidoidea (Hemiptera) of Fennoscandia and Denmark. V*, Vol. 28. Fauna Entomologica Scandinavica.

Heie, O. E. (1995) *The Aphidoidea (Hemiptera) of Fennoscandia and Denmark. VI*, Vol. 31. Fauna Entomologica Scandinavica.

Hemptinne, J.-L., Dixon, A. F. G. and Coffin, J. (1992) Attack strategy of ladybird beetles (Coccinellidae): factors shaping their numerical response. *Oecologia (Berlin)*, **90**, 238–245.

Hemptinne, J.-L., Dixon, A. F. G., Doucet, J.-L. and Petersen, J.-E. (1993) Optimal foraging by hoverflies (Diptera: Syrphidae) and ladybirds (Coleoptera: Coccinellidae): mechanisms. *European Journal of Entomology*, **90**, 451–455.

Hennig, E. (1966) Zur Histologie und Funktion von Einstichen der Schwarzen Bohnenlaus (*Aphis fabae* Scop.) in *Vicia faba*-Pflanzen. *Journal of Insect Phsyiology*, **12**, 65–76.

Herbert, P. D. N., Finston, T. L. and Foottit, R. (1991) Patterns of genetic diversity in the sumac gall aphid *Melaphis rhois*. *Genome*, **34**, 757–762.

Herger, P. (1975) Einfluss von Farbe und Nahrungszusammensetzung auf das Saugverhalten der kunstlich ernahrten Ampferblattlaus *Aphis rumicis* (Homoptera: Aphididae). *Entomologica Germanica*, **2**, 149–166.

Heriot, A. D. (1938) Biological and morphological differences between *Eriosoma crataegi* (Oestlund) and *Eriosoma lanigera* (Haus.). *Proceedings of the British Columbia Entomological Society*, **1937**, 22–32.

Higuchi, H. (1972) A taxonomic study of the sub-family Callipterinae in Japan. *Insecta Matsumurana*, **35(2)**, 19–126.

Hille Ris Lambers, D. (1938) Contributions to a monograph of the Aphididae of Europe. I. *Temminckia*, **3**, 1–44.

Hille Ris Lambers, D. (1939) Contributions to a monograph of the Aphididae of Europe. II. *Temminckia*, **4**, 1–134.

Hille Ris Lambers, D. (1945) De Bloedulekkenluis van appel, *Sappahis devecta* (Walker). *Tijdschrift over plantenzieleten*, **51**, 57–72.

Hille Ris Lambers, D. (1947) Contributions to a monograph of the Aphididae of Europe. III. *Temminckia*, **7**, 179–319.

Hille Ris Lambers, D. (1949) Contributions to a monograph of the Aphididae of Europe. IV. *Temminckia*, **8**, 182–323.

Hille Ris Lambers, D. (1950) Host plants and aphid classification. In *Proceedings of the 8th International Congress in Entomology*, 1948, pp. 141–148. Stockholm,

Hille Ris Lambers, D. (1953) Contributions to a monograph of the Aphididae of Europe. V. *Temminckia*, **9**, 1–176.

Hille Ris Lambers, D. (1979) Aphids on Botanists? *Symbolae Botanicae Uppsaliensis*, **22**, 114–119.

Hodek, I. (1973) *Biology of Coccinellidae*. Prague: Academia.

Holler, C., Borgemeister, C., Haardt, H. and Powell, W. (1993) The relationship between primary parasitoids and hyperparasitoids of cereal aphids: an analysis of field data. *Journal of Animal Ecology*, **62**, 12–21.

Holman, J. (1971) Factors influencing the host range of the aphids in the tropics. In *Proceedings of the 13th International Congress on Entomology*, **2**, pp. 339–340. Moscow.

Holt, R. D. and Lawton, J. H. (1994) The ecological consequences of shared natural enemies. *Annual Review Ecology and Systematics*, **25**, 495–520.

Honek, A. (1978) Trophic regulation of postdiapause ovariole maturation in *Coccinella septempunctata* (Col.: Coccinellidae). *Entomophaga*, **23**, 212–216.

Houk, E. J. and Griffiths, G. W. (1980) Intracellular symbiotes of the Homoptera. *Annual Review of Entomology*, **25**, 161–187.

Hussain, A., Forrest, J. M. S. and Dixon, A. F. G. (1974) Sugar, organic acid, phenolic acid and plant growth regulator content of extracts of honeydew of the aphids, *Myzus persicae* and its host plant, *Raphanus sativus*. *Annals of Applied Biology*, **78**, 65–73.

Huxley, T. H. (1858) On the agamic reproduction and morphology of *Aphis* – Part 1. *Transactions of the Linnean Society*, **22**, 193–219.

Ibbotson, A. and Kennedy, J. S. (1950) The distribution of aphid infestation in relation to leaf age II. The progress of *Aphis fabae* Scop. infestations on sugar beet in pots. *Annals of Applied Biology*, **37**, 680–696.

Inbar, M. and Wool, D. (1995) Phloem-feeding specialists sharing a host tree: resource partitioning minimizes interference competition among galling aphid species. *Oikos*, **73**, 109–119.

Ishikawa, H. (1984) Molecular aspects of intracellular symbiosis in the aphid mycetocyte. *Zoological Science*, **1**, 509–522.

Istock, C. A. (1967) The evolution of complex life history phenomena: an ecological perspective. *Evolution*, **21**, 211–234.

Istock, C. A. (1970) Natural selection in ecologically and genetically defined populations. *Behavioral Science*, **15**, 101–115.

Ito, Y. (1994) A new epoch in joint studies of social evolution: molecular and behavioural ecology of aphid soldiers. *Trends in Ecology and Evolution*, **9**, 363–365.

Ito, Y., Tanaka, S., Tukawa, J. and Tsuji, K. (1995) Factors affecting the proportion of soldiers in eusocial bamboo aphid, *Pseudoregma bambucicola*, colonies. *Ethology, Ecology and Evolution*, **7**, 335–345.

Jackson, D. L. and Dixon, A. F. G. (1996) Factors determining the distribution of the green spruce aphid, *Elatobium abietinum*, on young and mature needles of spruce. *Ecological Entomology*, **21**, 358–364.

James, A. C., Jakubchak, J., Riley, M. P. and Jaenike, J. (1988) On the causes of monophagy in *Drosophila quinaria*. *Evolution*, **42**, 626–630.

James, B. D. and Luff, M. L. (1982) Cold-hardiness and development of eggs of *Rhopalosiphum insertum*. *Ecological Entomology*, **7**, 277–282.

Janzen, D. H. (1977) What are dandelions and aphids? *American Naturalist*, **111**, 586–589.

Jenkins, R. L. (1991) *Colour and symbionts of aphids*. PhD Thesis, University of East Anglia.

Jeremy, T. (1985) Is there competition between phytophagous insects? *Zeitschrift für Zoologie System Evolutionsforsch*, **23**, 275–285.

Johnson, B. (1957) Studies on the degeneration of the flight muscles of alate aphids. I. A comparative study of the occurrence of muscle breakdown in relation to reproduction in aphids. *Journal of Insect Physiology*, **1**, 248–256.

Johnson, B. (1958) Factors affecting the locomotor and settling response of alate aphids. *Animal Behaviour*, **6**, 9–26.

Johnson, B. (1959) Studies on the degeneration of the flight muscles of alate aphids. II. Histology and control of muscle breakdown. *Journal of Insect Physiology*, **3**, 367–377.

Johnson, C. G. (1969) *Migration and Dispersal of Insects by Flight*. London: Methuen.

Johnson, C. G., Taylor, L. R. and Haine, E. (1957) The analysis and reconstruction of diurnal flight curves in alienicolae of *Aphis fabae* Scop. *Annals of Applied Biology*, **45**, 682–701.

Johnson, M. C., Dahlman, D. H., Siegel, M. R., Bush, L. P., Latch, C. G. M., Potter, D. A. and Varney, D. R. (1985) Insect feeding deterrents in endophyte-infected tall fescue. *Applied Environmental Microbiology*, **49**, 568–571.

Jones, A. T. (1988) The influence of cultivating new raspberry varieties on the incidence of viruses in raspberry crops in the UK. *Aspects of Applied Biology*, **17**, 179–186.

Jones, M. G. (1942a) A description of *Aphis (Doralis) rumicis*, L. and comparison with *Aphis (Doralis) fabae*, Scop. *Bulletin of Entomological Research*, **33**, 5–20.

Jones, M. G. (1942b) The summer hosts of *Aphis fabae* Scop. *Bulletin of Entomological Research*, **33**, 161–169.

Judge, F. D. (1967) Overwintering in *Pemphigus bursarius* (L.). *Nature*, **216**, 1041–1042.

Kan, E. (1988a) Assessment of aphid colonies by hoverflies. I. Maple aphids and *Episyrphus balteatus* (de Geer) (Diptera: Syrphidae). *Journal of Ethology*, **6**, 39–48.

Kan, E. (1988b) Assessment of aphid colonies by hoverflies. II. Pea aphids and 3 syrphid species; *Betasyrphus serarius* (Wiedemann), *Metasyrphus frequens* (Matsumura) and *Syrphus vitripennis* (Meigen) (Diptera: Syrphidae). *Journal of Ethology*, **6**, 135–142.

Kan, E. (1989) Assessment of aphid colonies by hoverflies. III. Pea aphids and *Episyrphus balteatus* (de Geer) (Diptera: Syrphidae). *Journal of Ethology*, **7**, 1–6.

Kan, E. and Sasakawa, M. (1986) Assessment of the maple aphid colony by the hoverfly *Episyrphus balteatus* (de Geer) (Diptera: Syrphidae) I. *Journal of Ethology*, **4**, 121–127.

Karczewska, M. (1976) Studies on the development of the aphid *Longiunguis pyrarius* Pass. (Homoptera, Aphididae). *Polskie Pismo Entomologiczne*, **46**, 319–341.

Karczewska, M. (1979) Bionomia i ekologia *Hyperomyzus lactucae* (L.) (Hom., Aphididae) na tle innych gatunków mszye Wystepujacych na *Ribes nigrum* L. *Roczniki Akademii Rolniczej w Poznaniu Rozprawy Naukowe*, **95**, 72 pp.

Kennedy, G. G. and Kishaba, A. N. (1977) Response of alate melon aphids to resistant and susceptible muskmelon lines. *Journal of Economic Entomology*, **70**, 407–410.

Kennedy, J. S. (1953) Host plant selection in Aphididae. In *Transactions of the 9th International Congress of Entomology*, **2**, pp. 106–113. Vienna.

Kennedy, J. S. (1958) Physiological conditions of the host plant and susceptibility to aphid attack. *Entomologia Experimentalis et Applicata*, **1**, 50–56.

Kennedy, J. S. (1975) Insect dispersal. In D. Pimentel (ed.), *Insects, Science and Society*, pp. 103–119. New York: Academic Press.

Kennedy, J. S. (1976) Host-plant finding by flying aphids. *Symposium Biologica Hungarica*, **16**, 121–123.

Kennedy, J. S. and Booth, C. D. (1951) Host alternation in *Aphis fabae* Scop. I. Feeding preferences and fecundity in relation to age and kind of leaves. *Annals of Applied Biology*, **38**, 25–64.

Kennedy, J. S. and Booth, C. O. (1954) Host alternation in *Aphis fabae* Scop. II. Changes in the aphids. *Annals of Applied Biology*, **41**, 88–106.

Kennedy, J. S. and Booth, C. D. (1963a) Free flight of aphids in the laboratory. *Journal of Experimental Biology*, **40**, 67–85.

Kennedy, J. S. and Booth, C. D. (1963b) Co-ordination of successive activities in an aphid. The effect of flight on the settling responses. *Journal of Experimental Biology*, **40**, 351–369.

Kennedy, J. S., Booth, C. O. and Kershaw, W. J. S. (1959) Host finding by aphids in the field. I. Gynoparae of *Myzus persicae* (Sulzer). *Annals of Applied Biology*, **47**, 410–423.

Kennedy, J. S., Booth, C. O. and Kershaw, W. J. S. (1961) Host finding by aphids in the field. III. Visual attraction. *Annals of Applied Biology*, **49**, 1–21.

Kennedy, J. S., Ibbotson, A. and Booth, C. O. (1950) The distribution of aphid infestation in relation to leaf age. I. *Myzus persicae* (Sulz.) and *Aphis fabae* Scop. on spindle trees and sugarbeet plants. *Annals of Applied Biology*, **37**, 651–679.

Kennedy, J. S. and Stroyan, H. L. G. (1959) Biology of aphids. *Annual Review of Entomology*, **4**, 139–160.

Kidd, N. A. C. (1990) Why is aphid flight density dependent? *Acta Phytopathologica et Entomologia Hungarica*, **25**, 105–112.

Kidd, N. A. C. and Cleaver, A. M. (1984) The relationship between pre-flight reproduction and migratory urge in alatae of *Aphis fabae* Scopoli (Hemiptera: Aphididae). *Bulletin of Entomological Research*, **74**, 517–527.

Kidd, N. A. C. and Cleaver, A. M. (1986) The control of migratory urge in *Aphis fabae* Scopoli (Hemiptera: Aphididae). *Bulletin of Entomological Research*, **76**, 77–87.

Kindlmann, P. and Dixon, A. F. G. (1989a) Developmental constraints in the evolution of reproductive strategies: telescoping of generations in parthenogenetic aphids. *Functional Ecology*, **3**, 531–537.

Kindlmann, P. and Dixon, A. F. G. (1989b) Role of population density in determining the sex ratios of species that show local mate competition, with aphids as a model group. *Functional Ecology*, **3**, 311–314.

Kindlmann, P. and Dixon, A. F. G. (1992) Optimum body size: effects of food quality and temperature, when reproductive growth rate is restricted, with examples from aphids. *Journal of Evolutionary Biology*, **5**, 677–690.

Kindlmann, P. and Dixon, A. F. G. (1993) Optimal foraging in ladybird beetles (Coleoptera: Coccinellidae) and its consequences for their use in biological control. *European Journal of Entomology*, **90**, 443–450.

Kindlmann, P. and Dixon, A. F. G. (1994) Evolution of host range in aphids. *European Journal of Entomology*, **91**, 91–96.

Kindlmann, P. and Dixon, A. F. G. (1996) Population dynamics of a tree-dwelling aphid: individuals to populations. *Ecological Modelling*, **89**, 23–30.

Kindlmann, P., Dixon, A. F. G. and Gross, L. J. (1992) The relationship between individual and population growth rates in multicellular organisms. *Journal of Theoretical Biology*, **157**, 535–542.

Kirk, W. D. J. (1991) The size relationship between insects and their hosts. *Ecological Entomology*, **16**, 351–359.

Kiss, A. (1981) Melezitose, aphids and ants. *Oikos*, **37**, 382.

Klingauf, F. (1971) Die Wirkung des Glucosids Phlorizin auf das Wirtswahlverhalten von *Rhopalosiphum insertum* (Walk.) und *Aphis pomi* De Geer (Homoptera: Aphididae). *Zeitschrift für Angewandte Entomologie*, **68**, 41–55.

Klingauf, F. (1972) Die Bedeutung von peripher vorliegenden Pflanzen substanzen für die Wirtswahl von phloemsaugenden Blattläusen (Aphididae). *Zeitschrifft für Pflanzenkrankheiten und Pflanzenschutz*, **79**, 471–477.

Kloft, W. J. (1977) Radioisotopes in aphid research. In K. F. Harris and K. Maramorosch (eds), *Aphids as Virus Vectors*, pp. 291–310. New York: Academic Press.

Knight, J. D., Bale, J. S., Franks, F., Mathias, S. F. and Baust, J. G. (1986) Insect cold hardiness: supercooling points and pre-freeze mortality. *Cyro-Letters*, **7**, 194–203.

Knight, R. L. and Alston, F. H. (1974) Pest resistance in fruit breeding. In D. Price Jones and M. E. Solomon (eds), *Biology in Pest and Disease Control*, pp. 73–86. Oxford: Blackwell.

Kobayashi, M. and Ishikawa, H. (1993) Breakdown of indirect flight muscles of alate aphids (*Acyrthosiphon pisum*) in relation to their flight, feeding and reproductive behaviour. *Journal of Insect Physiology*, **39**, 549–554.

Kobayashi, M. and Ishikawa, H. (1994a) Mechanisms of histolysis in indirect flight muscles of alate aphid (*Acyrthosiphon pisum*). *Journal of Insect Physiology*, **40**, 33–38.

Kobayashi, M. and Ishikawa, H. (1994b) Involvement of juvenile hormone and ubiquitin-dependent proteolysis in flight muscle breakdown of alate aphid (*Acyrthosiphon pisum*). *Journal of Insect Physiology*, **40**, 107–111.

Kolesova, D. A. (1974) Pear aphids of the genus *Dysaphis* C. B. (Homoptera, Aphididae) – two host species. *Revue d'Entomologie de l'URSS*, **53**, 38–53.

Komatsu, T. and Akimoto, S. (1995) Genetic differentiation as a result of adaptation to the phenologies of individual host trees in the galling aphid *Kaltenbachiella japonica*. *Ecological Entomology*, **20**, 33–42.

Kundu, R. and Dixon, A. F. G. (1995) Evolution of complex life cycles. *Journal of Animal Ecology*, **64**, 245–255.

Kunkel, H. and Mittler, T. E. (1971) Einfluss der Ernährung bei Junglarven von *Myzus persicae* (Sulz.) (Aphididae) auf ihre Entwicklung zu Geflügelten oder Ungeflügeten. *Oecologia (Berlin)*, **8**, 110–134.

Lai, C.-Y., Baumann, L. and Baumann, P. (1994) Amplification of *trpEG*: Adaptation of *Buchnera aphidicola* to an endosymbiotic association with aphids. *Proceedings of the National Academy of Science USA*, **91**, 3819–3823.

Lai, C.-Y., Baumann, P. and Moran, N. A. (1995) Genetics of the tryptophan biosynthetic pathway of the prokaryotic endosymbiont (*Buchnera*) of the aphid *Schlechtendalia chinensis*. *Insect Molecular Biology*, **4**, 47–59.

Lamarck, J. B. (1809) *Philosophie Zoologique* (1984 Translation, Trans.). Chicago: University of Chicago Press.

Lamb, K. P. (1959) Composition of the honeydew of the aphid *Brevicoryne brassicae* (K.) feeding on swedes (*Brassica napobrassica* DC). *Journal of Insect Physiology*, **3**, 1–13.

Lamb, R. J. and Mackay, P. A. (1979) Variability in migratory tendency within and among natural populations of the pea aphid, *Acyrthosiphon pisum*. *Oecologia (Berlin)*, **39**, 289–299.

Lamb, R. J. and Mackay, P. A. (1983) Micro-evolution of the migratory tendency, photoperiodic response and developmental threshold of the pea aphid, *Acyrthosiphon pisum*. In V. K. Brown, I. Hodek and W. Junk (eds), *Diapause and Life Cycle Strategies in Insects*, pp. 209–217. The Hague: Dr. W. Junk.

Lampel, G. (1969) Untersuchungen sur Morphenfolge von *Pemphigus spirothecae* Pass. 1860 (Homoptera, Aphidoidea). *Bulletin der Naturforschenden Gesellschaft Freiburg*, **58**, 56–71.

Latch, G. C., Christensen, M. J. and Gaynor, D. L. (1985) Aphid detection of endophyte infection in tall fescue. *New Zealand Journal of Agricultural Research*, **28**, 129–132.

Lathrop, F. H. (1928) The biology of apple aphids. *Ohio Journal of Science*, **28**, 177–204.

Lawson, C. A. (1941) The effect of temperature on longevity, reproduction and growth in aphids. *Genetics*, **26**, 159.

Lawton, J. H. (1978) Host-plant influences on insect diversity: the effects of time and space. *Symposium of the Royal Entomological Society of London*, **9**, 105–125.

Lawton, J. H. and Strong, D. R. (1981) Community patterns and competition in folivorous insects. *American Naturalist*, **118**, 317–338.

Leather, S. R. and Dixon, A. F. G. (1981) Growth, survival and reproduction of the bird-cherry aphid, *Rhopalosiphum padi* on its primary host. *Annals of Applied Biology*, **99**, 115–118.

Leather, S. R. and Dixon, A. F. G. (1984) Aphid growth and reproduction rates. *Entomologia Experimentalis et Applicata*, **35**, 137–140.

Leather, S. R. and Wellings, P. W. (1981) Ovariole number and fecundity in aphids. *Entomologia Experimentalis et Applicata*, **30**, 128–133.

Leather, S. R., Ward, S. A. and Dixon, A. F. G. (1983) The effect of nutrient stress on some life history parameters of the black bean aphid, *Aphis fabae* Scop. *Oecologia (Berlin)*, **57**, 156–157.

Lees, A. D. (1959) The role of photoperiod and temperature in the determination of parthenogenetic and sexual forms in the aphid *Megoura viciae* Buckton I. The influence of these factors on apterous viginoparae and their progeny. *Journal of Insect Phsyiology*, **3**, 92–117.

Lees, A. D. (1960) The role of photoperiod and temperature in the determination of parthenogenetic and sexual forms in the aphid *Megoura viciae* Buckton II. The operation of the 'internal timer' in young clones. *Journal of Insect Phsyiology*, **4**, 154–175.

Lees, A. D. (1964) The location of the photoperiodic receptors in the aphid *Megoura viciae* Buckton. *Journal of Experimental Biology*, **41**, 119–133.

Lees, A. D. (1966) The control of polymorphism in aphids. *Advances Insect Physiology*, **3**, 207–277.

Lees, A. D. (1967) The production of the apterous and alate forms in the aphid *Megoura viciae* Buckton, with special reference to the role of crowding. *Journal of Insect Phsyiology*, **13**, 289–318.

Lees, A. D. (1973) Photoperiodic time measurement in the aphid *Megoura viciae*. *Journal of Insect Phsyiology*, **19**, 2279–2316.

Lees, A. D. (1979) The maternal environment and the control of morphogenesis in insects. In D. R. Newth and M. Balls (eds), *Maternal Effects in Development*, pp. 221–239. British Society Developmental Biology Symposium 4.

Lees, A. D. and Hardie, J. (1981) The photoperiodic control of polymorphism in insects: neuroendocrine and endocrine components. In B. K. Follett and D. E. Follett (eds), *Biological Clocks in Seasonal Reproductive Cycles*, pp. 125–135. Bristol: Scientechnica.

Leszczynski, B. and Dixon, A. F. G. (1990) Resistance of cereals to aphids: The interaction between hydroxamic acids and glutathione S-transferase in the grain aphid *Sitobion avenae* (F.) (Hom., Aphididae). *Journal of Applied Entomology*, **113**, 61–67.

Leszczynski, B., Matok, H. and Dixon, A. F. G. (1992) Resistance of cereals to aphids: the interaction between hydroxamic acids and UDP-glucose transferases in the aphid *Sitobion avenae* (Homoptera: Aphididae). *Journal of Chemical Ecology*, **18**, 1189–1200.

Leszczynski, B., Matok, M. and Dixon, A. F. G. (1994) Detoxification of cereal plant allelochemicals by aphids: activity and molecular weights of glutathione S-transferase in three species of cereal aphids. *Journal of Chemical Ecology*, **20**, 387–394.

Leszczynski, B., Urbanska, A., Matok, H. and Dixon, A. F. G. (1993) Detoxifying enzymes of the grain aphid. *IOBC/WPRS Bulletin*, **16**, 165–172.

Liadouze, I., Febuay, G., Guillaud, J. and Bonnot, G. (1995) Effect of diet on the free amino acid pools of symbiotic and aposymbiotic pea aphids, *Acyrthosiphon pisum*. *Journal of Insect Physiology*, **41**, 33–40.

Llewellyn, M. J. (1970) *The ecological energetics of the lime aphid Eucallipterus tiliae* (L.) *and its effect on tree growth.* PhD Thesis, University of Glasgow.

Llewellyn, M. (1972) The effect of the lime aphid, *Eucallipterus tiliae* L. (Aphididae) on the growth of the lime *Tilia* × *vulgaris* Hayne. I. Energy requirements of the aphid populations. *Journal of Applied Ecology*, **9**, 261–282.

Llewellyn, M. (1975) The effects of the lime aphid (*Eucallipterus tiliae* L.) (Aphididae) on the growth of the lime *Tilia* × *vulgaris* Hayne. II. The primary production of saplings and mature trees, the energy drain imposed by the aphid populations and revised standard deviations of aphid population energy budgets. *Journal of Applied Ecology*, **12**, 15–23.

Llewellyn, M. (1982) The energy economy of fluid-feeding herbivorous insects. In *Proceedings of the 5th International Symposium on Insect–Plant Relationships*, pp. 243–251. Wageningen: Pudoc.

Llewellyn, M. and Brown, V. K. (1985) A general relationship between adult weight and the reproductive potential of aphids. *Journal of Animal Ecology*, **54**, 663–673.

Llewellyn, M., Rashid, R. and Leckstein, P. (1974) The ecological energetics of the willow aphid *Tuberolachnus salignus* (Gmelin): honeydew production. *Journal of Animal Ecology*, **43**, 19–29.

Long, B. J., Dunn, G. M., Bowman, J. S. and Routley, D. G. (1977) Relationship of hydroxamic acid content in corn and resistance to the corn leaf aphid. *Crop Science*, **17**, 55–58.

Losey, J. E., Ives, A. R., Harman, J., Ballantyne, F. and Brown, C. (1997) A polymorphism maintained by opposite patterns of parasitism and predation. *Nature*, **388**, 269–272.

Lowe, H. B. J. (1973) Variation in *Myzus persicae* (Sulz) (Hemiptera, Aphididae) reared on different host-plants. *Bulletin of Entomological Research*, **62**, 549–556.

Loxdale, H. D., Hardie, J., Halberts, S., Foottit, R., Kidd, N. A. C. and Carter, C. I. (1993) The relative importance of short- and long-range movement of flying aphids. *Biological Review*, **68**, 291–311.

Macarthur, R. H. (1955) Fluctuations of animal populations and a measure of community stability. *Ecology*, **36**, 533–536.

Macfadyen, A. (1964) Energy flow in ecosystems and its exploitation by grazing. In D. J. Crisp (ed.), *Grazing in Terrestrial and Marine Environments*, pp. 3–20. Oxford: Blackwell.

MacGibbon, D. B. (1975) An improved reagent for detecting glucosinolase activity in acrylamide gel electropherograms. *New Zealand Journal of Science*, **18**, 217–219.

MacGibbon, D. B. and Benzenberg, E. J. (1978) Location of glucosinolase in *Brevicoryne brassicae* and *Lipaphis erysimi* (Aphididae). *New Zealand Journal of Science*, **21**, 398–392.

Mackauer, M. (1973) Host selection and host suitability in *Aphidius smithi*. In A. D. Lowe (ed.), *Perspectives in Aphid Biology*, Vol. 2, pp. 20–29. Bulletin of the Entomological Society of New Zealand.

Mackauer, M. and Völkl, W. (1993) Regulation of aphid populations by aphidiid wasps: does parasitoid foraging behaviour or hyperparasitism limit impact? *Oecologia (Berlin)*, **94**, 339–350.

Mackenzie, A. (1992) The evolutionary significance of host-mediated conditioning. *Antenna*, **16**, 141–150.

Mackenzie, A. (1996) A trade-off for host plant utilization in the black bean aphid, *Aphis fabae*. *Evolution*, **50**, 155–162.

Mackenzie, A. and Dixon, A. F. G. (1990) Host alternation in aphids: constraint versus optimization. *American Naturalist*, **136**, 132–134.

Mackenzie, A., Dixon, A. F. G. and Kindlmann, P. (1994) The relationship between the regional number of aphid species and plant species diversity. *European Journal of Entomology*, **91**, 135–138.

Mackenzie, A. and Guldemond, J. A. (1994) Sympatric speciation in aphids. II. Host race formation in the face of gene flow. In S. R. Leather, A. D. Watt, N. J. Mills and K. F. A. Walters (eds), *Individuals, Populations and Patterns in Ecology*, pp. 379–396. Andover: Intercept.

Mackenzie, A., Thacker, J. I. and Dixon, A. F. G. (In press) Density dependence in insects: the role of life cycle velocity.

Malcolm, S. B. (1990) Chemical defence in chewing and sucking insect herbivores: plant-derived cardenolides in the monarch butterfly and oleander aphid. *Chemoecology*, **1**, 12–21.

Marchal, P. (1928) Etude Biologique et Morphologique du Puceron Lanigere du Pommier, (*Eriosoma lanigerum* Hausmann). *Annales des Epiphyties*, **14**, 1–106.

Marcovitch, S. (1924) The migration of the Aphididae and the appearance of the sexual forms as affected by the relative length of daily light exposure. *Journal of Agricultural Research*, **27**, 513–522.

Margulis, L. (1970) *Origin of Eukaryotic Cells*. New Haven: Yale University Press.

Margulis, L. (1981) *Symbiosis in Cell Evolution*. San Francisco: Freeman.

Markkula, M. (1963) Studies on the pea aphid, *Acyrthosiphon pisum* Harris (Hom., Aphididae) with special reference to the difference in biology of the green and red forms. *Annales Agriculture Fenniae*, **2** (Suppl. 1), 1–30.

Markkula, M. and Roukka, K. (1970) Resistance of plants to the pea aphid *Acyrthosiphon pisum* Harris (Hom., Aphididae). *Annales Agriculture Fenniae*, **9**, 127–132.

Marthavan, S. and Pandian, T. J. (1975) Effect of temperature on food utilization in the monarch butterfly *Danaus chrysippus*. *Oikos*, **26**, 60–64.

Massee, A. M. (1946) *The Pests of Fruits and Hops*. London: Crosby Lockwood.

Matsuka, M. and Mittler, T. E. (1978) Enhancement of alata production by an aphid, *Myzus persicae*, in response to increases in daylength. *Bulletin of the Faculty of Agriculture of Tamagawa University Tokyo*, **18**, 1–7.

Maudsley, M. J. (1993) *Regional differences in the abundance of cereal aphids*. PhD Thesis, University of East Anglia.

May, R. M. and Seger, J. (1985) Sex ratios in wasps and aphids. *Nature*, **318**, 408–409.

Maynard-Smith, J. (1969) Limitations on growth rate. *Symposium of the Society for General Microbiology*, **19**, 1–13.

Maynard-Smith, J. (1978) *The Evolution of Sex*. Cambridge: Cambridge University Press.

Maynard-Smith, J. (1984) The ecology of sex. In J. R. Krebs and N. B. David (eds), *Behavioural Ecology: An Evolutionary Approach*, 2nd ed., pp. 201–221. Oxford: Blackwell.

Maynard-Smith, J. (1988) The evolution of recombination. In R. E. Michod and B. R. Levin (eds), *The Evolution of Sex*, pp. 106. Sunderland, MA: Sinauer.

McLean, D. L. and Kinsey, M. G. (1964) A technique for electronically recording aphid feeding and salivation. *Nature*, **202**, 1358–1359.

McLean, D. L. and Kinsey, M. G. (1968) Probing behaviour of the pea aphid, *Acyrthosiphon pisum*. II. Comparisons of salivation and ingestion in host and non-host leaves. *Annals of the Entomological Society of America*, **61**, 730–739.

McLean, D. L. and Kinsey, M. G. (1984) The precibarial value and its role in the feeding behaviour of the pea aphid, *Acyrthosiphon pisum*. *Bulletin of the Entomological Society of America*, **30**, 26–31.

Mercer, D. R. (1979) *Flight behaviour of the sycamore aphid* Drepanosiphum platanoidis *Schr.* PhD Thesis, University of East Anglia.

Michel, E. (1942) Beiträge zur Kenntnis von *Lachnus (Pterochlorus) roboris* L., einer wichtigen Honigtauerzeugerin an der Eiche. *Zeitschrift Angewandte Entomologie*, **29**, 243–281.

Miles, P. W. (1968a) Studies on the salivary physiology of plant-bugs: Experimental induction of galls. *Journal of Insect Phsyiology*, **14**, 97–106.

Miles, P. W. (1968b) Insect secretions in plants. *Review of Phytopathology* **6**, 137–164.

Miles, P. W. (1978) Redox reactions of hemipterous saliva in plant tissues. *Entomologia Experimentalis et Applicata*, **24**, 334–339.

Miles, P. W. (1985) Dynamic aspects of the chemical relation between the rose aphid and rose buds. *Entomologia Experimentalis et Applicata*, **37**, 129–135.

Miles, P. W. (1987) Feeding process of Aphidoidea in relation to effects on their food plants. In A. K. Minks and P. Harrewijn (eds), *Aphids, their Biology, Natural Enemies and Control*, Vol. 2A, pp. 321–339. Amsterdam: Elsevier.

Mittler, T. E. (1957) Studies on the feeding and nutrition of *Tuberolachnus salignus* (Gmelin) (Homoptera, Aphididae). I. The uptake of phloem sap. *Journal of Experimental Biology*, **34**, 334–341.

Mittler, T. E. (1958a) Studies on the feeding and nutrition of *Tuberolachnus salignus* (Gmelin) (Homoptera, Aphididae). II. The nitrogen and sugar composition of ingested phloem sap and excreted honeydew. *Journal of Experimental Biology*, **35**, 74–84.

Mittler, T. E. (1958b) Studies on the nutrition of *Tuberolachnus salignis* (Gmelin) (Homoptera, Aphididae). III. The nitrogen economy. *Journal of Experimental Biology*, **35**, 626–638.

Mittler, T. E. (1971) Some effects on the aphid *Myzus persicae* of ingesting antibiotics incorporated into artificial diets. *Journal of Insect Phsyiology*, **17**, 1333–1347.

Mittler, T. E. and Kleinjan, J. E. (1970) Effect of artificial diet composition on wing-production by the aphid *Myzus persicae*. *Journal of Insect Phsyiology*, **16**, 833–850.

Miyazaki, M. (1971) A revision of the tribe of Macrosiphini of Japan (Homoptera: Aphididae, Aphidinae). *Insecta Matsumurana*, **34**, 1–247.

Moericke, V. (1955) Über die Lebensgewohnheiten der geflugelten Blattläuse (Aphidina) unter besonderer Berucksichtigung des verhalterns beim Landen. *Zeitschrift für Angewandte Entomologie*, **37**, 29–91.

Molyneux, R. J., Campbell, B. C. and Dreyer, D. L. (1990) Honeydew analysis for detecting phloem transport of plant natural products. Implications for host-plant resistance to sap-sucking insects. *Journal of Chemical Ecology*, **16**, 1899–1909.

Montllor, C. B. (1991) The influence of plant chemistry on aphid feeding behaviour. In E. Bernays (ed.), *Insect–Plant Interactions*, pp. 125–173. Boca Raton: CRC Press.

Moran, N. A. (1983) Seasonal shifts in host usage in *Uroleucon gravicorne* (Homoptera: Aphididae) and implications for the evolution of host alternation in aphids. *Ecological Entomology*, **8**, 371–382.
Moran, N. A. (1987) Evolutionary determinants of host specificity. In J. Holman, J. Pelikan, A. F. G. Dixon and L. Weismann (eds), *Population Structure, Genetics and Taxonomy of Aphids and Thysanoptera*, pp. 29–38. The Hague: SPB Academic Publishing.
Moran, N. A. (1988) The evolution of host-plant alternation in aphids: evidence for specialisation as a dead end. *American Naturalist*, **132**, 681–706.
Moran, N. A. (1990) Aphid life cycles: two evolutionary steps. *American Naturalists*, **136**, 135–138.
Moran, N. A. and Baumann, P. (1994) Phylogenetics of cytoplasmically inherited microorganisms of arthropods. *Trends in Ecology and Evolution*, **9**, 15–20.
Moran, N. A., Munson, M. A., Baumann, P. and Ishikawa, H. (1993) A molecular clock in endosymbiotic bacteria is calibrated using the insect hosts. *Proceedings of the Royal Society of London B*, **253**, 167–171.
Mordvilko, A. K. (1900) Zur Biologie und Morphologie der Pflanzenläuse (Fam. Aphididae Pass.). *Horae Société Entomologique Rossicae*, **31**, 1–84.
Mordvilko, A. K. (1901) Zur Biologie und Morphologie der Pflanzenläuse (Fam. Aphididae Pass.). *Horae Société Entomologique Rossicae*, **33**, 162–???.
Mordvilko, A. K. (1908) Beitrage zur Biologie der Pflanzenläuse, *Aphididae* Passerini. *Biologische Zentralblatt*, **28**, 631–639.
Mordvilko, A. K. (1928) The evolution of cycles and the origin of heteroecy (migrations) in plant-lice. *Annals and Magazine Natural History (Series 10)*, **2**, 570–582.
Mordvilko, A. K. (1934) On the evolution of aphids. *Archiv für Naturgeschichte (NF)*, **3**, 1–60.
Mordvilko, A. K. (1935) Aphids: their generation cycles and evolution. *Priroda*, **11**, 35–44.
Moritsu, M. (1982) *Aphids of Japan in Colors*. Tokyo: Zenkoku Noson.
Morse, D. R., Lawton, J. H., Dodson, M. M. and Williamson, M. H. (1985) Fractal dimension of vegetation and the distribution of arthropod body lengths. *Nature*, **314**, 731–733.
Mosbacher, G. C. (1963) Über die Nahrungswahl bei *Dactynotus* RaF. (Aphididae) I. Die Wirtsspektren der Gruppe *D. jaceae* (L.) s. lat. und *D. cichorii* (Koch) s. lat. *Zietschrift für Angewandte Entomologie*, **51**, 378–428.
Müller, F. P. (1961) Der fakultative Wirtswechsel der Blattläuse (Homoptera: Aphididae). In *Proceedings of the 11th International Congress of Entomology*, **2**, pp. 100–102. Vienna.
Müller, F. P. (1962) Biotypen and Unterarten der 'Erbsenlaus' *Acyrthosiphon pisum* Harris. *Zeitschrift für Pflanzenkrankheiten und Pflanzenschutz*, **69**, 129–136.
Müller, F. P. (1966) Formen des Wirtswechsels der Blattläuse. *Forschung und Fortschritt*, **40**, 353–357.
Müller, F. P. (1971a) Isolationmechnismen zwischen sympatrischen bionomischen Rassen am Beispiel der Erbsenblattlaus *Acyrthosiphon pisum* (Harris) (Homoptera, Aphididae). *Zoologisches Jahrbuch Abteilung Systematiek und Oecologische Geographie der Tiere*, **98**, 131–152.
Müller, F. P. (1971b) Bisher unbekannte Überwinterungs-formen bei anholozyklischen Aphiden. *Wissenschaftliche der Zeitschrift Universität Rostock*, **20**, 91–96.
Müller, F. P. (1976) Host and non-host in subspecies of *Aulacorthum solani* (Kaltenbach) and intraspecific hybridisations (Homoptera; Aphididae). *Symposium Biologia Hungarica*, **16**, 187–190.
Müller, F. P. (1980) Wirtzpflanzen, Generationsfolge und reproductive Isolationintraspecifiker Formen von *Acyrthosiphum pisum*. *Entomologica Experimentalis et Applicata*, **28**, 145–157.
Müller, F. P. (1982) Das Problem *Aphis fabae*. *Zeitschrift für Angewandte Entomologie*, **94**, 432–446.
Müller, F. P. (1983) Untersuchungen zur Blattläuse der Gruppe *Acyrthosiphon pelargonii* im Freiland-Insektarium. *Zeitschrift für Angewandte Zoologie*, **70**, 351–367.
Müller, F. P. (1985a) Das Problem *Acyrthosiphon pisum* (Homoptera: Aphididae). *Zeitschrift für Angewandte Zoologie*, **72**, 317–334.
Müller, F. P. (1985b) Genetic and evolutionary aspects of host choice in phytophagous insects, especially aphids. *Biologische Zentralblatt*, **104**, 225–237.
Müller, F. P. and Steiner, H. (1985) Das Problem *Acyrthosiphon pisum* (Homoptera: Aphididae). *Zeitschrift für Angewandte Zoologie*, **72**, 317–334.
Müller, H. J. (1964) The relation of recombination to mutational advance. *Mutation Research*, **1**, 2–9.

Müller, H. J. (1966a) Über die Ursachen des unterschiedlichen Resistenz von *Vicia faba* L. gegenüber der Bohnenblattlaus. *Aphis (Doralis) fabae* Scop. IX. Der Einfluss okologischer Faktoren auf das Wachstum von *Aphis fabae* Scop. *Entomologia Experimentalis et Applicata*, **9**, 42–66.

Müller, H. J. (1966b) Über Mehrjährige Coccinelliden-Fänge auf Ackerbohnen mit hohem *Aphis fabae*-Besatz. *Zeitschrift für Morphologie und Okologie der Tiere*, **58**, 144–161.

Munson, M. A. and Baumann, P. (1993) Molecular cloning and nucleotide sequence of a putative trpDC(F)BA operon in *Buchnera aphidicola* (Endosymbiont of the aphid *Schizaphis graminum*). *Journal of Bacteriology*, **175**, 6426–6432.

Munson, M. A., Baumann, P., Clark, M. A., Baumann, L., Moran, N. A., Voegtlin, D. J. and Campbell, B. C. (1991a) Evidence for the establishment of aphid-eubacterium endosymbiosis in an ancestor of four aphid families. *Journal of Bacteriology*, **173**, 6321–6324.

Munson, M. A., Baumann, P. and Kinsey, M. G. (1991b) *Buchneria* gen nov. and *Buchneria aphidicola* sp nov., a taxon consisting of the mycetocyte-associated, primary endosymbionts of aphids. *International Journal of Systematics and Bacteriology*, **41**, 566–568.

Murdie, G. (1969a) Some causes of size variation in the pea aphid, *Acyrthosiphon pisum* Harris. *Transactions of the Royal Entomological Society of London*, **121**, 423–442.

Murdie, G. (1969b) The biological consequences of decreased size caused by crowding or rearing temperatures in apterae of the pea aphid, *Acyrthosiphon pisum* Harris. *Transactions of the Royal Entomological Society of London*, **121**, 443–455.

Nault, L. R. and Montgomery, M. E. (1977) Aphid pheromones. In K. F. Harris and K. Maramorosch (eds), *Aphids as Virus Vectors*, pp. 527–545. New York: Academic Press.

Nault, L. R. and Styer, W. E. (1972) Effects of sinigrin on host selection by aphids. *Entomologia Experimentalis et Applicata*, **15**, 423–437.

Newton, C. and Dixon, A. F. G. (1988) The cost of switching from asexual to sexual reproduction in an aphid. *Entomologia Experimentalis et Applicata*, **46**, 283–287.

Newton, C. and Dixon, A. F. G. (1990) Pattern of growth in weight of alate and apterous nymphs of the English grain aphid, *Sitobion avenae*. *Entomologia Experimentalis et Applicata*, **55**, 231–238.

Nicholson, A. J. (1933) The balance of animal populations. *Journal of Animal Ecology*, **2**, 132–178.

Nicholson, A. J. (1954) An outline of the dynamics of animal populations. *Australian Journal of Zoology*, **2**, 9–65.

Nielson, M. W. and Don, H. (1974) Probing behaviour of biotypes of the spotted alfalfa aphid on resistant and susceptible alfalfa clones. *Entomologia Experimentalis et Applicata*, **17**, 477–486.

Nishida, R. and Fukami, H. (1989) Host plant iridoid-based chemical defence of an aphid, *Acyrthosiphon nipponicus*, against ladybird beetles. *Journal of Chemical Ecology*, **15**, 1837–1845.

Noda, I. (1960) The emergence of winged viviparous female in aphid. VI. Difference in the rate of development between the winged and the unwinged forms. *Japanese Journal of Ecology*, **10**, 97–102.

Nottingham, S. F. and Hardie, J. (1989) Migratory and targeted flight in seasonal forms of the black bean aphid, *Aphis fabae*. *Physiological Entomology*, **14**, 451–458.

Nottingham, S. F. and Hardie, J. (1993) Flight behaviour of the black bean aphid, *Aphis fabae*, and the cabbage aphid, *Brevicoryne brassicae*, in host and non-host plant odour. *Physiological Entomology*, **18**, 389–394.

Nottingham, S. F., Hardie, J., Dawson, G. W., Hick, A. J., Pickett, J. A., Wadhams, L. J. and Woodcock, C. M. (1991) Behavioural and electrophysiological responses of aphids to host and non-host plant volatiles. *Journal of Chemical Ecology*, **17**, 1231–1242.

Nottingham, S. F., Hardie, J. and Tatchel, G. M. (1991) Flight behaviour of the bird cherry aphid, *Rhopalosiphum padi*. *Physiological Entomology*, **16**, 223–229.

Orlando, E. (1974) Sex determination in *Megoura viciae* Buckton (Homoptera, Aphididae). *Monitore Zoologica Italiana (NS)*, **8**, 61–70.

Osawa, N. (1992) The effect of shoot growth of *Spiraea thunbergii* (Rosaceae) on colonisation of *Aphis spiraecola* (Homoptera: Aphididae). *Applied Entomology and Zoology*, **27**, 557–563.

Owen, D. F. (1978) Why do aphids synthesize melezitose? *Oikos*, **31**, 264–267.

Owen, D. F. (1980a) How plants may benefit from the animals that eat them? *Oikos*, **35**, 230–235.

Owen, D. F. (1980b) Response to Petelle's comments. *Oikos*, **35**, 128.

Owen, D. F. and Wiegert, R. G. (1976) Do consumers maximise plant fitness? *Oikos*, **27**, 488–492.

Owen, D. F. and Wiegert, R. G. (1981) Mutualism between grasses and grazers: an evolutionary hypothesis. *Oikos*, **36**, 376–378.

Owen, R. (1849) *On Parthenogenesis as the Successive Production of Procreating Individuals from a Single Ovum*. London: John van Voorst.

Paik, W. H. (1965) *Aphids of Korea*. Seoul: Seoul National University.

Paik, W. H. (1972) Illustrated Encyclopedia of Fauna and Flora of Korea. *Insecta* **5, 13**.

Palmer, M. A. (1952) *Aphids of the Rocky Mountain Region*. Vol. 5. Denver Colorado: The Thomas Soy Foundation.

Parrish, W. B. (1967) The origin, morphology, and innervation of aphid stylets (Homoptera). *Annals of the Entomological Society of America*, **60**, 273–276.

Parrish, W. E. G. and Bale, J. S. (1991) Effect of low temperatures on the intracellular symbionts of the grain aphid, *Sitobion avenae* (F.) (Hem., Aphididae). *Journal of Insect Physiology*, **37**, 339–345.

Parry, W. H. (1979a) Factors affecting low temperature survival of *Cinara pilicornis* eggs on Sitka spruce. *International Journal of Biometeorology*, **23**, 185–193.

Parry, W. H. (1979b) Acclimatization in the green spruce aphid, *Elatobium abietinum*. *Annals of Applied Biology*, **92**, 299–306.

Parry, W. H. (1985) The overwintering of *Euceraphis punctipennis* eggs on birch in NE Scotland. *Cryo-Letters*, **6**, 5–12.

Peng, Z. and Miles, P. W. (1988) Acceptability of catechin and its oxidative condensation products to the rose aphid, *Macrosiphum rosae*. *Entomologia Experimentalis et Applicata*, **47**, 255–265.

Pergande, T. (1912) The life history of the alder blight aphis. *US Department of Agriculture Bureau Entomology Technical Series*, **24**, 1–28.

Perry, J. N., Woiwood, I. P. and Hanski, I. (1993) Using response-surface methodology to detect chaos in ecological time series. *Oikos*, **68**, 329–339.

Petelle, M. (1980) Aphids and melezitose: a test of Owen's 1978 hypothesis. *Oikos*, **35**, 127–128.

Pettersson, J. (1970) Studies on *Rhopalosiphum padi* (L.) I. Laboratory studies on olfactometric responses to winter host, *Prunus padus*. *Lantbrukshogskolans Annaler*, **36**, 381–399.

Pettersson, J. (1971) An aphid sex attractant. II. Histological, ethological and comparative studies. *Entomologica Scandinavica*, **2**, 81–93.

Pfeiffer, D. G., Brown, M. W. and Varn, M. W. (1989) Incidence of spirea aphid (Homoptera: Aphididae) in apple orchards in Virginia, West Virginia and Maryland. *Journal Entomological Society*, **24**, 145–149.

Pickett, J. A. and Griffiths, D. C. (1980) Composition of aphid alarm pheromones. *Journal of Chemical Ecology*, **6**, 349–360.

Pickett, J. A., Wadhams, L. J. and Woodcock, C. M. (1994) Attempts to control aphid pests by integrated use of semiochemicals. *Brighton Crop Protection Conference – Pests and Diseases*, pp. 1239–1246.

Pickett, J. A., Wadhams, L. J., Woodcock, C. M. and Hardie, J. (1992) The chemical ecology of aphids. *Annual Review of Entomology*, **37**, 67–90.

Pimm, S. L. (1991) *The Balance of Nature*. Chicago: University of Chicago Press.

Polaszek, A. (1987) Studies on the comparative anatomy of aphid reproductive systems. In J. Holman, J. Pelikan, A. F. G. Dixon and L. Weismann (eds), *Population Structure, Genetics and Taxonomy of Aphids and Thysanoptera*, pp. 261–266. The Hague: SPB Academic Publishing.

Pollard, D. G. (1973) Plant penetration by feeding aphids (Hemiptera, Aphidoidea): a review. *Bulletin of Entomological Research*, **62**, 631–714.

Pollard, D. G. (1977) Aphid penetration of plant tissues. In K. F. Harris and K. Maramorosch (eds), *Aphids as Virus Vectors*, pp. 105–118. New York: Academic Press.

Ponsen, M. B. (1972) The site of potato leafroll virus multiplication in its vector, *Myzus persicae*, an anatomical study. *Wageningen: Mededlingen Landbouwhogeschool*, *72-16*, 147 pp.

Ponsen, M. B. (1979) The digestive system of *Subsaltusaphis ornata* (Homoptera: Aphididae). *Mededelingen Landbouwhogeschool Wageningen*, *79–17*, **30** pp.

Ponsen, M. B. (1990) Phylogenetic implications of the structure of the alimentary tract of the Aphidoidea. I and II. *Wageningen Agricultural University Papers*, *90–4*, **52** pp.

Ponsen, M. B. (1991) Structure of the digestive system of aphids, in particular *Hyalopterus* and *Coloradoa*, and its bearing on the evolution of filter chambers in the Aphidoidea. *Wageningen Agricultural University Papers, 91–5*, **61** pp.

Pontin, A. J. (1978) The numbers and distribution of subterranean aphids and their exploitation by the ant *Lasius flavus* (Fabr.). *Ecological Entomology*, **3**, 203–207.

Powell, W. and Parry, W. H. (1976) Effects of temperature on overwintering populations of the green spruce aphid *Elatobium abietinum*. *Annals of Applied Biology*, **82**, 209–219.

Prado, E. and Tjallingii, W. F. (1994) Aphid activities during sieve element punctures. *Entomologia Experimentalis et Applicata*, **72**, 157–165.

Prosser, W. A., Simpson, S. J. and Douglas, A. E. (1992) How an aphid (*Acyrthosiphon pisum*) symbiosis responds to variation in dietary nitrogen. *Journal of Insect Physiology*, **38**, 301–307.

Puterka, G. J. and Peters, D. C. (1990) Sexual reproduction and inheritance of virulence in the green bug, *Schizaphis graminum* (Rondani). In R. K. Campbell and R. D. Eikenbary (eds), *Aphid Plant Genotype Interactions*, pp. 289–318. Amsterdam: Elsevier Press.

Rabinowitz, D. (1981) Seven forms of rarity. In H. Synge (ed.), *The Biological Aspects of Rare Plant Conservation*, pp. 205–217. New York: Wiley.

Rahbé, Y., Delobel, B., Febvay, G. and Chantegrel, B. (1994) Aphid-specific triglycerides in symbiotic and aposymbiotic *Acyrthosiphon pisum*. *Insect Biochemistry and Molecular Biology*, **24**, 95–101.

Rahbé, Y., Delobel, B., Febvay, G., Nardon, C. and Nardon, P. (1993) Are aphid bacterial symbionts involved in the biosynthesis of aphid peculiar triglyceriders? In *Proceedings of Endocytobiology V*, 1992. Kyoto.

Rahbé, Y., Sauvion, N., Febvay, G., Peumans, W. J. and Gatehouse, A. M. R. (1995) Toxicity of lectins and processing of ingested proteins in the pea aphid *Acyrthosiphon pisum*. *Entomologia Experimentalis et Applicata*, **76**, 143–155.

Rausher, M. D. (1984) Trade-offs in performance on different hosts: evidence from within- and between-site variation in the beetle *Deloyala guttata*. *Evolution*, **38**, 582–595.

Redfearn, A. and Pimm, S. L. (1988) Population variability and polyphagy in herbivorous insect communities. *Ecological Monographs*, **58**, 39–55.

Remaudière, G. (1953) Nutrition et variations du cycle évolutif des Aphidoidea. *Revue de Pathologie Végétale et d'Entomologie Agricole de France*, **32**, 190–207.

Remaudière, G. and Stroyan, N. L. G. (1984) Un *Tamalia* nouveau de Californie (USA) discussion sur les *Tamaliinae* subfam. Nov. (Hom. Aphididae). *Annales de la Société Entomologique de France (NS)*, **20**, 93–103.

Rhode, K. (1979) A critical evaluation of intrinsic and extrinsic factors in parasites. *American Naturalist*, **114**, 648–671.

Rhodes, J. D., Croghan, P. R. and Dixon, A. F. G. (1996) Uptake, excretion and respiration of sucrose and aminoacids in the pea aphid *Acyrthosiphon pisum*. *Journal of Experimental Biology*, **199**, 1269–1276.

Richards, W. R. (1960) A synopsis of the genus *Rhopalosiphum* Koch in Canada. *Canadian Entomologist*, **92**, 1–51.

Richards, W. R. (1963) The myzaphidines of Canada (Homoptera: Aphididae). *Canadian Entomologist*, **95**, 680–704.

Richards, W. R. (1965) The Callaphidini of Canada (Homoptera: Aphididae). *Memoirs of the Entomological Society of Canada*, **44**.

Richards, W. R. (1972) The Chaitophorinae of Canada (Homoptera: Aphididae). *Memoirs of the Entomological Society of Canada*, **87**.

Roberti, D. (1939) Contributi alla conoscenza degli afidi d'Italia II. *Bollettino del Laboratorio di zoologia generale e agraria della Facoltà agraria* in *Portici*, **31**, 137–157.

Rodriguez, A. E., Tingey, W. M. and Mutschler, M. A. (1993) Acylsugars of *Lycopersicon pennellii* deter settling and feeding of the green peach aphid (Homoptera: Aphididae). *Journal of Economic Entomology*, **86**, 34–39.

Roff, D. A. (1986) The evolution of wing polymorphism and its impact on life cycle adaptation in insects. In F. Taylor and R. Karban (eds), *The Evolution of Insect Life Cycles*, pp. 209–221. Berlin: Springer.

Rohfritsch, O. (1966) Dehiscence de la Galle d'*Adelges abietis* Kalt. *Marcellia*, **33**, 149–158.

Rosengren, R. and Sundström, L. (1991) The interaction between wood ants, *Cinara* aphids, and pines. A ghost of mutualism past? In C. R. Huxley and D. F. Cutler (eds), *Ant–Plant Interactions*, pp. 80–91. Oxford: Oxford University Press.

Ross, J. C. (1828) Appendix: Zoology. In W. E. Parry (ed.), *Narrative of an Attempt to Reach the North Pole in Boats fitted for the Purpose and Attached to His Majesty's Ship* Hecla, *in the year 1827*, pp. 187–206. London: Murray.

Rothschild, M. J., von Euw, J. and Reichstein, T. (1970) Cardiac glycosides in the Oleander aphid *Aphis nerii*. *Journal of Insect Phsyiology*, **16**, 1141–1145.

Royama, T. (1977) Population persistence and density dependence. *Ecological Monographs*, **47**, 1–35.

Royama, T. (1981) Fundamental concepts and methodology for the analysis of animal population dynamics, with particular reference to univoltine species. *Ecological Monographs*, **51**, 473–493.

Royama, T. (1992) *Analytical Population Dynamics*. London: Chapman & Hall.

Ruzicka, Z. (1994) Oviposition deterring pheromone in *Chrysopa oculata* (Neuroptera: Chrysopidae). *European Journal of Entomology*, **91**, 361–370.

Ryti, R. T. (1992) Relationship between density of aphid host plants and an associated aphid-tending ant (*Formica altipetens*). *American Midland Naturalists*, **127**, 190–197.

Sakata, H. (1994) How an ant decides to prey on or attend aphids. *Researches on Population Ecology*, **36**, 45–51.

Sakata, K., Hô, Y., Yukawa, J. and Yamane, S. (1991) Ratio of sterile soldiers in the bamboo aphid, *Pseudoregma bambucicola* (Homoptera: Aphididae), colonies in relation to social and habitat conditions. *Applied Entomology and Zoology.*, **26**, 463–468.

Salyk, R. P. and Sullivan, D. J. (1982) Comparative feeding behaviour of two aphid species: bean aphid (*Aphis fabae* Scopoli) and pea aphid (*Acyrthosiphon pisum* (Harris)) (Homoptera: Aphididae). *Journal of the New York Entomological Society*, **90**, 87–93.

Sanders, C. J. and Knight, F. B. (1968) Natural regulation of the aphid *Pterocomma populifoliae* on the big-tooth aspen in northern lower Michigan. *Ecology*, **49**, 234–244.

Sasaki, T., Aoki, T., Hayashi, H. and Ishikawa, H. (1990) Amino acid composition of the honeydew of symbiotic and aposymbiotic pea aphids *Acyrthosiphon pisum*. *Journal of Insect Physiology*, **36**, 35–40.

Sasaki, T., Fukuchi, N. and Ishikawa, H. (1993) Amino acid flow through aphid and its symbiont: studies with N-labelled glutamine. *Zoological Science*, **10**, 787–791.

Sasaki, T., Hayashi, H. and Ishikawa, H. (1991) Growth and reproduction of the symbiotic and aposymbiotic pea aphid, *Acyrthosiphon pisum* maintained on artificial diets. *Journal of Insect Physiology*, **37**, 749–756.

Sasaki, T. and Ishikawa, H. (1993) Nitrogen recycling in the endosymbiotic system of the pea aphid, *Acyrthosiphon pisum*. *Zoological Science*, **10**, 779–785.

Savary, A. (1953) Le puceron cendré du poirier (*Sappaphis pyri* Fonsc.) en Suisse Romande. *Annales Agricole Suisse*, **2**, 249–314.

Schaefers, G. A. and Judge, F. D. (1971) Effects of temperature, photo-period and host plant on alary polymorphism in the aphid, *Chaetosiphon fragaefolii*. *Journal of Insect Phsyiology*, **17**, 365–379.

Schültze, M. and Maschwitz, U. (1991) Enemy recognition and defence within trophobiotic associations with ants and by the soldier caste of *Pseudoregma sundanica* (Homoptera: Aphidoidea). *Entomologia Generalis*, **16**, 1–12.

Schwarzbach, E. (1962) Die Bildung von Fluglöchern an Gallen von *Pemphigus spirothecae* Pass. *Naturwissenschaften*, **49**, 91.

Sen Gupta, G. C. and Miles, P. W. (1975) Studies on the susceptibility of varieties of apple to the feeding of two strains of woolly aphis (Homoptera) in relation to the chemical content of the tissues of the host. *Australian Journal of Agricultural Research*, **26**, 157–168.

Sequeira, R. and Dixon, A. F. G. (1996) Life history responses to host quality changes and competition in the Turkey-oak aphid, *Myzocallis boerneri* (Hemiptera: Sternorrhyncha: Callaphididae). *European Journal of Entomology*, **93**, 53–58.

Sequeira, R. and Dixon, A. F. G. (1997) Population dynamics of tree-dwelling aphids: the importance of seasonality and time scale. *Ecology* (in press).

Setzer, R. W. (1980) Intergall migration in the aphid genus *Pemphigus*. *Annals of the Entomological Society of America*, **73**, 327–331.

Shambaugh, G. F., Frazier, J. L., Castell, A. E. M. and Coons, L. B. (1978) Antennal sensilla of seventeen aphids (Homoptera: Aphidinae). *International Journal of Insect Morphology and Embryology*, **7**, 389–404.

Shaposhnikov, G. C. (1951) Evoljucija nekotorych grupp tlej v svyazi sz evoljuciej rozocvetnych (Evolution of some groups of aphids associated with the Rosaceae). *Izdatel'stuo Akadamii Nauk SSR*, **1**, 23–60.

Shaposhnikov, G. C. (1956) Filogeneticeskoe obosnovane sistemy korolkochvostnych tlej (Anuraphidina) ucetom ich sujazej s rasteniami (The phylogenetic basis of the systematics of the short-tailed aphids (Anuraphidina) with reference to their relationships with hosts). *Trudy Zoologiceskogo Instituta Akademii Nauk SSSR*, **23**, 215–320.

Shaposhnikov, G. K. (1964) Suborder Aphidinae – plant lice. In G. Y. Bei-Benko (ed.), *Keys to the Insects of the European USSR*, pp. 616–799. Moscow and Leningrad.

Shaposhnikov, G. K. (1977) The trend of evolution. *Zhurnal Obshchei Biologii*, **38**, 649–655.

Shaposhnikov, G. K. (1981) *Populations and Species in Aphids and the Need for a Universal Species Concept.* Special Publication, Research Branch, Agriculture Canada.

Shaposhnikov, G. K. H. (1985) The main features of the evolution of aphids. In *Proceedings of the International Aphidological Symposium at Jablonna*, pp. 19–99. Warsaw: Polska Akademia Nauk, Instytut Zoologii.

Shaposhnikov, G. C. (1987) Evolution of aphids in relation to evolution of plants. In A. K. Minks and P. Harrewijn (eds), *Aphids, their Biology, Natural Enemies and Control*, Vol. 2A, pp. 409–414. Amsterdam: Elsevier.

Shaw, M. J. P. (1970a) Effects of population density on alienicolae of *Aphis fabae* Scop. I. The effect of crowding on the production of alatae in the laboratory. *Annals of Applied Biology*, **65**, 191–196.

Shaw, M. J. P. (1970b) Effects of population density on alienicolae of *Aphis fabae* Scop. I. The effects of crowding on the expression of migratory urge in the laboratory. *Annals of Applied Biology*, **69**, 197–203.

Shaw, M. J. P. (1970c) Effects of population density on alienicolae of *Aphis fabae* Scop. III. The effect of isolation on the development of form and behaviour of alatae in a laboratory clone. *Annals of Applied Biology*, **65**, 205–212.

Shearer, J. W. (1976) *Polymorphism and population ecology of European maple aphid*, Periphyllus testudinaceus *(Fernie)*. PhD Thesis, University of Glasgow.

Shinji, O. (1941) *Monograph of Japanese Aphididae.* Tokyo: Shinkyo Sha Shoin.

Sibly, R. M. and Calow, P. (1986) *Physiological Ecology of Animals. An Evolutionary Approach.* Oxford: Blackwell Scientific Publications.

Simon, J.-C., Martinez-Torres, D., Latorre, A., Moya, A. and Herbert, P. D. N. (1996) Molecular characterisation of cyclic and obligate parthenogens in the aphid *Rhopalosiphum padi* (L.). *Proceedings of the Royal Society of London B*, **263**, 481–486.

Singer, M. C. (1983) Determinants of multiple host use by a phytophagous insect population. *Evolution*, **37**, 389–403.

Skinner, G. J. and Whittaker, J. E. (1981) An experimental investigation of inter-relationships between the wood-ant (*Formica rufa*) and some tree-canopy herbivores. *Journal of Animal Ecology*, **50**, 313–326.

Slade, N. A. and Wassersug, R. J. (1975) On the evolution of complex life cycles. *Evolution*, **29**, 568–571.

Slansky, F. and Feeny, P. (1977) Stabilisation of the rate of nitrogen accumulation by larvae of the cabbage butterfly on wild and cultivated food plants. *Ecological Monographs*, **47**, 209–228.

Sluss, R. R. (1967) Population dynamics of the walnut aphid *Chromaphis juglandicola* (Kalt.) in northern California. *Ecology*, **48**, 41–58.

Smith, C. F. and Denmark, H. A. (1984) Life history and synchrony of *Grylloprociphilus imbricator* (Fitch) (Homoptera: Aphididae). *Florida Entomologist*, **67**, 430–434.

Smith, J. D. (1948) Symbiotic micro-organisms of aphids and fixation of atmospheric nitrogen. *Nature*, **162**, 930–931.

Smith, L. M. (1936) Biology of the mealy plum aphid, *Hyalopterus pruni* (Geoffroy). *Hilgardia*, **10**, 167–209.

Smith, M. A. H. and MacKay, P. A. (1990) Latitudinal variation in the photoperiodic responses of populations of pea aphid (Homoptera: Aphididae). *Environmental Entomology*, **19**, 618–624.

Sokal, R. R., Unnasch, R. S. and Thomson, B. A. (1991) *Pemphigus* revisited: changes in geographical variation but constancy in variation and covariation. *Evolution*, **45**, 1585–1605.

Sömme, L. (1969) Mannitol and glycerol in overwintering aphid eggs. *Norsk Entomologisk Tidsskrift*, **16**, 107–111.

Southwood, T. R. E. (1977) Habitat, the templet for ecological strategies. *Journal of Animal Ecology*, **46**, 337–345.

Srivastava, P. N. and Auclair, J. L. (1963) Characteristics and nature of proteases from the alimentary canal of the pea aphid, *Acyrthosiphon pisum* (Harr.) (Homoptera: Aphididae). *Journal of Insect Physiology*, **9**, 469–474.

Stam, P. (1983) The evolution of reproductive isolation in closely adjacent plant populations through differential flowering time. *Heredity*, **50**, 105–118.

Stanley, F. M. (1973) An explanation for Cope's rule. *Evolution*, **27**, 1–26.

Stearns, S. C. (1976) Life-history tactics: a review of the ideas. *Quarterly Review of Biology*, **51**, 3–47.

Stearns, S. C. (1982) The role of development in the evolution of life histories. In J. T. Bonner (ed.), *Evolution and Development*, pp. 237–248. Berlin: Springer Verlag.

Stearns, S. C. and Koella, J. C. (1986) The evolution of phenotypic plasticity in life-history traits: predictions of reaction norms for age and size at maturity. *Evolution*, **40**, 893–913.

Steel, C. G. H. and Lees, A. D. (1977) The role of neurosecretion in the photoperiodic control of polymorphism in the aphid *Megoura viciae*. *Journal of Experimental Biology*, **67**, 117–135.

Steffan, A. W. (1983) Zur pheromonalem kommunikation bei der Geschlechterfindung von Blattläusen (Homoptera: Aphidinea). *Verhandlungen der Deutschen Zoologisch Gesellschaft*, 168.

Steffan, A. W. (1987) Fern- und Nahorientierung geflügelter Gynoparae und Sexualis – Männchen bei Blattläusen (Homoptera: Aphidinea: Aphididae). *Entomologia Generalis*, **12**, 235–258.

Steffan, A. W. (1990) Courtship behaviour and possible pheromone spread by hind leg raising in sexual females of aphids (Homoptera: Aphididae). *Entomologia Generalis*, **15**, 33–49.

Stern, D. L. (1994) A phylogenetic analysis of soldier evolution in the aphid family Hormaphididae. *Proceedings of the Royal Society of London B*, **256**, 203–209.

Stern, D. L. and Foster, W. A. (1996) The evolution of soldiers in aphids. *Biological Reviews*, **71**, 27–79.

Stern, D. L., Aoki, S. and Kurosu, U. (1994) A test of geometric hypotheses for soldier investment patterns in the gall producing tropical aphid *Cerataphis fransoeni* (Homoptera, Hormaphididae). *Insectes Sociaux*, **41**, 457–460.

Stiling, P. (1987) The frequency of density dependence in insect host–parasitoid systems. *Ecology*, **68**, 844–856.

Stiling, P. (1988) Density-dependent processes and key factors in insect populations. *Journal of Animal Ecology*, **57**, 581–593.

Strathdee, A. T., Bale, J. S., Block, W. C., Webb, N. R., Hodkinson, I. D. and Coulson, S. J. (1993) Extreme adaptive life-cycle in a high arctic aphid, *Acyrthosiphon svalbardicum*. *Ecological Entomology*, **18**, 254–258.

Strathdee, A. T., Howling, G. G. and Bale, J. S. (1995a) Cold hardiness of overwintering aphid eggs. *Journal of Insect Physiology*, **41**, 653–657.

Strathdee, A. T., Bale, J. S., Strathdee, F. C., Block, W. C., Coulson, J. C., Webb, N. R. and Hodkinson, I. D. (1995b) Climatic severity and the response to temperature elevation of arctic aphids. *Global Change Biology*, **1**, 23–28.

Strong, D. R., Lawton, J. N. and Southwood, T. R. E. (1984) *Insects on Plants*. Oxford: Blackwell.

Stroyan, H. L. G. (1958) Contribution to the taxonomy of some British species of *Sappaphis* Matsumura 1918 (Homoptera, Aphidoidea). *Journal of the Linnean Society*, **43**, 644–713.

Stroyan, H. L. G. (1963) *The British Species of Dysaphis Börner Part II. The Subgenus Dysaphis sensa stricto*. London: HMSO.

Stroyan, H. L. G. (1977) Homoptera Aphidoidea: Chaitophoridae and Callaphidae. *Royal Entomological Society (London) Handbooks for the Identification of British Insects*, **2**(4a).

Stroyan, H. L. G. (1984) Aphids-Pterocommatinae and Aphidinae (Aphidini) Homoptera, Aphididae. *Royal Entomological Society (London) Handbooks for the Identification of British Insects*, **2**(6).

Sunose, T. S., Yamane, K., Tsuda, K. and Takasu, K. (1991) What do the soldiers of *Pseudoregma bambucicola* (Homoptera, Aphidoidea) defend? *Japanese Journal of Entomology*, **59**, 141–148.

Suomalainen, E. (1950) Parthenogenesis in animals. *Advances in Genetics*, **3**, 193–253.

Suomalainen, E., Saura, A. and Lokki, J. (1976) Evolution of parthenogenetic insects. *Evolutionary Biology*, **9**, 209.

Suomalainen, E., Saura, A. and Lokki, J. (1987) *Cytology and Evolution in Parthenogenesis*. Boca Raton: CRC Press.

Suomalainen, E., Saura, A., Lokki, J. and Teeri, T. (1980) Genetic polymorphism and evolution in parthenogenetic animals. Pt. 9: Absence of variation within parthenogenetic aphid clones. *Theoretical and Applied Genetics*, **57**, 129–132.

Sutherland, O. R. W. (1968) Dormancy and lipid storage in the pemphigine aphid *Thecabius affinis*. *Entomologia Experimentalis et Applicata*, **11**, 348–354.

Sutherland, O. R. W. (1969a) The role of crowding in the production of winged forms by two strains of the pea aphid, *Acyrthosiphon pisum*. *Journal of Insect Physiology*, **15**, 1385–1410.

Sutherland, O. R. W. (1969b) The role of the host plant in the production of winged forms by two strains of the pea aphid, *Acyrthosiphon pisum*. *Journal of Insect Physiology*, **15**, 2179–2201.

Sutherland, O. R. W. and Mittler, T. E. (1971) Influence of diet composition and crowding on wing production by the aphid *Myzus persicae*. *Journal of Insect Physiology*, **17**, 321–328.

Szelegiewicz, H. (1978) Roznodomnsc (heteroecja) u mszyc, jej pochodzenie i ewolucja. *Zeszyty Problemowe Postepów Nauk Rolniczych*, **208**, 19–31.

Taimr, L., Kudelova, A. and Kriz, J. (1978) Diurnal periodicity in the flight activity of migrant alatae of *Phorodon humuli* Schrank (Hom. Aphididae). *Zeitschrift für Angewandte Entomologie*, **36**, 373–380.

Takada, H. (1979) Characteristics of forms of *Myzus persicae* (Sulzer) (Homoptera: Aphididae) distinguished by colour and esterase differences and their occurence in populations on different host plants in Japan. *Applied Entomology and Zoology*, **14**, 370–375.

Takada, H. (1981) Inheritance of body colours in *Myzus persicae* (Sulzer) (Homoptera: Aphididae). *Applied Entomology and Zoology*, **16**, 242–246.

Takeda, S., Kinomura, K. and Sakurai, H. (1982) Effects of ant-attendance on the honeydew excretion and larviposition of the cowpea aphid, *Aphis craccivora*. *Applied Entomology and Zoology*, **17**, 133–135.

Tamaki, G. and Allen, W. W. (1969) Competition and other factors influencing the population dynamics of *Aphis gossypii* and *Macrosiphoniella sanborni* on greenhouse chrysanthemums. *Hilgardia*, **39**, 447–505.

Tanaka, S. and Ito, Y. (1995) Interrelationships between the eusocial aphid, *Pseudoregma bambicicola*, and its syrphid predator, *Eupeodes confrater*. *Japanese Journal of Entomology*, **63**, 221–228.

Tarn, T. R. and Adams, J. B. (1982) Aphid probing and feeding, electronic monitoring, and plant breeding. In K. F. Harris and K. Maramorosch (eds), *Pathogens, Vectors and Plant Diseases: Approach to Control*, pp. 221–246. New York: Academic Press.

Tashev, D. and Markova, E. (1982) Morphofunktionelle Grundlagen der Fruchtbarkeit bei den Blattläusen (Aphidina, Viviovipara). I. Anzahl Der Eirohen. *Godishnik Na Sofiiskiya Universitet Kliment Okhridski Biologicheski Fakultet (Zoologiya)*, **70**, 19–28.

Tatchell, G. M. and Parker, S. J. (1990) Host plant selection by migrant *Rhopalosiphum padi* in autumn and the occurrence of an intermediate morph. *Entomologia Experimentalis et Applicata*, **54**, 237–244.

Taylor, L. R. (1961) Aggregation, variance and the mean. *Nature*, **189**, 732–735.

Taylor, L. R. (1965) Flight behaviour and aphid migration. *Proceedings North Central Branch Entomological Society of America*, **20**, 9–19.

Taylor, L. R. (1973a) Monitor surveying for migrant insect pests. *Outlook on Agriculture*, **7**, 109–116.

Taylor, L. R. (1973b) Monitoring change in the distribution and abundance of insects. *Reports of the Rothamsted Experimental Station*, **1973**, 202–239.

Taylor, L. R. (1974) Insect migration, flight periodicity and the boundary layer. *Journal of Animal Ecology*, **43**, 225–238.

Taylor, L. R. (1977) Migration and the spatial dynamics of an aphid, *Myzus persicae*. *Journal of Animal Ecology*, **46**, 411–423.

Taylor, L. R. (1979) The Rothamsted Insect Survey – an approach to the theory and practice of synoptic pest forecasting in agriculture. In R. L. Rabb and G. G. Kennedy (eds), *Movement of Highly Mobile Insects*, pp. 148–185. Raleigh: North Carolina State University.

Taylor, L. R. and Taylor, R. A. J. (1977) Aggregation, migration and population mechanics. *Nature*, **265**, 415–421.

Taylor, L. R. and Taylor, R. A. J. (1978) The dynamics of spatial behaviour. In F. J. Ebling and D. M. Stoddart (eds), *Population Control by Social Behaviour*, pp. 181–212. London: Institute of Biology.

Taylor, L. R. and Taylor, R. A. J. (1983) Insect migration as a paradigm for survival by movement. In I. R. Swingland and P. J. Greenwood (eds), *The Ecology of Animal Movement*, pp. 181–214. Oxford: Oxford University Press.

Taylor, L. R. and Woiwood, I. P. (1980) Temporal stability as a density-dependent species characteristic. *Journal of Animal Ecology*, **49**, 209–224.

Taylor, L. R., Taylor, R. A. J. and Perry, J. N. (1983) Behavioural dynamics. *Nature*, **303**, 801–804.

Taylor, L. R., Woiwod, I. P. and Perry, J. N. (1978) The density-dependence of spatial behaviour and the rarity of randomness. *Journal of Animal Ecology*, **47**, 383–406.

Taylor, L. R., Woiwod, I. P. and Perry, J. N. (1980) Variance and the large scale spatial stability of aphids, moths and birds. *Journal of Animal Ecology*, **49**, 831–854.

Taylor, L. R., Woiwod, I. P. and Taylor, R. A. J. (1979) The migratory ambit of the hop aphid and its significance in aphid population dynamics. *Journal of Animal Ecology*, **48**, 955–972.

Taylor, R. A. J. and Taylor, L. R. (1979) A behavioural model for the evolution of spatial dynamics. In R. M. Anderson, D. D. Turner and L. R. Taylor (eds), *Population Dynamics*, pp. 1–27. Oxford: Blackwell.

Thacker, J. I. (1995) *The role of density dependence and weather in determining aphid abundance.* PhD Thesis, University of East Anglia.

Thacker, J. I., Thieme, T. and Dixon, A. F. G. (1997) Forecasting of periodic fluctuations in annual abundance of the bean aphid: the role of density dependence and weather. *Journal of Applied Entomology.*

Thieme, T. (1988) Zur Biologie von *Aphis fabae mordvilkowi* Börner und Janisch 1922 (Hom., Aphididae). *Zeitchrift für Angewandte Entomologie*, **105**, 510–515.

Thieme, T. and Dixon, A. F. G. (1996) Mate recognition in the *Aphis fabae* complex: daily rhythm of release and specificity of sex pheromones. *Entomologia Experimentalis et Applicata*, **79**, 85–89.

Thomas, K. H. (1968) Die Blattläuse aus der engere Verwandtschaft von *Aphis gossypii* Glover und *A. frangulae* Kaltenbach unter besonderer Berücksichtigung ihres Vorkommens an Kartoffel. *Entomologische Abhandlunges staatliches Museum für Tierkunde in Dresden*, **35**, 337–389.

Thornback, N. (1983) *The factors determining the abundance of* Metopolophium dirhodum *(Walk.), the rose grain aphid.* PhD Thesis, University of East Anglia.

Thurston, R. (1970) Toxicity of trichome exudates of *Nicotiana* and *Petunia* species to tobacco hornworm larvae. *Journal of Economic Entomology*, **63**, 272–274.

Thurston, R., Smith, W. T. and Cooper, B. P. (1966) Alkaloid secretion by trichomes of *Nicotiana* species and resistance to aphids. *Entomologia Experimentalis et Applicata*, **9**, 428–432.

Tingey, W. M. and Laubengayer, J. E. (1981) Defense against the green peach aphid and potato leafhopper by glandular trichomes of *Solanum berthaultii. Journal of Economic Entomology*, **74**, 721–725.

Tingey, W. M. and Sinden, S. L. (1982) Glandular pubescence, glycoalkaloid composition, and resistance to the green peach aphid, potato leaf hopper, and potato flea beetle in *Solanum berthaultii. American Potato Journal*, **59**, 95–106.

Tjallingii, W. F. (1978a) Electronic recording of penetration behaviour by aphids. *Entomologia Experimentalis et Applicata*, **24**, 521–530.

Tjallingii, W. F. (1978b) Mechanoreceptors of the aphid labium. *Entomologia Experimentalis et Applicata*, **24**, 531–537.

Tjallingii, W. F. (1985) Electrical nature of recorded signals during stylet penetration by aphids. *Entomologia Experimentalis et Applicata*, **38**, 178–186.

Tjallingii, W. F. (1986) Wire effects on aphids during electrical recording of stylet penetration. *Entomologia Experimentalis et Applicata*, **40**, 89–98.

Tjallingii, W. F. and Hogen Esch, T. (1993) Fine structure of aphid stylet routes in plant tissues in correlation with EPG signals. *Physiological Entomology*, **18**, 317–328.

Toba, H. H., Paschke, J. D. and Friedman, S. (1967) Crowding as the primary factor in the production of the agamic alate form of *Therioaphis maculata* (Homoptera: Aphididae). *Journal of Insect Physiology*, **13**, 381–396.

Tomiuk, J. (1990) Genetic stability in aphid clones and its implication for host-plant interactions. In R. K. Campbell and R. D. Eikenbary (eds), *Aphid–Plant Genotype Interactions*, pp. 273–288. Amsterdam: Elsevier Press.

Tomiuk, J. and Wöhrmann, K. (1980) Enzyme variability in populations of aphids. *Theoretical and Applied Genetics*, **57**, 125–127.

Tomiuk, J. and Wöhrmann, K. (1981) Changes of the genotype frequencies at the MDH-Locus in populations of *Macrosiphum rosae* (L.) (Hemiptera, Aphididae). *Biologisches Zentralblatt*, **100**, 631–640.

Tomiuk, J. and Wöhrmann, K. (1982) Comments on the genetic stability of aphid clones. *Experientia*, **38**, 320–321.

Tóth, L. (1933) Über die Frühembryonale entwicklung der viviparen Aphiden. *Zeitschrift für Morphologie und Okologie der Tiere*, **27**, 692–731.

Tóth, L. (1937) Entwicklungszyklus und Symbiose von *Pemphigus spirothecae* Pass. (Aphidina). *Zeitschrift für Morphologie und Okologie der Tiere*, **33**, 412–437.

Tóth, L. (1940) The protein metabolism of the aphids. *Annales du Musée d'histoire naturelle part Zoologie*, **33**, 167–171.

Trager, W. (1970) *Symbiosis*. New York: Van Nostrand.

Tsumuki, H., Nagatsuka, H., Kawada, K. and Kanehisa, K. (1990) Comparison of nutrient reserves in apterous and alate pea aphids, *Acyrthosiphon pisum* (Harris). I. Developmental time and sugar content. *Applied Entomology and Zoology*, **25**, 215–221.

Turchin, P. (1990) Rarity of density dependence or population regulation with lags? *Nature*, **344**, 660–663.

Turchin, P. and Taylor, A. D. (1992) Complex dynamics in ecological time series. *Ecology*, **73**, 289–305.

Ueda, N. and Takada, H. (1977) Differential relative abundance of green-yellow and red forms of *Myzus persicae* (Sulzer) (Homoptera: Aphididae) according to host plant and season. *Applied Entomology and Zoology*, **12**, 124–133.

Uichanco, L. B. (1924) Studies on the embryology and postnatal development of the Aphididae with special reference to the history of the 'symbiotic organ' or 'mycetome'. *Philippine Journal of Science*, **24**, 143–247.

Unterman, B. M., Baumann, P. and McLean, D. L. (1989) Pea aphid symbiont relationships established by analysis of 16S rRNA. *Journal of Bacteriology*, **171**, 2970–2974.

van den Heuvel, J., Verbeek, M. and van der Wilk, F. (1994) Endosymbiotic bacteria associated with circulative transmission of potato leafroll virus by *Myzus persicae*. *Journal of General Virology*, **75**, 2559–2565.

van Emden, H. F. (1972) Aphids as phytochemists. In J. B. Harbourne (ed.), *Phytochemical Ecology*, pp. 25–43. London: Academic Press.

van Emden, H. F. (1978) Insects and secondary plant substances – an alternative viewpoint with special reference to aphids. In J. B. Harbourne (ed.), *Biochemical Aspects of Plant and Animal Coevolution*, pp. 309–323. London: Academic Press.

van Emden, H. F. and Bashford, M. A. (1971) The performance of *Brevicoryne brassicae* and *Myzus persica* in relation to plant age and leaf amino-acids. *Entomologia Experimentalis et Applicata*, **14**, 349–360.

van Emden, H. F., Eastop, V. F., Hughes, R. D. and Way, M. J. (1969) The ecology of *Myzus persicae*. *Annual Review of Entomology*, **14**, 197–270.

Varley, G. C. (1967) The effects of grazing by animals on plant productivity. In K. Petrusewicz (ed.), *Secondary Productivity of Terrestrial Ecosystems*, Vol. 2, pp. 773–778. Warsaw: Kraków.

Varley, G. C., Gradwell, G. R. and Hassell, M. P. (1973) *Insect Population Ecology: An Analytical Approach*. Oxford: Blackwell.

Vehrs, S. L. C., Walker, G. P. and Parrella, M. P. (1992) Comparison of population growth rate and within-plant distribution between *Aphis gossypii* and *Myzus persicae* (Homoptera: Aphididae) reared on potted chrysanthemums. *Journal of Economic Entomology*, **85**, 799–807.

Verbeek, M. and van den Heuvel, J. F. J. M. (1994) Localisation of symbionin, a protein abundantly produced by the primary endosymbiont of *Myzus persicae*. *Proceedings of Experimental and Applied Entomology*, **5**, 91–92.

Vereijken, B. H. (1979) Feeding and multiplication of three cereal aphid species and their effect on yield of winter wheat. *Agricultural Research Report, Wageningen*, 888.

Via, S. (1984) The quantitative genetics of polyphagy in an insect herbivore. II. Genetic correlations in larval performance within and among hosts. *Evolution*, **38**, 896–905.

Via, S. (1991) The genetic structure of host plant adaptation in a spatial patchwork: Demographic variability among reciprocally transplanted pea aphid clones. *Evolution*, **45**, 827–852.

Visser, J. H. and Piron, P. G. M. (1994) Perception of plant odour components by the vetch aphid *Megoura viciae*: shape characteristics of electroantennogram responses. *Proceedings of Experimental and Applied Entomology*, **5**, 85–90.

Visser, J. H. and Piron, P. G. M. (1995) Olfactory antennal responses to plant volatiles in apterous virginoparae of the vetch aphid *Megoura viciae*. *Entomologia Experimentalis et Applicata*, **77**, 37–46.

Visser, J. H. and Taanman, J. W. (1987) Odour-conditioned anemotaxis of apterous aphids (*Cryptomyzus korschelti*) in response to host plants. *Physiological Entomology*, **12**, 473–479.

Vπlkl, W. (1989) Resource partitioning in a guild of aphids species associated with creeping thistle *Cirsium arvense*. *Entomologia Experimentalis et Applicata*, **51**, 41–47.

von Dohlen, C. D. and Gill, D. E. (1989) Geographic variation and evolution in the life cycle of the witch-hazel leaf gall aphid *Hormaphis hamamelidis*. *Oecologia (Berlin)*, **78**, 165–175.

Vrieling, K., Smit, W. and van der Meijden, E. (1991) Tritrophic interactions between aphids (*Aphis jacobaeae* Shrank), ant species, *Tyria jacobaeae* L. and *Senecis jacobaea* L. lead to maintenance of genetic variation in pyrrolizidine alkaloid concentration. *Oecologia (Berlin)*, **86**, 177–182.

Wagner, G. P. and Gabriel, W. (1990) Quantitative variation in finite parthenogenetic populations: what stops Muller's ratchet in the absence of recombination. *Evolution*, **44**, 715–.

Walker, F. (1848) Remarks on the migration of Aphides. *Annals of the Magazine of Natural History*, **1**, 372–373.

Walker, F. (1850) Description of aphids. *Annals of the Magazine of Natural History*, **2**, 14–28.

Walsh, B. D. (1864) On phytophagic varieties and polyphagous species. *Proceedings of the Entomological Society of Philadelphia*, **3**, 403–430.

Walters, F. S. and Mullin, C. A. (1988) Sucrose dependent increase in oligosaccharide production and associated glycosidase activities in the potato aphid *Macrosiphum euphorbiae* (Thomas). *Archives of Insect Biochemistry and Physiology*, **9**, 35–46.

Walters, K. F. A., Brough, C. and Dixon, A. F. G. (1988) Habitat quality and reproductive investment in aphids. *Ecological Entomology*, **13**, 337–345.

Walters, K. F. A. and Dixon, A. F. G. (1982) Effect of host quality and crowding on the settling and take-off of cereal aphids. *Annals of Applied Biology*, **101**, 211–218.

Walters, K. F. A. and Dixon, A. F. G. (1983) Migratory urge and reproductive investment in aphids: variation within clones. *Oecologia (Berlin)*, **58**, 70–75.

Walters, K. F. A. and Dixon, A. F. G. (1984) The effect of temperature and wind on the flight activity of cereal aphids. *Annals of Applied Biology*, **104**, 17–26.

Walters, K. F. A., Dixon, A. F. G. and Eagles, G. (1984) Non-feeding by adult gynoparae of *Rhopalosiphum padi* and its bearing on the limiting resource in the production of sexual females in host alternating aphids. *Entomologia Experimentalis et Applicata*, **36**, 9–12.

Ward, S. A. (1987a) Cyclical parthenogenesis and the evolution of host range in aphids. In J. Holman, J. Pelikan, A. F. G. Dixon and L. Weismann (eds), *Population Structure, Genetics and Taxonomy of Aphids and Thysanoptera*, pp. 39–44. The Hague: SPB Academic Publishing.

Ward, S. A. (1987b) Optimal habitat selection in time-limited dispersers. *American Naturalist*, **129**, 568–579.

Ward, S. A. (1991a) Reproduction and host selection by aphids: the importance of 'rendezvous' hosts. In W. Bailey and J. Ridsdill-Smith (eds), *Reproductive Behaviour in Insects*, pp. 202–226. London: Chapman & Hall.

Ward, S. A. (1991b) Theoretical perspectives on sympatric speciation in aphids (Homoptera: Aphidinea: Aphididae). *Entomologia Generalis*, **16**, 177–192.

Ward, S. A. (1992a) Assessing functional explanations of host-specificity. *American Naturalist*, **139**, 883–891.
Ward, S. A. (1992b) Environmental uncertainty and polyphagy in herbivorous insects. *Oikos*, **53**, 506–512.
Ward, S. A. and Dixon, A. F. G. (1982) Selective resorption of aphid embryos and habitat changes relative to life-span. *Journal of Animal Ecology*, **51**, 859–864.
Ward, S. A. and Dixon, A. F. G. (1984) Spreading the risk, and the evolution of mixed strategies: seasonal variation in aphid reproductive biology. *Advances in Invertebrate Reproduction*, **3**, 367–386.
Ward, S. A. and Dixon, A. F. G. (In Prep.) Seasonal sex allocation. I. Mating and fitness.
Ward, S. A., Dixon, A. F. G. and Wellings, P. W. (1983b) The relation between fecundity and reproductive investment in aphids. *Journal of Animal Ecology*, **52**, 451–461.
Ward, S. A., Leather, S. R. and Dixon, A. F. G. (1984) Temperature prediction and the timing of sex in aphids. *Oecologia (Berlin)*, **62**, 230–233.
Ward, S. A., Leather, S. R., Pickup, J. and Harrington, R. (1998) Mortality during dispersal and the cost of host-specificity in parasites: how many aphids find hosts? *Journal of Animal Ecology* (submitted).
Ward, S. A. and Wellings, P. W. (1994) Deadlines and delays as factors in aphid sex allocation. *European Journal of Entomology*, **91**, 29–36.
Ward, S. A., Wellings, P. W. and Dixon, A. F. G. (1983a) The effect of reproductive investment on pre-reproductive mortality in aphids. *Journal of Animal Ecology*, **52**, 305–313.
Watson, M. A., Heathcote, G. D., Lauckner, F. B. and Sowray, P. A. (1975) The use of weather data and counts of aphids in the field to predict the incidence of yellowing viruses of sugar-beet crops in England in relation to the use of insecticides. *Annals of Applied Biology*, **81**, 181–198.
Watson, S. J. (1983) *Effects of weather on the numbers of cereal aphids*. PhD Thesis, University of East Anglia.
Watt, A. D. and Dixon, A. F. G. (1981) The role of cereal growth stages and crowding in the induction of alatae in *Sitobion avenae* and its consequences for population growth. *Ecological Entomology*, **6**, 441–447.
Watt, K. E. F. (1964) Comments on fluctuations of animal populations and measure of community structure. *Canadian Entomologist*, **96**, 1434–1442.
Way, M. J. (1963) Mutualism between ants and honeydew-producing Homoptera. *Annual Review of Entomology*, **8**, 307–344.
Way, M. J. (1967) The nature and causes of annual fluctuations in numbers of *Aphis fabae* Scop. on field beans (*Vicia faba*). *Annals of Applied Biology*, **59**, 175–188.
Way, M. J. and Banks, C. J. (1967) Intraspecific mechanisms in relation to the natural regulation of numbers of *Aphis fabae* Scop. *Annals of Applied Biology*, **59**, 189–205.
Way, M. J. and Banks, C. J. (1968) Population studies on the active stages of the black bean aphid, *Aphis fabae* Scop., on its winter host *Euonymus europaeus* L. *Annals of Applied Biology*, **62**, 177–197.
Way, M. J. and Cammell, M. (1970) Aggregation behaviour in relation to food utilisation by aphids. In A. Watson (ed.), *Animal Populations in Relation to their Food Resources*, pp. 229–247. Oxford: Blackwell.
Way, M. J. and Heathcote, G. D. (1966) Interactions of crop density of field beans, abundance of *Aphis fabae* Scop., virus incidence and aphid control by chemicals. *Annals of Applied Biology*, **57**, 409–423.
Way, M. J., Cammell, M. E., Taylor, L. R. and Woiwood, I. P. (1981) The use of egg counts and suction trap samples to forecast the infestation of spring-sown field beans, *Vicia faba*, by the black bean aphid, *Aphis fabae*. *Annals of Applied Biology*, **98**, 21–34.
Weber, G. (1985a) On the ecological genetics of *Sitobion avenae* (F.) (Hemiptera, Aphididae). *Zeitschrift für Angewandte Entomologie*, **100**, 100–110.
Weber, G. (1985b) On the ecological genetics of *Metopolophium dirhodum* (Walker) (Hemiptera, Aphididae). *Zeitschrift für Angewandte Entomologie*, **100**, 451–458.
Weber, G. (1985c) Genetic variability in host plant adaptation of the green peach aphid, *Myzus persicae*. *Entomologia Experimentalis et Applicata*, **38**, 49–56.
Weed, C. M. (1896) The hibernation of aphids. *Psyche*, **7**, 351–362.
Wegorek, W. and Krzymanska, J. (1971) Further studies on the resistance of lupine to the pea aphid (*A. pisum* Harris). *Prace Naukovo Institute Ochrony Roslin*, **13**, 7–23.

Wellings, P. W. (1981) The effect of temperature on the growth and reproduction of two closely related aphid species on sycamore. *Ecological Entomology*, **6**, 209–214.

Wellings, P. W. and Dixon, A. F. G. (1983) Physiological constraints on the reproductive activity of the sycamore aphid: the effect of development experience. *Entomologia Experimentalis et Applicata*, **34**, 227–232.

Wellings, P. W. and Dixon, A. F. G. (1987a) The role of weather and natural enemies in determining aphid outbreaks. In P. Barbosa and J. C. Schultz (eds), *Insect Outbreaks*, pp. 313–346. New York: Academic Press.

Wellings, P. W. and Dixon, A. F. G. (1987b) Sycamore aphid numbers and population density. III. The role of aphid-induced changes in plant quality. *Journal of Animal Ecology*, **56**, 161–170.

Wellings, P. W., Chambers, R. R. and Dixon, A. F. G. (1985) Sycamore aphid numbers and population density. I. Some patterns. Journal of Animal Ecology, **54**, 411–424.

Wellings, P. W., Leather, S. R. and Dixon, A. F. G. (1980) Seasonal variation in reproductive potential: a programmed feature of aphid life cycles. *Journal of Animal Ecology*, **49**, 975–985.

Wellington, W. S. (1983) Biometerology of dispersal. *Bulletin of the Entomological Society of America*, **29**, 24–29.

Wensler, R. J. D. (1962) Mode of host selection by an aphid. *Nature*, **195**, 830–831.

Wensler, R. J. D. and Filshie, B. K. (1969) Gustatory sense organs in the food canal of aphids. *Journal of Morphology*, **129**, 473–492.

Wertheim, G. (1954) Studies on the biology and ecology of the gall-producing aphids of the tribe Fordini (Homoptera: Aphidoidea) in Israel. *Transactions of the Royal Entomological Society of London*, **105**, 79–96.

White, P. L. (1970) *The effect of aphids on tree growth.* PhD thesis, University of Glasgow.

Whitehead, L. F., Wilkinson, T. L. and Douglas, A. E. (1992) Nitrogen recycling in the pea aphid (*Acythosiphon pisum*) symbiosis. *Proceedings of the Royal Society of London B*, **250**, 115–117.

Whitham, T. G. (1978) Habitat selection by *Pemphigus* aphids in response to resource limitation and competition. *Ecology*, **59**, 1164–1176.

Whitham, T. G. (1979) Territorial behaviour of *Pemphigus* gall aphids. *Nature*, **279**, 324–325.

Whitham, T. G. (1980) The theory of habitat selection: examined and extended using *Pemphigus* aphids. *American Naturalist*, **115**, 449–466.

Whittaker, J. B. (1991) Effects of ants on temperate woodland trees. In C. R. Huxley and D. F. Cutler (eds), *Ant–Plant Interactions*, pp. 69–79. Oxford: Oxford University Press.

Wiktelius, S. (1987) The role of grassland in the yearly life-cycle of *Rhopalosiphum padi* (Homoptera: Aphididae) in Sweden. *Annals of Applied Biology*, **110**, 9–15.

Wilbur, H. M. (1980) Complex life cycles. *Annual Review of Ecological Systems*, **11**, 67–93.

Williams, C. B. (1964) *Patterns in the Balance of Nature*. London: Academic Press.

Williams, C. T. (1980) Low temperature mortality of cereal aphids. *IOBC/WPRS Bull.* **111**, **4**, 63–66.

Williams, G. C. (1966) *Adaptation and Natural Selection*. Princeton: Princeton University Press.

Williams, G. C. (1975) *Sex and Evolution*. Princeton: Princeton University Press.

Williams, G. C. (1992) *Natural Selection: Domains, Levels and Challenges*. New York: Oxford University Press.

Williams, W. G., Kennedy, C. G., Yamamoto, R. T., Thacker, J. D. and Bordner, J. (1980) 2-Tridecanone: a naturally occurring insecticide from the wild tomato *Lycopersicon hirsutum f. glabratum*. *Science*, **207**, 888.

Wink, M., Hartmann, T., Witte, L. and Rheinheimer, J. (1982) Interrelationship between quinolizidine alkaloid producing legumes and infesting insects: exploitation of the alkaloid-containing phloem sap of *Cytisus scoparius* by the broom aphid *Aphis cytisorum*. *Zeitschrift für Naturforschung*, **37**, 1081–1086.

Wink, M. and Romer, P. (1986) Acquired toxicity – the advantages of specialising on alkaloid-rich lupins to *Macrosiphum albifrons* (Aphidae). *Naturwissenschaften*, **73**, 210–212.

Wink, M. and Witte, L. (1991) Storage of quinolizidine alkaloids in *Macrosiphum albifrons* and *Aphis genistae* (Homoptera: Aphididae). *Entomologia Generalis*, **15**, 237–254.

Woiwood, I. P. and Hanski, I. (1992) Patterns of density dependence in moths and aphids. *Journal of Animal Ecology*, **61**, 619–629.

Wood, B. W., Tedders, W. L. and Thompson, J. M. (1985) Feeding influence on three pecan aphid species on carbon exchange and phloem integrity of seedling pecan foliage. *Journal of the American Society for Horticultural Science*, **110**, 393–397.

Wool, D. (1984) Gall forming aphids. In T. N. Ananthakrishnan (ed.), *Biology of Gall Insects*, pp. 11–58. Oxford and IBH, New-Delhi.

Wool, D. and Manheim, O. (1986) Population ecology of the gall-forming aphid, *Aploneura lentisci* (Pass.) in Israel. *Research in Population Ecology*, **28**, 151–162.

Wratten, S. D. (1977) Reproductive strategy of winged and wingless morphs of the aphids *Sitobion avenae* and *Metapolophium dirhodum*. *Annals of Applied Biology*, **85**, 319–331.

Yamaguchi, H. (1976) Biological studies on the todo-fir aphid *Cinara todicola* Inouye with special reference to its population dynamics and morph determination. *Bulletin of the Government Forestry Experimental Station*, **283**, 1–102.

Yamaguchi, Y. (1985) Sex ratios of an aphid subject to local mate competition with variable maternal condition. *Nature*, **318**, 460–462.

Young, J. P. W. (1981) Sib competition can favour sex in two ways. *Journal of Theoretical Biology*, **88**, 755–756.

Zehavi, A. and Rosen, D. (1987) Population trends of the spirea aphid, *Aphis citricola* van der Goot, in a citrus grove in Israel. *Journal of Applied Entomology*, **104**, 271–277.

Zhang, G. X. and Zhong, T. S. (1990) Experimental studies on some aphid life-cycle patterns and the hybridisation of two sibling species. In R. K. Campbell and R. D. Eikenbary (eds), *Aphid–Plant Genotype Interactions*, pp. 37–50. Amsterdam: Elsevier.

Zhou, X., Honek, A., Powell, W. and Carter, N. (1995) Variations in body length, weight, fat content and survival in *Coccinella septempunctata* at different hibernation sites. *Entomologia Experimentalis et Applicata*, **75**, 99–107.

Ziegler, H. (1975) Nature of transported substances. In M. H. Zimmerman and J. A. Milburn (eds), *Transport in Plants. I. Phloem Transport*, pp. 59–100. Berlin: Springer.

Ziolko, M. and Kozlowski, J. (1983) Evolution of body size: an optimization model. *Mathematical Bioscience*, **64**, 127–143.

Zucker, W. V. (1982) How aphid choose leaves: the roles of phenolics in host selection by a galling aphid. *Ecology*, **63**, 972–981.

Index

MIX
Papier aus verantwortungsvollen Quellen
Paper from responsible sources
FSC® C105338

If you have any concerns about our products,
you can contact us on
ProductSafety@springernature.com

In case Publisher is established outside the EU,
the EU authorized representative is:
Springer Nature Customer Service Center GmbH
Europaplatz 3, 69115 Heidelberg, Germany

Printed by Libri Plureos GmbH
in Hamburg, Germany